Time, Law, and Change

An Interdisciplinary Study

Edited by
Sofia Ranchordás
and
Yaniv Roznai

•HART•
OXFORD • LONDON • NEW YORK • NEW DELHI • SYDNEY

HART PUBLISHING

Bloomsbury Publishing Plc

Kemp House, Chawley Park, Cumnor Hill, Oxford, OX2 9PH, UK

1385 Broadway, New York, NY 10018, USA

HART PUBLISHING, the Hart/Stag logo, BLOOMSBURY and the Diana logo are trademarks of Bloomsbury Publishing Plc

First published in Great Britain 2020

A catalogue record for this book is available from the British Library.

Library of Congress Cataloging-in-Publication data

Names: International Conference on Law and Time (2017 : Israel). | Ranchordás, Sofia, editor. | Roznai, Yaniv, editor. | Bet sefer Radziner le-mishpaṭim (Hertseliyah, Israel), sponsoring body. | Israel. Keneset, sponsoring body. | Universiteit Leiden, sponsoring body.

Title: Time, law, and change : an interdisciplinary study / edited by Sofia Ranchordás and Yaniv Roznai.

Description: Oxford ; New York : Hart, 2020. | Includes papers presented as a "international academic conference that took place on 6–7 December 2017 at IDC Herzliya and the Knesset (the Israeli Parliament), and was co-organised by Leiden University"—ECIP acknowledgements. | Includes bibliographical references and index.

Identifiers: LCCN 2019052998 (print) | LCCN 2019052999 (ebook) | ISBN 9781509930937 (hardback) | ISBN 9781509930944 (Epub)

Subjects: LCSH: Time (Law)—Congresses. | Legislation—Congresses. | Justice, Administration of—Congresses. | BISAC: LAW / Courts

Classification: LCC K579.T5 I58 2017 (print) | LCC K579.T5 (ebook) | DDC 340/.1—dc23

LC record available at https://lccn.loc.gov/2019052998

LC ebook record available at https://lccn.loc.gov/2019052999

ISBN: HB: 978-1-50993-093-7
ePDF: 978-1-50993-059-1
ePub: 978-1-50993-094-4

Typeset by Compuscript Ltd, Shannon
Printed and bound in Great Britain by CPI Group (UK) Ltd, Croydon CR0 4YY

MIX
Paper from responsible sources
FSC® C013604

We dedicate this book to our grandparents (Candida and Rodrigo, SR; Dina and Tsila, YR), because they taught and teach us about how precious time is.

ACKNOWLEDGEMENTS

This book project was born out of an international academic conference that took place on 6–7 December 2017 at IDC Herzliya and the Knesset (the Israeli Parliament), and was co-organised by Leiden University. In this conference, international scholars gathered to discuss issues of law and temporality, time and the effect of law, the timing of law, temporary legislation, the legislative cycles and practical issues of time in legislation. Apart from the discussion, from this academic gathering the Israeli Association of Legislation was established, as a branch of the International Association of Legislation, so we are delighted that this international collaboration has produced two exciting products: the Association and this edited collection.

We would like to thank all the book's participants, who facilitated the work, in addition to those who are not included in the book but who took part in the conference: Alon Harel, Daniel Greenberg, Daphna Barnai, Eyal Zandberg, Gali Ben-Or, Gur Bligh, Joshua Shoffman, Keren Weinshall-Margel, Malcolm Feeley, Pauline Westerman and Rivka Weill. We would like to thank Eyal Yinon, the Legal Advisor of the Knesset, for co-organising this event and for hosting us in the Knesset and Amnon Lehavi, the Dean of the Harry Radzyner Law School at IDC Herzliya for hosting the conference at IDC. We would also like to extend our gratitude to the conference coordinators, Michal Dibner-Carmel and Tali Gai.

We would like to thank Kate Whetter and Rosamund Jubber of Hart Publishing for their guidance, support and patience, Victoria Broom for her careful editing, and – finally – Angie Sánchez Medina for her tremendous work and efforts in bringing this collection to completion. Her assistance was more than valuable.

Sofia Ranchordás and Yaniv Roznai

August 2019

TABLE OF CONTENTS

PART III
LEGISLATORS AND TIME

PART IV
TECHNOLOGY AND TIME

LIST OF CONTRIBUTORS

Enrico Albanesi, Associate Professor of Constitutional Law at the Faculty of Law, University of Genoa and Associate Research Fellow at the Institute of Advanced Legal Studies, University of London.

Richard Albert, William Stamps Farish Professor in Law and Professor of Government at the University of Texas, United States.

Yifat Aran, JSD Candidate, Stanford University, United States.

Ittai Bar-Siman-Tov, Associate Professor at Bar-Ilan University Faculty of Law, Israel.

Lior Barshack, Associate Professor of Social and Legal Theory at the Harry Radzyner Law School, Interdisciplinary Center, Herzliya, Israel.

Lyria Bennett Moses, Director of the Allens Hub for Technology, Law and Innovation and Professor in the Faculty of Law at the University of New South Wales, Sydney.

Steven G Calabresi, Clayton J & Henry R Barber Professor of Law at Northwestern Pritzker School of Law, United States.

Gijs van Dijck, Professor of Private Law at the Faculty of Law, Maastricht University.

Catalina Goanta, Assistant Professor in Private Law at the Faculty of Law, Maastricht University.

Gaya Harari-Heit, Law Clerk on the Supreme Court of Israel.

Antonios Kouroutakis, Assistant Professor at IE University in Madrid, Spain.

Moran Ofir, Assistant Professor at the Harry Radzyner Law School, Interdisciplinary Center Herzliya, Israel.

Patricia Popelier, Full Professor of Constitutional Law at the University of Antwerp, Belgium.

Zoltán Pozsár-Szentmiklósy, Assistant Professor of Constitutional Law at the ELTE Eötvös Loránd University, Budapest.

Sofia Ranchordás, Professor of European and Comparative Public Law at the Faculty of Law of the University of Groningen.

Yaniv Roznai, Senior Lecturer at the Harry Radzyner Law School, Interdisciplinary Center, Herzliya, Israel.

Guy Seidman, Professor at the Harry Radzyner Law School, Interdisciplinary Center Herzliya, Israel.

Gerasimos Spanakis, Assistant Professor of Computer Science at the Department of Data Science and Knowledge Engineering, Maastricht University.

Mattis van 't Schip, Junior Researcher at the Department of Law and ICT at the University of Groningen, the Netherlands.

Luc Verhey, Member of the Advisory Division of the Dutch Council of State and Professor of Constitutional Law, University of Leiden, the Netherlands.

Wim Voermans, Full Professor of Constitutional Law, University of Leiden, Faculty of Law, the Netherlands.

Helen Xanthaki, Professor of Law, University College of London, Faculty of Law.

Monika Zalnieriute, Research Fellow, Lead of 'Technologies and Rule of Law' Stream, Allens Hub for Technology, Law and Innovation, Faculty of Law, University of New South Wales, Sydney.

Introduction

SOFIA RANCHORDÁS AND YANIV ROZNAI

I. How Time Shapes Law and Regulation

'Time is always necessary in the law', Rebecca R French wrote.[1] Time is indeed an important element in law, as law changes over time and raises many challenges for the conception and status of time in law, such as inter-temporal conflicts and the retroactive application of law,[2] the entrenchment and eternity of laws[3] and so forth. That said, law is often taught, researched and applied as a *linear reality* that is not subject to significant alterations. Several constitutions, for example, include various procedural, temporal and substantive limitations to constitutional change, and yet empirical data suggests that flexible constitutions (up to a certain point) actually tend to endure longer than extremely rigid constitutions.[4] The vast majority of lawyers, particularly in civil law countries, rarely confronts temporary or experimental legislation and regards laws as stable and enduring legal instruments. Empirical evidence shows, nonetheless, that although statutes tend indeed to last decades, legislators have been devoting more attention to temporary and experimental dispositions in the last decade.[5] Temporality is

[1] Rebecca R French, 'Time in the Law' (2001) 72 *University of Colorado Law Review* 663.

[2] See, eg FC Savigny, *A Treatise on the Conflict of Laws and the Limits of their Operation in Respect of Place and Time* (translated by W Guthrie 1880) (T & T Clark, Law Publishers; Stevens & Sons, 1869); Paul Roubier, *Le Droit Transitoire (Conflits des lois dans le Temps)*, 2nd edn (Dalloz, 2008); JR Trahan, 'Time for a Change: A Call to Reform Louisiana's Intertemporal Conflicts Law (Law of Retroactivity of Laws)' (1998–1999) 59 *Louisiana Law Review* 661.

[3] See, eg Richard Albert and Bertil Emrah Oder (eds), *An Unamendable Constitution? Unamendability in Constitutional Democracies* (Springer, 2018); Yaniv Roznai, *Unconstitutional Constitutional Amendments – The Limits of Amendment Powers* (Oxford University Press, 2017); Christoph Bezemek, 'Constitutional Core(s): Amendments, Entrenchments, Eternities and beyond Prolegomena to a Theory of Normative Volatility' (2011) 11 *Journal Jurisprudence* 517; Melissa Schwartzberg, *Democracy and Legal Change* (Cambridge University Press, 2007).

[4] Zachary Elkins, Tom Ginsburg and James Melton, *The Endurance of National Constitutions* (Cambridge University Press, 2009). On the flexibility and rigidity of constitutions, see Richard Albert, *Constitutional Amendments: Making, Breaking, and Changing Constitutions* (Oxford University Press, 2019) 95–174.

[5] Ittai Bar-Siman-Tov, 'Temporary Legislation, Better Regulaton, and Experimentalist Governance: An Experimentalist Study' (2018) 12 (2) *Regulation & Governance* 192. See also Sofia Ranchordás, *Constitutional Sunset Clauses and Experimental Legislation* (Edward Elgar, 2014); Antonios Kouroutakis, *The Constitutional Value of Sunset Clauses: An Historical and Normative Analysis* (Routledge, 2016);

becoming a feature not only of regulations and ordinary legislation, but also of constitutions. Temporary constitutional rules can function as a mechanism for relaxing the 'dead hand' and counter-majoritarian difficulties, allowing flexibility and reducing cognitive biases or error costs that are inherent in permanent constitutions.[6] Such mechanism of constitutional temporality may also be misused or abused by incumbents, for advancing narrow political goals and entrenching their power against political opponents, while avoiding both public accountability and judicial review of their actions.[7] Constitutional temporality thus requires more careful attention than one might prima facie suspect.

The idea of time as a given variable in law or a linear or reactive view of law tend to be oversimplifications of reality. First, time dimensions have a chronological impact on law and its instruments. Legislators legislate for the future in order to avoid injustice and protect legitimate expectations. Judges make decisions based on past events, and a wide discussion exists as for the temporal – retrospective or prospective – application of judicial rulings.[8] Administrative decision-makers seek to translate present needs into policies, balancing past and future actions, and trying to achieve policy coherence.[9] Likewise, judges attempt to bridge the gap between laws enacted in the past and changing reality by adopting a dynamic interpretation (in the midst of a battle between 'originalist' and 'living instrument' theories of interpretation).[10]

Second, time as a natural or physical phenomenon (hours, days, months, day/night) influences the *effectiveness* of laws and sometimes the contents of legal decisions and legislation. Laws and judicial decisions that fail to respond swiftly to timely debates (eg, authorisation of emergent technology) may stifle innovation and offer less attractive conditions for companies. Laws that become obsolete put at stake the trust of citizens in the ability of laws to solve the problems for which they were enacted. While legislative and policy persistence, or the tendency of laws and policies to change slowly, if at all, is often regarded as the outcome of

Madeleine Martinek, *Experimental Legislation in China between Efficiency and Legality: The Delegated Legislative Power of the Shenzhen Special Economic Zone* (Springer, 2018).

[6] Ozan O Varol, 'Temporary Constitutions' (2014) 102 *California Law Review* 409. See also Tom Ginsburg and Eric Alston, 'Playing for Constitutional Time: Interim Constitutions and Transitional Provisions' in Frank Fagan and Saul Levmore (eds), *The Timing of Lawmaking* (Edward Elgar, 2017) 110.

[7] See, eg Nadav Dishon, 'Temporary Constitutional Amendments as a Means to Undermine the Democratic Order: Insights from the Israeli Experience' (2018) 51(3) *Israel Law Review* 389.

[8] On the temporal effect of judicial decisions see, eg Patricia Popelier, Sarah Verstraelen, Dirk Vanheule, Beatrix Vanlerberghe (eds), *The Effects of Judicial Decisions in Time* (Intersentia, 2013); Eva Steiner (ed), *Comparing the Prospective Effect of Judicial Rulings Across Jurisdictions* (Springer, 2015); Richard HS Tur, 'Time and Law' (2002) 22(3) *Oxford Journal of Legal Studies* 463; ML Friedland, 'Prospective and Retrospective Judicial Lawmaking' (1974) 24(2) *The University of Toronto Law Journal* 170.

[9] For a reflection on administrative rulemaking and its evolution, see Christopher DeMuth, 'Can the Administrative State Be Tamed?' (2016) 8 *Journal of Legal Analysis* 121.

[10] See Gert-Fredrik Malt, 'Dynamic Interpretation, Spatial and Temporal Aspects in Interpretation' in Jes Bjarup and Mogens Blegvad (eds), *Time, Law, and Society* (Franz Steiner Verlag Stuttgart, 1995) 85.

political inertia, this persistence becomes problematic when there are no rational reasons to maintain a law in place and when it impedes the enactment of more effective solutions.[11] This occurs, for example, when suboptimal policies are not terminated in a timely manner or when they become superfluous because the problem they aimed to address no longer exists.[12]

At the same time, legislators seek to improve the effectiveness of law by taking into account the impact of time and change on legal subjects: individuals that work night shifts may require additional legal protection because of the biological effect of sleep deficiency, crimes committed at night may be more severely punished as victims are left more exposed to dangers.

Third, time is 'visible' in our legal instruments and influences our legal design and doctrines for resolving legal conflicts. When deciding a judicial case, we may look to past precedents, and when interpreting legislation we may go back to the legislative debates (or the *travaux préparatoires* when interpreting a treaty) or the lawmakers' intentions. When drafting a constitution, constitution-makers look to the future, in an attempt and aspiration to design a better one for the polity, but they also, inevitably, look to the past and react to crucial histories and past events in an attempt to address – and never return to – the previous regime's failures, atrocities or abuses.[13] Thus, the past informs our current lawmaking and influences our future. Our legal design not only addresses the past, but also seeks to preserve it or part thereof, very often in order to allow the advancement of law, as paradoxically as it may sound. For instance, in evolutionary patterns of constitution-making, many civil and common law countries use 'saving' or 'preservation' clauses, shielding past laws from judicial review.[14] This preservation of the past is often a compromise that allows the enactment of a new constitution or a bill of rights. Moreover, sometimes law not only preserves or saves the past, but aims to shape how it would be remembered, for instance through 'memory laws' that enshrine state-approved interpretations of crucial historical events.[15] Here law governs – or attempts to govern – time.

[11] Iris Geva-May, 'When the Motto is "Till Death Do Us Part": The Conceptualization and the Craft of Termination in the Public Policy Cycle' (2001) 24 *International Journal of Public Administration* 263; Jinhua Zhao and Catherine L Kling, 'Policy Persistence in Environmental Regulation' (2003) 25(3) *Resource and Energy Economics* 255 (analysing the optimal emission standards under uncertain pollution damages and transaction costs associated with policy changes; and arguing that in this case policy persistence might be a more rational response to forward-looking policymakers due to the existence of future transaction costs).

[12] See Mark R Daniels, *Terminating Public Programs: An American Political Paradox* (Routledge, 1997) 31.

[13] Kim Lane Scheppele, 'A Constitution Between Past and Future' (2008) 49 *William & Mary Law Review* 1377.

[14] Rivka Weill, 'Bills of Rights with Strings Attached: Protecting Death Penalty, Slavery, Discriminatory Religious Practices and the Past from Judicial Review' in Goeffrey Sigalet, Grégoire Webber and Rosalind Dixon (eds), *Constitutional Dialogue: Rights, Democracy, Institutions* (Cambridge University Press, 2019) 308.

[15] Uladzislau Belavusau and Aleksandra Gliszczyńska-Grabias (eds), *Law and Memory: Towards Legal Governance of History* (Cambridge University Press, 2017).

Consequently, 'time is ontologically and epistemologically involved in the most basic questions of law', as Nousiainen puts it.[16] Nonetheless, it appears that only recently, the relationship between time and law is receiving greater interdisciplinary attention.[17]

II. Time and Law: From Reaction to Prediction

Time imperatives and social change have permeated law throughout the centuries: constitutional and legal interpretation, constitutional amendment procedures, legislative techniques and judicial review regularly seek to accommodate the impact of time on law, shape it or avoid it.[18] For example, while eternity clauses and other entrenchment instruments aim to isolate constitutions and quasi-constitutional statutes from future changes, sustainability dispositions in environmental treaties, aspirational constitutions and sunrise clauses are forward-looking instruments that seek to spark change.[19]

Although law students still learn that legal rules are a response to social change that offers primarily *ex post* regulations,[20] technology is reshaping the relationship between law and time.[21] As Neuwirth puts it,

> technology … has a profound impact on law and vice versa. This, in particular, as both influence each other in the boost they provide to change. Technology and law also mutually influence each other in terms of changes in perception, as the cinematograph and subsequent inventions like the computer and digital technologies prove …. Law and

[16] Kevät Nousiainen, 'Time of Law – Time of Experience' in Bjarup and Blegvad (n 10) 23.

[17] See Emily Grabham, *Brewing Legal Times: Things, Form, and the Enactment of Law* (University of Toronto Press, 2016) (describing how 'things' change our understanding of law and time and how time arises in and through the interaction between humans and 'things' (objects or other non-human actors)); Siân M Beynon-Jones and Emily Grabham (eds), *Law and Time* (Routledge, 2019) (bringing together socio-legal scholars, anthropologists, sociologists, geographers and historians to explore the impact of law's relationship with time; how concepts of time influence law and regulation and how law influences our idea of time; Fagan and Levmore (n 6) (explore the diverse interactions between time and law, and how the design and architecture of law can be understood through time); Luigi Corrias and Lyana Francot (eds), *Temporal Boundaries of Law and Politics: Time Out of Joint* (Routledge, 2018) (examines the tensions between temporal possibilities and limitations in the intersections between law, time and politics).

[18] For a thorough reflection on the role of time in law, see Francois Ost, *Le temps du droit* (Odile Jacob, 1999).

[19] On constraining constitutional change and entrenchment, see, eg, NW Barber, 'Why Entrench?' (2016) 14(2) *International Journal of Constitutional Law* 325; David Landau and Rosalind Dixon, 'Constraining Constitutional Change' (2015) 50 *Wake Forest Law Review* 859; Richard Albert, Constitutional Handcuffs' (2010) 42 *Arizona State Law Journal* 663; Douglas A Schmeiser, 'The Entrenchment of a Bill of Rights' (1981) 19 *Alberta Law Review* 381; Jerome B Elkind, 'A New Look at Entrenchment' (1987) 50 *Modern Law Review* 158. On sunrise clauses, see: Sofia Ranchordás, 'Constitutional Sunrise', in Richard Albert, Xenophon Contiades, and Alkmini Fotiadou (eds), *The Foundations and Traditions of Constitutional Amendment* (Hart, 2017) 177.

[20] Yehezkel Dror, 'Law and Social Change' (1958–1959) 33 *Tulane Law Review* 787.

[21] See, eg Jerry Jie Hua, *Toward A More Balanced Approach: Rethinking and Readjusting Copyright Systems in the Digital Network Era* (Springer, 2014); Emilie Cloatre and Martyn Pickersgill (eds), *Knowledge, Technology and Law* (Routledge, 2014).

technology may converge and merge to combine its task, namely when technological standards become or enforce regulatory norms.[22]

Until recently, legal literature on law and technology primarily criticised the 'pacing gap' between law and technology: law lagged behind technological developments, imposed regulatory burdens that were disadvantageous to innovators, and impeded the free circulation of information and knowledge in the digital society.[23]

Nowadays law, with the help of data science and digital technology, is often expected to fulfil an anticipatory function, predict rather than punish. Automated systems currently allow law enforcers to predict crime before it happens, rank citizens according to the probability that they will commit a crime or social welfare fraud, and better allocate police and emergency services to the areas most affected by crime or natural disasters.[24] Technology helps law enforcers and public authorities detect patterns and future misconduct based on past data and statistics, creating a new dimension in the complex relationship between time and law. While predictive systems do not immediately alter the legal sphere of individuals, this novel forward-looking aspect of law-making and law enforcement is already producing stigmatising effects on underrepresented communities who feel profiled and watched. Predictive policing and other anticipatory legal mechanisms also often result in profiling and the public deterioration of the image of many minorities due to the frequent disclosure of information on the groups of citizens who are the most likely to commit fraud or crimes. This focus on preventing future misconducts is also visible in other areas of law where new applications of artificial intelligence ('AI') are giving rise to novel ethical questions and a shift towards an anticipatory approach: shouldn't legislators provide guidelines for the ethical development of AI? What should be the future limits for its application in decision-making? Can law anticipate change and offer future-proof rules for what is yet to come in an algorithmic society?[25]

This book seeks to address the different temporal dimensions experienced by lawyers in the roles of legislators, judges and regulators, as well as the impact of technology and the creation of a new time dimension. This book is about how law, in different fields and through varied instruments resists, accommodates or promotes change. It engages with different strands of literature from several

[22] Rostam J Neuwirth, *Law in the Time of Oxymora – A Synaesthesia of Language, Logic and Law* (Routledge, 2018) 198–99.

[23] Gary Marchant, Braden R Allenby and Joseph Herkert (eds), *The Growing Gap between Emerging Technologies and Legal Ethical Oversight* (Springer, 2011).

[24] See, eg, Virginia Eubanks, *Automating Inequality: How High-Tech Tools Profile, Police, and Punish the Poor* (St Martin's Press, 2018); Andrew Guthrie Ferguson, 'Predictive Policing and Reasonable Suspicion' (2012) 62 *Emory Law Journal* 259; Samir Chopra and Laurence F White, *A Legal Theory for Autonomous Artificial Agents* (University of Michigan Press, 2011).

[25] See, eg, Donghan Jin (ed), *Reconstructing Our Orders: Artificial Intelligence and Human Society* (Springer, 2018); Woodrow Barfield and Ugo Pagallo (eds), *Research Handbook on the Law of Artificial Intelligence* (Edward Elgar Publishing, 2018).

fields of law (eg, constitutional law, business law, legal theory, legal history, legisprudence, private law, law and technology) in order to provide a thorough understanding of the different tensions experienced by public authorities when trying to accommodate or predict change in their laws and decisions.

Time considerations permeate different fields of law on countless dimensions; law evolves with time and time influences how law evolves; law aims to adapt to changing circumstances but also aspires to shape the way society changes; law – whether made by legislators, judges, extra-legal actors, public officials or others – is simultaneously backward, present and forward looking. Questions concerning the temporal enactment, application and validity of law and judicial decisions affect the very fundamental principles of a legal system: certainty, stability, resilience, rule of law, separation of powers and sovereignty. The very question of ultimate foundations of legal validity (*Grundnorm*) may yield a conception of an 'essentially temporal' category, in that 'legal validity must be understood as an essentially temporal category, in the sense that legally valid norms must always necessarily have a past … and also open the horizon for a possible future'.[26] Consider, as another example, Eule's famous ideas of temporal limits on the legislative mandate, according to which arguably, the mandate of lawmaker is temporally limited so that it cannot legislate retroactively, invading into the mandates of past legislatures, just as it cannot bind future legislators.[27] Alternatively, consider how the very notion of constituent power is influenced when emergency measures suspending constitutional norms, which are supposedly temporary in nature, become permanent.[28] Time, law and politics are intertwined.[29]

Social, political, and technological change is also an element inherent to this overarching analysis of time: when societies and their customs evolve, laws become ineffective or even obsolete and citizens contest existing legal frameworks. Nevertheless, thus far, notwithstanding recent important contributions,[30] the impact of time considerations on law and vice versa is still under-studied. Instead, these phenomena have been examined in a segmented way (retroactive legislation, temporary laws, constitutional change, etc). This book fills this gap by providing a more complete analysis of how time considerations as well as social, political and technological change affect the legislative process, the interpretation of laws, the definition of the powers of the executive, and the ability of legal orders to promote innovation.

[26] Carlo Invernizzi Accetti, 'The Temporality of Normativity: Hans Kelsen's Overcoming of the Problem of the Foundation for Legal Validity' (2016) 42(1) *Philosophy and Social Criticism* 25, 34.

[27] Julian N Eule, 'Temporal Limits on the Legislative Mandate: Entrenchment and Retroactivity' (1987) 12(2/3) *American Bar Foundation Research Journal* 379.

[28] See Alan Greene, *Permanent States of Emergency and the Rule of Law – Constitutions in an Age of Crisis* (Hart Publishing, 2018).

[29] Corrias and Francot (n 17).

[30] See n 17.

III. A Roadmap to the Book

This book analyses the relationship between time, law and change from four perspectives: that of *scholars* who have studied the theoretical impact of time on legal interpretation, constitutional change and lawmaking as such; *judges* who try to balance the need to accommodate change with the principle of legal certainty and legitimate expectations; *legislators* who are often required to legislate fast or adopt temporary legislation to respond to social demands or under the pretext of improving the quality of legislation; and from the perspective of *technology* that is giving rise to new challenges in the legal context. This book is hence divided into four parts.

Part I presents a number of theoretical and historical perspectives on the relationship between time and law, and how time influences legal interpretation, constitutional change, and how time as a physical or natural reality (ie, day or night) has reshaped law as such. In the first chapter, Lior Barshack offers a reflection on the work of several authors on the object of legal interpretation. The chapter starts with a discussion of Dworkin's position that judges are bound exclusively by the past. The object of interpretation is thus the entire history of the legal system, seen as a single text written by a single fictional enunciator. Judges would have to look for a moral message inscribed in that past, a message that is so evasive that it is bound to be misunderstood by many, possibly most, judges. Steven Calabresi shows a different perception in historical constitutional discussions regarding time in US constitutional law. Calabresi describes in chapter two the experience of the US Constitution over time, taking the readers back to the letters exchanged between Thomas Jefferson and James Madison and the position that constitutions should always be temporary.

The question of constitutional revision is also addressed in chapter three, in which Richard Albert analyses an overlooked dimension of constitutional amendment: time. It is well-known that formal amendment rules are designed to fragment power, whether among political parties or branches of government, or along ethnic, cultural, linguistic, geographic, subnational or other divisions. However, rarely have scholars delved into the temporal dimension of constitutional amendments. Democratic constitutions sometimes entrench rules requiring political actors to adhere to certain specifications as to the timing of various steps in the formal amendment process. This temporal fragmentation of power entails unique implications for how we understand the formation of constitutional consensus and how we evaluate contemporaneity in amendment ratification. In this chapter, Albert adopts a comparative perspective to explore and evaluate the use of time in fragmenting the amending power. In chapter four, Guy Seidman discusses the impact of physical time (the night) on legislation. It describes multiple fields of law (eg, labour law), professions and measures (eg, curfew) that were regulated in specific ways so as to take into account the perils of a specific time of day. In this chapter, Seidman discusses how 'night laws' have shaped the narrative around certain professions and inquires whether it still makes sense nowadays to take this time dimension into account.

Part II shows the different ways in which courts approach the impact of time on law. Patricia Popelier's chapter (chapter five) discusses the principle of legal certainty and legitimate expectations in the case law of the Court of Justice of the European Union. Through the analysis of case law, Popelier discusses the two-time dimensions of laws: the interval of time between laws coming alive and subsequently being abolished; and the interval of time during which the individuals affected by laws plan and act. Popelier shows how tensions arise when one dimension interferes with the other. At the intersection of both dimensions are 'legitimate expectations'. This chapter inquires how courts employ this legal concept to deal with tensions of time.

The issue of certainty is also raised in chapter six, where Zoltán Pozsár-Szentmiklósy and Yaniv Roznai examine the timing of judicial review of constitutional amendments, an extreme exercise of judicial power that raises perhaps the ultimate counter-majoritarian difficulty. The legitimacy of the judicial invalidation of constitutional norms, which are positioned at the top of a country's legal pyramid, is even more questionable when the judicial decision arrives many years after the constitutional norm was constituted and relied upon. Against the backdrop of the Moldovan Constitutional Court's decision on the modality of electing the President, the authors ask whether the time element should be taken into consideration by courts when reviewing constitutional amendments. They answer this question affirmatively, and after analysing the decision on the modality of presidential elections, they propose a 'time sensitivity test' be taken into the consideration of the courts when reviewing constitutional amendments.

Chapter seven by Yifat Aran and Moran Ofir analyses the effects of specialised economic courts over time. This chapter develops a research design that reveals empirical evidence of whether or not – and in what ways – a specialised court offers litigants increased efficiency and better quality of judicial decision-making in terms of accuracy and uniformity. Aran and Ofir also examine how specialised courts stimulate private enforcement of corporate law. This contribution shows that judicial specialisation fosters the fast development of a coherent and consistent body of law. This effect is driven by the fast adjudication procedures and the judicial reliance on past decisions and joint effort to develop the case law. This chapter also shows that specialisation can contribute to an efficient management of particularly time-consuming dockets.

Part III explains how legislators address time and change imperative with legislation. Enrico Albanesi's chapter eight describes the use of so-called 'provisional emergency decrees' in Italy. This type of emergency legislation was only originally meant to be used in exceptional circumstances. These legislative instruments come into force immediately and are thus extremely attractive to the executive. In the last decades, the Italian legislature has frequently used these emergency laws even for situations that were not necessarily characterised by any type of emergency. Helen Xanthaki's contribution (chapter nine) delves into the nature and drafting of sunset clauses, a type of temporary disposition that expires on a set date. Sunset clauses have been primarily analysed from a constitutional law perspective by scholars who seek to define the constitutional framework of these dispositions. By studying sunset clauses from the perspective of legislative studies, this chapter fills an important gap in the literature. This chapter offers a definition and discussion of what sunset clauses are from a legislative drafting perspective. Following this definition of sunset clauses, Ittai Bar-Siman-Tov and Gaya Harari-Heit present in chapter ten an interdisciplinary analysis of the political rationale behind temporary legislation. This chapter describes from a normative and empirical perspective two primary functions of temporary legislation: legisprudential functions such as the use of temporary legislation as a regulatory tool to deal with situations in which the nature of the law, the policy or circumstances require the temporality of the norm; and political uses, for example, using temporary legislation as a political tool for gathering political support for a legislative change.

In chapter 11, Wim Voermans elaborates on a different dimension of time: the timing of the legislative process. In his chapter, Voermans explains the challenges of accelerating the legislative process to meet social demands. Voermans discusses the difficulty in striking a balance in practice between the demands for expedient legislation and administrative decisions and the requirements of transparency and openness of government. Although efficiency and transparency are not mutually exclusive goals, their practical realisation is often underestimated.

This part is concluded with chapter 12, in which Yaniv Roznai seeks to provocatively challenge the distinction between law's treatment of retroactive legislation in the criminal sphere vis-à-vis the civil law sphere. In the former, most jurisdictions are hostile towards the retroactive application of criminal law, while in contrast, in the civil law era, constitutions and courts are usually more permissible. In his chapter, Roznai seeks to revisit this differential treatment and argues that the basic rationales underpinning the objections towards retroactivity apply in civil legislation, at least to some extent and often even more, than in criminal legislation.

Part IV focuses on the relationship between technology, law and time. In chapter 13, Antonios Kouroutakis analyses the definition of disruptive innovation, a well-explored concept in business literature but still rather unknown in legal scholarship, which requires a different legislative approach. In this context, several authors have argued that temporary legislation would be more suitable to promote disruptive innovation than permanent legislation. In this contribution, Kouroutakis applies this analysis to the regulation of on-demand transportation networks such as Uber. This chapter aims to examine the utility of sunset clauses in the regulation of on-demand transportation and it uses the progressive abolition of roaming charges as a pilot case in order to show how such clauses may regulate this topic by creating a predictable business environment for all parties involved.

Lyria Bennett Moses and Monika Zalnieriute (chapter 14) delve into the time dimension of law and technology. By drawing on theories of law and technology, this chapter discusses how attempts to regulate a socio-techno-legal landscape (including design-based regulation) can fall out of alignment. This chapter illustrates this narrative with a case study on automated driving systems and data protection rules in Australia and the European Union. By exploring how law and technology interact with each other across time, this chapter considers how the adaptability of socio-techno-legal systems can be improved.

In chapter 15, Catalina Goanta, Gijs van Dijck and Gerasimo Spanakis offer a meta-analysis of legal scholarship on AI. They show that while modern scholarship portrays ethical questions on AI as a novel topic, many of the concerns raised by recent scholarship have been present for decades. This chapter offers a thorough analysis of scholarship on AI and places legal debate on law and technology in a broad time frame in an attempt to identify patterns and questions which are at the heart of legal research AI. This chapter delves into the history of AI and shows how 'old became new' in the regulation of AI.

In chapter 16, Sofia Ranchordás and Mattis van 't Schip explore the possibility of 'legislating future-proof'. In the last few years, several European governments have discussed the concept of 'future-proof legislation' as a way to ensure that legislation could be resilient to future changes, particularly considering the fast pace of technological change. The concept of 'future-proof' refers to an interdisciplinary approach often used in design and architecture that aims to test the resilience of a product against the ravages of time. This chapter discusses the challenges of designing legislation according to this model and suggests the adoption of experimental legislation and the inclusion of future-thinking considerations in regulatory impact assessments as steps towards future-proofing law.

This book concludes with the reflections of Luc Verheij who presents the view of a practitioner who is often faced with the social and political pressure to quickly update legislation, regardless of the lack of information regarding the effectiveness of these legislative measures. In his concluding remarks, Verheij highlights the time-consuming character of the legislative process and the particular challenge of legislating in the context of emergencies. Under exceptional circumstances, legislators may have to give priority to the time dimension (urgency) of the legislative process and set aside constitutional guarantees that typically delay the quick adoption or amendment of legislation. Nevertheless, as Enrico Albanesi's contribution to this book explained, once a law is introduced as a temporary or emergency measure, it is unlikely to be terminated. While the temporary character of these legislative measures is typically used to help convince the opponents of legislative change that the proposed reform will be limited and transitory, the reality is nevertheless that once an important legislative change is introduced, it tends to persist. This book shows that despite the existence of different time dimensions, our legal order continues to have a natural bias towards the path of dependence and stability.

PART I

Historical, Comparative and Theoretical Perspectives on Law and Time

1

Interpretation and the Legal Fabrication of Time

LIOR BARSHACK

Dworkin's account of interpretation is cast in mythological terms, with Judge Hercules as its protagonist. For Dworkin, judges are bound exclusively by the past, they speak in the name of the dead. The object of interpretation is the entire history of the legal system, seen as a single text written by a single fictional enunciator.[1] The dead, then, are not so much the individual drafters of laws and constitutions, but rather a big 'Other' who looms large behind the historical accumulation of legal texts. Judges have to decipher the moral message treasured in the mass of old legal documents, a message that is so evasive that it is bound to be misunderstood by many, possibly most, judges. In addition to being controversial and elusive, the message of the dead can be interpreted differently at any moment because the one right answer is right only for the moment in which it is given. The violence of the law can be legitimate if it is based on principles that can be attributed to the legal system from its foundation. The mythical dimension of Dworkin's account is enhanced by his theoretical preference to speak about interpretative practices from the perspective of their own internal self-understanding, the influence of Gadamer's account of the authority of tradition in the self-understanding of interpreters, and Dworkin's own somewhat metaphysical concept of 'community of principle'.[2] In fact, Dworkin's theory anticipated the contemporary interest in political theology and in the theological foundations of the constitution. Dworkin articulates the theological premises of interpretation, which include commitment to a fictional, retrospectively projected, unity of the legal system from its foundational moment onwards, and denial of society's ability to ascertain the content of the system's everlasting principles.

I would like to support Dworkin's claims about law's continuity/perpetuity, contestability and legitimacy on the basis of an account of the relations between law and time that I have outlined in earlier work.[3] The following argument

[1] Ronald Dworkin, *Law's Empire* (Harvard University Press, 1986) 225, 242.
[2] *Ibid* 214–28.
[3] Lior Barshack, 'Time and the Constitution' (2009) 7(4) *International Journal of Constitutional Law* 553.

remains in an outline form. Its purpose is to draw the implications of a certain view of the legal construction of time on the theory of interpretation. I will argue that the law fabricates historical time by propagating representations of corporate perpetuity, of historical changes and of the contestability of the system's everlasting principles. Alongside representations of legal permanence and change, the law represents controversy over what is and what is not changeable. Dworkin's accounts of interpretation and essentially contested concepts foreground and knit together these basic elements of society's representation of time. The same threefold representation – of perpetuity through change, and of the unknowability of the message transmitted from generation to generation – can also be found in other practices, such as art, that play a role in the fabrication of historical time. As part of their representation of time, interpretive practices such as law and art separate the figures of historically situated authors (artists, lawmakers, and so on) from those of immortal collective bodies that author eternal principles. At the same time, they entangle these two authorial instances – the historically situated and the immortal/corporate – in a way that complicates in each practice the interpretive task of sorting out the changeable and the unchangeable. My account of the legal construction of time will draw on some concepts and ideas of Hannah Arendt and on insights from art history and social anthropology. I would like to illustrate in particular the enduring relevance of anthropological theories of descent to theories of social time.

I. The Collapse of Social Space and Social Time in Constitutional Moments

My starting point is the idea of constitutional moments as moments in which the flow of historical time is interrupted. Constitutional moments bring us back to the moment of foundation. They approximate to the state of nature, because they are moments in which the rule of law and the separation of powers are relaxed. From the perspective of modern social thought, such moments are no longer perceived as they were by the philosophers of the social contract. Even Hobbes's account of the state of nature seems benign in comparison to modern accounts. For Hobbes, the self-preserving rationality that governs social life is perfectly in place in the state of nature. In the aftermath of the atrocities of the twentieth century, states of lawlessness, which are not always identifiable, are conceived in more sinister and apocalyptic terms. In Hannah Arendt's accounts of totalitarianism, such moments of bare life seem as breakdowns of law, space, time, language, work – not only action – and plurality. In other words, the very existence of the human world is at stake when the rule of law breaks down. Sociological and psychoanalytic theories of atrocities echo of course numerous myths that invoke the recurrence of apocalyptic destruction at the beginning, end, and occasional breakdowns of historical time. Arendt, however, does not endorse a view of constitutional moments as cyclical breakdowns of historical time. For Arendt, constitutional beginnings and transformations need not involve collapse of the human world and interruptions of time. In *On Revolution*, Arendt seeks to demythologise constitutional beginnings and place them within historical time. While her notion of a 'gap between past and future' suggests an interruption of time that is akin to mythical time,[4] Arendt seems to resist any sharp and easy contrast between continuity and beginning, at least in the life of institutions.

In a similar vein, oppositions that have been drawn between ordinary, continuous time and the sanctity and intensity of mythical time can be either moderated or viewed as inevitable paradoxes. An important example is Victor Turner's distinction between social structure and *communitas*, which readily suggests itself as a theoretical framework for elucidating the social dynamics of constitutional moments.[5] If constitutional moments are more or less intense re-enactments of the moment of foundation, then theories of ritual are likely to shed light on their nature. For Turner, *communitas* and social structure are two modalities of social life between which every society alternates. *Communitas* is characterised by dissolution of social roles and statuses; reversal of hierarchies; the collapse of fundamental social dichotomies between private and public, the local and the universal, the human and the animal, the sacred and the profane; passage from traditional to charismatic authority, and from pragmatic to symbolic use of language; violation of the most fundamental social prohibitions; and collapse of social space and social time. These are just some of the main aspects of *communitas*. There are multiple types of *communitas*, including carnivals, happenings, wars, natural disasters, economic crises, official rituals and a variety of new 'media events' and constitutional moments. Bakhtin's work on carnival and ethnographic studies of different kinds of *liminal* events fit into and enrich the general account proposed by Turner.[6]

In earlier work, I have characterised *communitas* and social structure respectively in terms of the place that the collective body occupies in them. In *communitas*, the collective body of the group is enacted in a way that brings about the dissolution of interpersonal boundaries. The individual disappears in the collective body. I refer to the collective body as it is incarnated in constitutional moments as the *communal body*. When society returns from *communitas* to social structure, the collective body is projected outside of the social, and assumes the form of an absent, transcendent body to which I refer as the *corporate body*.

[4] The theory of constitutional beginnings can benefit from studies of the structure of beginnings in literature, cinema, theatre and music. See, eg, the discussion of beginnings 'before the beginning' in Steven Vande Moortele, *The Romantic Overture and Musical Form from Rossini to Wagner* (Cambridge University Press, 2017) 108.

[5] Victor Turner, *The Ritual Process: Structure and Anti-Structure* (Cornell University Press, 1969).

[6] Bakhtin's work of course antedated Turner's. See, Mikhail Bakhtin, *Rabelais and his World* (Indiana University Press, 1984); Elihu Katz and Daniel Dayan, *Media Events: The Live Broadcasting of History* (Harvard University Press, 1994).

The collective body assembles not only the individuals that make up society in a particular moment, but all generations. It is a multigenerational body. In the corporate body, generations succeed each other in time and temporal distances between them are observed. In moments of *communitas* all the generations assemble as the horizons of past and future collapse into an everlasting present.[7] *Communitas* is experienced as an arrest of time because it occasions the simultaneous presence of all generations. Lawless regimes aspire to enact the communal body because its presence liberates the living from the authority of the dead. By contrast, Dworkin's concept of community of principle subordinates the living to the authority of the dead and arguably evokes the image of a corporate subject/body. The arrest of time in *communitas* can also be couched in terms of repetition. The temporal experience in *communitas* is one of repetition because it involves the dissolution of individual agency, and because it consists in the re-enactment of founding, mythical events. As a reenactment of something that assumes the character of an origin, repetition shapes both dramatic historical moments and the everyday satisfaction of the necessities of life. The separation between everyday life and the realm of great historical events makes room for linear historical time between the repetitive cycles of private life and those of political beginnings and transformations.

In his celebrated book, *The Nuer*, Evans-Pritchard characterised time in terms of 'constant structural relationship between two points, the first and last persons in a line of agnatic descent'. Time is equated by Evans-Pritchard with intergenerational distance. Evans-Pritchard referred to time-reckoning of that sort as 'structural time', as opposed to deep historical time, because it always serves to explain and legitimise actual structural relations between different contemporary groups in terms of their genealogical origins. As Evans-Pritchard notes, '... time-reckoning is a conceptualization of the social structure, and the points of reference are a projection into the past of actual relations between groups of persons'. Before the beginning of the genealogical line, lies the realm of mythical time. As Evans-Pritchard writes, 'Beyond tradition lies the horizon of pure myth ...'[8]

Evans-Pritchard's notion of structural time as an ordering of past generations along a temporal axis by means of projected temporal distances applies, in its general outline, also to modern notions and institutions of historical time. Despite the fact that historical time is longer than structural time and in principle unlimited in length, linear historical time also places generations – past,

[7] Vast historical and anthropological literature documents the multiple forms of presence of the dead among the living. To cite just one recent study that considers various forms of such presence, see Nancy Caciola, *Afterlives: The Return of the Dead in the Middle Ages* (Ithaca, Cornell, 2016).

[8] E Evans-Pritchard, *The Nuer* (Oxford University Press, 1940) 108. On Evans-Pritchard's concept of 'structural' time, see Mary Douglas, *Edward Evans-Pritchard* (Routledge, 1980) 88; Max Gluckman, 'The Utility of the Equilibrium Model in the Study of Social Change' (1968) 70(2) *American Anthropologist* 219–37.

present and future – in imagined relations to each other, in ways that structure distances and relations among the living. Like structural time, historical time is bracketed and interrupted by episodes of mythical time in which all generations become simultaneously present. When the generations assemble in constitutional moments to take momentous decisions, and in other instances of mythical time, historical time stands still.

In order to return from *communitas* to social structure, spatial/horizontal boundaries among the living, and temporal/vertical distances between the generations, have to be re-established. Social statuses, roles, categories and hierarchies are reaffirmed. Fundamental dichotomies between spheres of life and basic prohibitions are enforced. In Arendt's terms, the passage from *communitas* to social structure amounts to a construction of a world. Arendt's account in *The Human Condition* of the work (as opposed to labour and action) of world-making describes how a habitable world is built through the construction of fences and walls; the division between different spheres of life; the creation of objects and masks that separate and mediate between individuals, and enable everyday interaction between individuals engaging in work and action; the confrontation between different perspectives on reality that endows experience with a sense of objectivity. Arendt refers also to the construction of artefacts – monuments, artworks, laws – that survive their usage and the death of their creators, and thereby create links between the generations, as one of the conditions for the existence of a habitable world. Narratives that pass from generation to generation ensure the 'earthly immortality' of glorious deeds, and thereby endow human affairs with a modest sort of realness, and nourish the human commitment to the taxing work of world-making.[9] Arendt's account of world-making in *The Human Condition* can be supplemented by her observations on totalitarianism. In *The Origins of Totalitarianism*, she cites some of the factors that caused the destruction of human worlds under totalitarianism: the destruction of social relations, the denial of a 'place in the world' and of a 'right to have rights' to numerous – if not, arguably, all – citizens. These and other totalitarian forms of destruction are characteristic of *communitas* in general, and have to be repaired with the return to social structure.

The renewed recognition of social categories in social structure disperses the collective body. In this process, interpersonal and intergenerational boundaries are reaffirmed. The former produce social space, the latter – social time. The concept of the collective body explains how the construction of social time is linked to the construction of social space as the building-blocks of the world shared by humans. The interdependence of social time and social space has been suggested

[9] The creation of a historical world exacts efforts and sacrifices. Arendt discussed different forms of rejection of worldliness, such as the choice of solitary life of intellectual or spiritual pursuit, or the attempts of certain religious and political groups to repudiate the burdens of historicity and plurality, and realise here and now a communion of individuals and generations.

in different theoretical and historical studies. Panofsky, for example, remarked that the thinkers and artists of the Renaissance who developed the principles of spatial perspective, like Alberti, also conceptualised the temporal distance between historical periods. These figures of the Renaissance, often celebrated as champions of modern worldliness, changed medieval culture by treating the distance between historical epochs as objective and unmalleable. Thus, spatial separation between different perspectives and separation in time go hand in hand.[10]

II. The Representation of Corporate Perpetuity as a Premise of Historical Time

In order to disentangle the communion of the generations, society stages representations of the temporal distance between the generations in every possible media, including mythical, official and popular narratives, legal records, a diversity of monuments and sites of commemoration that position the living in relation to past generations. Certain generations enjoy a privileged place in the collective memory and the mythical representation of the past as generations that founded the group or took part in important turning points. Representations of temporal distances between generations always postulate intergenerational continuity. They amount to representations of a collective body that encompasses multiple generations. In contrast to *communitas*, in the course of ordinary historical time the generations are united not through simultaneous enactment, but by succeeding each other as organs of an immortal body. As indicated above, I refer to this form of the collective body as the corporate body, following the work of Henry Maine, who greatly influenced Kantorowicz (through Maitland), and the central figures in the British school of social anthropology, such as Radcliffe-Brown, Evans-Pritchard and Fortes.[11]

[10] Erwin Panofsky, *Renaissance and Renascences in Western Art* (Almqvist, 1960) 108. Anthony Grafton's essay 'Panofsky, Alberti, and the Ancient World' sheds light on the conditions for the appearance of historical time by considering Alberti's attitude to the classics, an attitude that combined an emphasis on temporal distance with an affirmation of both cultural continuity and change. See, 'Panofsky, Alberti, and the Ancient World' in A Grafton, *Bring out your Dead: The Past as Revelation* (Harvard University Press, 2001) 19. Evans-Pritchard argued that the correlation between temporal and spatial distances is quantitative. Greater temporal depth corresponds to greater spatial spread, because the more ancient the ancestors the more numerous and widespread their descendants. See, Evans-Pritchard (n 8). Such an analysis may seem entirely irrelevant to contemporary society unless we are willing to admit, following Godelier and other anthropologists, that descent remains all-important in contemporary Western society. See, Maurice Godelier, *The Metamorphoses of Kinship* (London, 2011) 127: '… in Europe the institution of marriage is crumbling while descent stands fast'.

[11] Fortes explicates the notion of the corporate body as follows: 'The notion of a descent group as constituting "one person" takes many forms. The essential idea is that the living plurality of persons constitutes a single body by reason of being the current representation and continuation of a single founder. Whether this is conceptualized and expressed in beliefs about being "the children of so and so," or "of one womb," or "of one blood," or "of one penis," or – more metaphysically – of one spiritual

The concept of the multigenerational corporate body consists of various fictions that concern time, such as the fictions of intergenerational continuity through temporal distance, and the fiction of corporate perpetuity. Humans engage in work and action on the basis of fictions of perpetuity – of human glory, of the state, the constitution, the arts – even though we are well aware of the fact that none of these is immortal. In an essay on ancestor worship, Fortes proposed that the importance attached to genealogical relations to ancestors has to do with anxieties in the face of the ephemerality of human existence.[12] Annette Weiner explained in these terms the preoccupation of all cultures with intergenerational transmission, and the universal obligation to keep certain things inalienable in order to transmit them to future generations. According to Weiner, partaking in something immortal is our way to authenticate our existence, to anchor ephemeral human existence in the cosmic order. 'Cosmological authentication' is achieved by 'material resources and social practices' that 'link individuals and groups with an authority that transcends present social and political action'.[13] A similar claim was made about the continuity of legal systems. In his essay, *The Ancestral Constitution*, Finley observed that the idea of an eternal constitution is rooted in the fear of death.[14]

According to a related but somewhat different explanation, the fiction of the immortality of the corporate body gives a sense of realness to human affairs. This explanation rests on the assumption that in the eyes of humans ephemeral things are not really real. The idea has been elaborated by Arendt in *The Human Condition*. For Arendt, the reality of the world inhabited by humans depends on it being shared not only by multiple individuals and groups, but also by multiple generations. Therefore, as Arendt repeatedly argues, the durability of things conditions the realness of the world.[15]

essence or totemic origin, or of common ritual allegiance to ancestors or other supernatural agencies, the implications are the same. The group is one by physical perpetuation and moral identity.' See, Meyer Fortes, *Kinship and the Social Order* (Aldine, 1969) 304. See also Meyer Fortes, *The Concept of the Person, in Religion, Morality and the Person: Essays on Tallensi Religion* (Cambridge University Press, 1987) 283: '… [A] lineage is a collective person because it is the perpetuation of its founding ancestor in each of his descendants'. For Kantorowicz's theory of the corporate body politic, see EH Kantorowicz, *The King's Two Bodies* (Princeton University Press, 1957).

[12] Meyer Fortes, 'An Introductory Commentary' in William H Newell (ed), *Ancestors* (Mouton, 1976).

[13] Annette B Weiner, *Inalienable Possessions: The Paradox of keeping-while-giving* (University of California Press, 1992) 4–7.

[14] Moses I Finley, *The Ancestral Constitution: An Inaugural Lecture* (Cambridge University Press, 1971). Such explanations are inconsistent with my assumptions. Society devises representations of perpetuity, in my view, in order to resist the seductions of presence, rather than overcome the fear of death. Only once presence is dispelled do the prospects of individual death become abhorrent. In a highly speculative vein, it can be argued that corporate perpetuity overcomes the love, not fear, of death. Individual life becomes valuable only when the communal body is transformed into a corporate body.

[15] On profane time as lack of realness, see Mircea Eliade, *Cosmos and History: The Myth of the Eternal Return* (Harper, 1959) 34.

I prefer a different explanation, according to which the fiction of perpetuity is necessary because it substitutes for the immediate access to the absolute in *communitas*. Perpetuity offers an indirect, postponed access to the absolute, mediated by and dependent upon the cooperation of other generations. Once direct participation in the collective body becomes impossible, we must content ourselves with the position of links in an eternal chain of generations.[16] This accounts for the capacity of the eternal to bestow not only a sense of realness but also legitimacy. The eternity of a tradition, a law, a myth confers ultimate authority, because it confirms their authenticity as manifestations of corporate identity. This traditional notion of legitimacy can be found in Dworkin's theory of law as integrity, where the law is seen as legitimate only if it is grounded in principles that can be attributed to a multigenerational 'community of principle' from its beginning.

The anthropological literature on descent is full of illuminating insights about corporate perpetuity and other temporal aspects of corporate bodies. In the British school of social anthropology, descent groups were conceptualised as corporate groups. I would like to cite two characteristic passages from the work of the anthropologist Meyer Fortes, who was greatly influenced by Maine, concerning the temporality of corporate bodies. In my view, such observations remain indispensable to theories of social time, especially those centred around juridical notions of corporateness and corporate perpetuity. As Fortes explains, the notion of intergenerational continuity has a forward-looking orientation:

> [C]orporate groups … are kept in existence by mobilizing the succession of generations regulated by the principle of filiation … The dead are … thought of as having a stake in the continuity, i.e. in the future persistence of the society to which they belonged in life. In ancestor worship this is accepted as just and natural; it is this that, paradoxical as it may sound, gives ancestor worship a future orientation, rather than, as might superficially be thought to be the case, a fixation on the past.[17]

Thus, contrary to appearances, descent does not place an emphasis on the past. While 'regimes of historicity' differ in the respective significance they attach to past, present and future, the theory of descent suggests a basic model in which intergenerational continuity stretches between the poles of past and future as two sources of orientation and value. If we agree with Maurice Godelier that descent remains a fundamental category in contemporary kinship, then this basic structure of temporality remains intact.

In the following passage, Fortes illustrates the mythical dimension of the idea of corporate perpetuity:

> … fission and accretion are processes inherent in lineage structure. However, it is a common experience to find an informant who refuses to admit that his lineage or even

[16] See Barshack (n 3).

[17] Fortes (n 12) 6. Like ancestor worship, it is often argued that interpretation is future-oriented, despite the fact that it seems focused on the past. See, eg, Bruce Ackerman, *We the People: Foundations* (Harvard University Press, 1993) 264; Dworkin (n 1) 228.

> his branch of a greater lineage did not at one time exist. Myth and legend, believed, naturally, to be true history, are quickly cited to prove the contrary. But investigation shows that the stretch of time, or rather of duration, with which perpetuity is equated varies according to the count of generations needed to conceptualize the internal structure of the lineage and link it on to an absolute, usually mythological origin for the whole social system in a first founder.[18]

These and similar observations drawn from theories of descent highlight recurrent aspects of collective representations of linear time and corporate perpetuity, such as the role of founding fathers and mothers as symbols of continuity.[19] The continuity of public institutions enjoys, however, a certain primacy in the representation of time. In order to loosen the hold of communal presence over individuals, there has to be an *official, public* representation of the continuity of the corporate personality of the group, of the constitution, and of society's system of institutions and statuses. As Pocock notes:

> History – in all but a few, rather esoteric, senses of the term – is public time … it is time experienced by the individual as public being, conscious of a framework of public institutions in and through which events, processes and changes happen to the society of which he perceives himself to be part. The public realm, unlike the social realm, must be thought of as institutionalized and formalized, since otherwise the distinction between public and private cannot be maintained; and the institutionalization of the public realm leads to the institutionalization of social experience and of modes of apprehending it, and consequently to the institutionalization and differentiation of apprehended time.[20]

The representation of legal continuity is not limited to the continuity of the corporate subject, of the constitution and of institutions.[21] It typically includes, for example, ideas of sacred legal text, an everlasting interpretive tradition, and an eternal body of law. Some of these elements feature, for example, in the ideological representation of the common law as an eternal and exhaustive ancient constitution that perpetuates itself in the form of custom and whose origins go back to times immemorial.[22]

These and other representations of legal continuity are inseparable from a wealth of non-legal representations of corporate perpetuity, which have been

[18] Meyer Fortes, 'The Structure of Unilineal Descent Groups' in J Goody (ed), *Kinship* (Penguin, 1971) 264. See also Meyer Fortes, 'Pietas in Ancestor Worship' (1961) 2 *The Journal of the Royal Anthropological Institute of Great Britain and Ireland* 91, 166–91.

[19] On ancestors as symbols of the continuity of the social structure and of the law, see Meyer Fortes, 'Some Reflections on Ancestor Worship in Africa' in M Fortes and G Dieterlen (eds), *African Systems of Thought* (Oxford University Press, 1972) 122–44.

[20] JGA Pocock, *Politics, Language and Time: Essays on Political Thought and History* (Chicago, 1989) 87.

[21] On the constitution as collecting the generations into a collective subject, see Jed Rubenfeld, *Freedom and Time* (Yale University Press, 2001). In contrast to the present argument, Rubenfeld views the constitution as connecting rather than disconnecting, and is not dealing with the social construction of time.

[22] JGA Pocock, *The Ancient Constitution and the Feudal Law: A Study of English Historical Thought in the Seventeenth Century* (Cambridge University Press, 1987).

studied by anthropologists and sociologists. They include a large variety of images, symbols, myths and ideologies that express the perpetuity of the constituent subject, such as the mythology of founding fathers and mothers, the ideology of an eternal homeland, representations of the groups' traditional ways of life, and varying cultural representations of bodily substances that are eternally transmitted from generation to generation and distinguish the group from other groups.[23] The representation of corporate immortality is also built into the very structure of narration in works of fiction, in 'low' and 'high' culture alike. Fictional worlds in cinema and literature are premised on the fiction of a transcendent enunciator who guarantees the objectivity of (fictional) space and time.[24] The immortality of the enunciator vouches for the flow of linear time within literary and cinematic fictional worlds. Its transcendent gaze guarantees the objectivity of fictional events and situations.

The conceptualisation of time in terms of intergenerational continuity often rests on the intuition that linear time is the product of a social effort to *connect* the generations in the face of the ephemerality of human existence. I have suggested that representations of corporate perpetuity and intergenerational continuity mainly entrench and anchor the separation and distance between generations. Before intergenerational continuity overcomes the passage of time, it produces time by placing the generations along a linear temporal axis and creating a relation of succession between them.

III. Time and Transcendence

Arendt's account of the construction of public space, and more generally her view of the human condition, do not take very seriously the separation between immanent and transcendent spheres.[25] From the point of view of today's 'post-secular' intellectual climate, a critique of Arendt along these lines readily suggests itself. Arendt did pay attention to the otherworldly pursuits of the *vita contemplativa*, but

[23] As Fortes and Godelier showed, on the basis of different models of descent, in every society individuals receive from preceding generations both life-giving substances and a legal inheritance consisting of status, rights and obligations. See Fortes, *Kinship and the Social Order* (n 11); Godelier (n 10).

[24] On objectivity (in real and fictional worlds) as dependent on a transcendent point of view, see, eg, Mikhail Bakhtin, *Problems of Dostoevsky's Poetics* (Minnesota, 1984) 52: '… the construction of [the objective] authorial world with its points of view and finalizing definitions presupposes a fixed external position, a fixed authorial field of vision. The self-consciousness of the hero is inserted into this rigid framework, to which the hero has no access from within and which is part of the authorial consciousness defining and representing him – and is presented against the firm background of the external world'. For Bakhtin, the external point of view belongs to the author, not to a fictional enunciator. On the 'implied author' in fiction and cinema, see Seymour Chatman, *Coming to Terms* (Cornell University Press, 1990).

[25] Kristeva noted that Arendtian public space lacks transcendent foundations; Julia Kristeva, *Hannah Arendt: Life is a Narrative* (translated by Frank Collins)(Columbia University Press, 2001) 56.

it seems that for her the relation between the world and transcendence was basically a relation of mutual exclusion. It can be argued that when society succeeds in repelling the expansionist attempts of the communal body – of which Arendt was well aware – and institutes plurality, it preserves its unity by relegating that unity to a separate sphere. In this way, individuals and groups engage in political, ideological, economic and legal battles, while preserving their unity on a mythical sphere outside of the social, and transforming it into some form of a 'generalised Other'. In the same way that social space is created through the relegation of interpersonal unity to a separate sphere, social time is constructed through the projection of intergenerational communion. The separation between immanent and transcendent spheres allows the institution of social space and social time. It allows for the existence of a mundane realm of interpersonal and intergenerational separation, of autonomy of the living vis-a-vis other generations and other individuals, a realm in which we can distinguish ourselves from the dead and leave our mark on history.

The principle of intergenerational continuity in its various formulations – for example, the universal 'obligation to keep and transmit' postulated by Maurice Godelier[26] – is instituted in 'immanent' social life. It makes possible the separation between the generations and the flow of historical time. It recruits individuals to the taxing work of world-making. It ties the generations to each other and postulates the existence of an essential and immortal common denominator to all generations, but it first and foremost separates the generations from each other and places them along the axis of historical time. The worldliness of the principles of permanence and immortality is emphasised by Arendt: these principles are anchored in the world and count among its building-blocks. They substitute for promises of otherworldly eternity in metaphysical spheres unimpaired by the relativity and fragility of worldly realms. However, and this point was perhaps missed or rejected by Arendt, the world-making fictions of earthly immortality and corporate perpetuity depend on the institution of otherworldliness. The collective body itself, the communal Thing, the ultimate meaning of the message that is handed down from generation to generation, remain always ungraspable and elusive. The content of the objects to which perpetuity is attributed, eg, laws and artworks and other tokens of the immortal identity of the group, must remain a matter of controversy among individuals and generations.

IV. The Affirmation of Work and Action

Representations of intergenerational continuity, while necessary for the linear flow of time, can prove inimical to time. They endanger historical time if they are not accompanied by representations of individual agency and social change, and of the

[26] Godelier (n 10) 470.

limited autonomy of the living in relation to other generations. In Arendt's terms, representations of intergenerational continuity, as embedded in a diversity of cultural monuments, must be accompanied by representations of work and action. As Kristeva pointed out, Arendt's account of public space celebrates both continuity and the singularity of the moment.[27] Representations of intergenerational continuity cannot prevent on their own the enactment of the communal body. Unless the message of past generations is represented as essentially contested, lawless leaders will appropriate and monopolise the power to speak in the name of the dead, in a way that will deny the past any power over the living. Similarly, if continuity is not counterbalanced by representations of human action, it deteriorates into stagnation and dissolution of time. While it may be the case that representations of historical processes and transformations, and the enhanced sense of historicity, are modern phenomena,[28] representations that celebrate the human capacity for work and action in unpredictable circumstances are not. By means of such representations, the realms of action and work affirm themselves against dangers identified by Arendt: the aspirations of the communal body, the repetitiveness of labour,[29] and attempts to subordinate the public world to the principle of the family. In order to display and affirm the human capacities to work and act on the level of collective representation, every generation inscribes in the collective memory new historical dates and narratives of courage and defiance. This is necessary in order to safeguard the limited autonomy of the living and keep the collective body in its removed dwelling place outside the social.[30]

Practices, such as law and art, are constantly engaged in the representation of continuity through change. I cannot offer even a preliminary account of how law represents action and adaptation. The practice of legislation embodies commitment to both ideals of permanence and change in law. Interpretation affirms time

[27] '... the political realm ... is a space filled with a temporality that is at once eternal and fragmented ...' Kristeva (n 25) 204.

[28] Historicity has multiple meanings. It can refer to changing collective representations of the horizons of past and future, and to the collective imagination of how the past flows into the future. A richer notion of historicity, in social practices such as law and art, would involve a reflective response to one's position in a historical continuum, typically an affirmative attitude towards the historicity of any form of participation in the practice. On the nature and history of historicity, see Reinhart Koselleck, *Future's Past: On the Semantics of Historical Time* (Columbia University Press, 2004); Christian Delacroix, Francois Dosse, Patrick Garcia (eds), *Historicités* (Paris, 2009); François Hartog, *Regimes of Historicity: Presentism and Experiences of Time* (Columbia University Press, 2015).

[29] On labour and repetition, see Hannah Arendt, *The Human Condition* (Chicago, 1958) 125; Hannah Arendt, *Between Past and Future* (Viking, 1961) 41–42.

[30] It is clear that representations of action and continuity challenge each other. But they are also mutually-dependent. As much as action prevents continuity from deteriorating into stagnation, historical continuity tames catastrophe and sets the stage – the institutional frameworks and the public spaces – for the exercise of the capacities for work and action. Arendt's work on action calls for elaboration of detailed and refined accounts of the place of actions and beginnings within continuity and tradition. See, eg, Adam Lindsay, 'Hannah Arendt, the Problem of the Absolute and the Paradox of Constitutionalism, or: "How to Restart Time within an Inexorable Time Continuum" (2017) 43(10) *Philosophy and Social Criticism* 1022; Christian Volk, *Arendtian Constitutionalism* (Hart Publishing, 2015).

by representing law's changeability and unchangeability. Interpretive dispute is an important instance of the simultaneous affirmation of dependence and independence in relation to the dead. Law represents human capacities for action in the numerous narratives it produces and the unofficial narratives (literary, popular, cinematic, journalistic) it inspires. Legal narratives on the deeds of citizens, lawyers, judges and lawmakers attach great importance to detailed and precise dating of events. Laws and court sessions are identified by means of reference to the dates they were given or held. Every legal document places itself in relation to preceding and future legal materials within a legal *process* that forms part of an endless sequence of similar processes. The legal preoccupation with dating cannot be reduced to a matter of bureaucratic utility. It has to do with the production of time. It reconciles continuity and change by subsuming endlessly variable sets of facts and institutional responses under the formal categories of the legal process. The terminology of process confirms society's hold over time and its capacity to contain risks of discontinuity and disintegration. In these and other ways, the law affirms its continuity with the legal past and legal future, but also the world-making powers of the living and the significance of human action in response to changing circumstances.[31]

V. Architecture, Art and Time

The historicity of art, its reflexive commitment to the historical process of its evolution, is analogous to the historicity of law, but it has been more profoundly theorised. Works of art, like constitutional principles, embody the collective body of the group: their permanence enacts the continuity of historical time and the continuous existence of society.[32] Art, as opposed to design, belongs in the public realm. It plays an important role in the fabrication of public time by materialising

[31] As Pocock writes, the continuity of custom was always perceived as continuity through change: 'The past is a continuity, but a continuity of adaption; the image of a body changing its cells, of a river changing its water, is more than once used; and we are being told that all we can ever know of the past is that it was unlike the present and yet continuous with it. Even this we cannot know so much as presume, for we cannot know what the past was like at any single moment. Such moments are inapprehensible. The tradition has become a flux, Hale's river that of Heraclitus and Cratylus. No less than the radical, than, the conservative can think both historically and unhistorically. The radical asserts that the past can be known, but that some elements of it can be known out of their traditional context; the conservative asserts that all knowledge must be knowledge in a continuous social context, but denies to that knowledge any status more positive than that of presumption.' See Pocock (n 20) 264.

[32] As Arendt writes in *The Human Condition* (n 29): 'In this permanence, the very stability of the human artifice, which, being inhabited and used by mortals, can never be absolute, achieves a representation of its own. Nowhere else does the sheer durability of the world of things appear in such purity and clarity, nowhere else therefore does this thingworld reveal itself so spectacularly as the non-mortal home for mortal beings. It is as though worldly stability had become transparent in the permanence of art, so that a premonition of immortality, not the immortality of the soul or of life but of something immortal achieved by mortal hands, has become tangibly present, to shine and to be seen, to sound and to be heard, to speak and to be read.'

corporate perpetuity. In order to fabricate time, interpretive practices like art and law embody change and adaptation as much as they represent continuity. Artworks stand both for tradition and continuity and for the unique circumstances of creation that conditioned them. The difference between 'conservative' and 'revolutionary' artistic movements in so far as their relation to the past is concerned is often overstated.[33] Every work of art rests on established landmarks and authoritative models in the history of its genre, not only because they serve as conventional reference points that define the identity of the genre, but because they draw a linear temporal axis that new works of art extend into the future.

These aspects of the artwork – its fictional permanence and its rootedness in the historical moment of its creation – are conspicuous. The ultimate meaning of an artwork must remain less accessible. An artwork must make room for competing interpretations and for a multiplicity of perspectives. In this way, artworks affirm both plurality among the living and changes in their reception throughout the generations, and keep social unity outside of the social. Art and law constitute time by embodying and announcing permanence and change, and by calling for competing interpretations.[34]

Such a view of the historicity of art is largely supported by the work of Nagel and Wood on the passage from medieval to renaissance art. Nagel and Wood describe the modern concept of art as premised on a new understanding of the historicity of the artwork.[35] In contrast to the modern understanding of art's relationship to historical time, the medieval conception of art did not recognise the distinction between an original and a copy. No importance was attached to the continuous existence of the original. There was no negative notion of forgery, and the accurate dating and attribution of an artwork were not considered important. Modern works could be regarded as ancient. In the absence of distinction between original and copy an unlimited number of authentic objects could exist, all enjoying the aura of an original. According to Nagel and Wood, the unbridgeable gap between copy and original developed in the renaissance together with the interest in the historical circumstances of creation. The renaissance celebrated the figure of the artist/author as the main protagonist of an evolving genre of art history that formed an integral part of the social discourse on art.

It can be speculated, on the basis of the work of Nagel and Wood and other works in art history and the history of literature, that the function of the author in practices such as law and art, as a figure that is both mythical and historically

[33] Different studies demonstrate how modernism in art and architecture came to define itself not in terms of a break with the past, but by reference to historical *precedents*. See, eg (on architecture), Anthony Vidler, *Histories of the Immediate Present* (MIT Press, 2008).

[34] On art as constitutive of time, see also, M Gubser, *Time's Visible Surface: Alois Riegl and the Discourse on History and Temporality in fin-de-siècle Vienna* (Wayne State University Press, 2006).

[35] Alexander Nagel and Christopher S Wood, *Anachronic Renaissance* (Zone Books, 2010). See also, Christopher Wood, *Forgery, Replica, Fiction: Temporalities of German Renaissance Art* (Chicago, 2008).

situated, forms a structural component of society's organisation of time. The social structuring of time may explain why the presence of the author remains so powerful for us despite the constant rumours about its demise. The work of Nagel and Wood demonstrates how art embodies permanence in the form of permanence of the original, and embodies change through the importance attached to the attribution of every artwork and to stories about the historical circumstances of creation. The emphasis on the uniqueness of the original postulates, in my view, the transcendent realm of hidden meaning of the artwork: the artwork cannot be simply copied because it calls for interpretation. It cannot be reduced to what is immediately visible. It has to be continually present in order to allow access to its absent dimension and compel the spectators to engage in interpretation.

Architecture is perhaps more effective than other arts in embodying historical time because of its unavoidability. The theory of architectural conservation illustrates the role of art in the social fabrication of time. Cities that were built from scratch or cities that were destroyed and had to be completely rebuilt seem to exist outside of time. There is neither permanence nor change in them. While demonstrating the world-making powers of the living, they do not offer their inhabitants a sense of realness and worldliness. Cities that do not exhibit a multiplicity of historical styles, and threads of continuity that link them, seem to exist without roots and anchors in real time and space. As Kevin Lynch noted, the built environment can enhance society's comprehension of time by displaying successive changes of style, in a way that sensibly links past periods to each other, leads to the present and anticipates a future.[36] As in the case of law, the ultimate meaning of each architectural specimen and of the process of architectural evolution must remain contested, elusive and open to interpretation.

Fascist art and architecture entertain a different relation to historical time. While art in general plays an important role in the projection of the collective body outside the social, and in setting historical time in motion, fascist art seeks to incarnate the communal body and arrest the flow of time. True art, like true law, invites the subject to contemplate an absent sovereign body through the interpretive engagement with hidden meanings. Fascist art, by contrast, draws no line between the overt and the covert, surface and essence, society and its body, and relieves the subject of the burden of interpretation. It seeks to render the sovereign body, the common body of all generations, immediately present and refrain from any reference to an absent ideal or goal. It is perceived uniformly by all members of the political community finding their oneness immediately present in it.

Because fascism rejects the transcendence of the collective body, it rejects linear time and the historicity of the artwork. Fascist art searched for perfect beauty that is not the product of the historical process of development,[37] and

[36] Kevin Lynch, *What Time is this Place?* (MIT Press, 1972) 163.

[37] Under fascism, Neoclassicism formed part of an aesthetics which denies time, confining beauty to self-evident and self-identical timeless forms. It was devised to liberate society from history through

which does not leave room for interpretation, a beauty that exists outside historical time and which is fully immanent. In a lecture entitled 'What National Socialism has done to the Arts', Adorno described the tendency of Nazi art to rest content with the level of appearances in terms of a 'taboo of expressing the essence, the depth of things' and a 'compulsion of keeping to the visible, the fact, the datum and accepting it unquestionably'.[38] Hitler's notion of artistic clarity amounted to a requirement that meaning in art should be transparent and unambiguous.[39] The imperative of superficiality derived from the goal of enacting the oneness of the group and engulfing the individual in it.

Despite its often colossal dimensions, totalitarian architecture fails to embody corporate perpetuity and support the corporate structure of society. Fascist architecture seeks to precipitate the simultaneous presence of all generations in an everlasting present. In an essay on Albert Speer, Elias Canetti has argued that Speer's architecture was designed to propel the formation of vast crowds and to prevent them from falling apart.[40] The concept of the corporate body allows us to observe in the totalitarian gathering space the enactment of a single body which comprises the dead and the living alike. In order to stimulate the immediate presence of sovereignty and generate an all-encompassing oneness, Nazi architecture refrains from cultivating distinctions between the overt and the latent. It aims at utter explicitness and outwardness. Because its meaning is exhausted by its surface, totalitarian architecture has often been qualified as scenographic.[41] Since complexity entails the existence of multiple layers of meaning and dispels communal presence, it cannot play a genuine role in totalitarian architecture. Complexity and dialectics were eliminated in order to engulf the individual in an immediately present collective body. Totalitarian architecture cannot be the object of individual contemplation and interpretive controversy. Its utter explicitness

the construction of a timeless Valhalla. Wilhelm Kreis's view that Neoclassicism was the architectural expression of German soil is instructive in that it reveals the quest for an ahistorical architecture that is validated not as a stage in the evolution of style but by reference to nature. Quoted by Kenneth Frampton in *Modern Architecture: A Critical History*, 4th edn (Thames and Hudson, 2007) 218. On Wilhelm Kreis, see Holger H Herwig, 'The Cult of Heroic Death in Nazi Architecture' in H Herwig and M Keren (eds), *War Memory and Popular Culture* (Jefferson, NC: McFarland, 2009) 105. On the distortion of classical architecture in the proto-totalitarian architecture of Tsarist St Petersburg, see Uriel Procaccia, *Russian Culture, Property Rights, and the Market Economy* (Cambridge University Press, 2007) 140.

[38] Theodor Adorno, 'What National Socialism has done to the Arts' [1945] in Richard Leppert (ed), *Essays on Music by Theodor W Adorno* (University of California Press, 2002) 373 at 381.

[39] On Hitler's notion of clarity, see Barbara Miller Lane, *Architecture and Politics in Germany, 1918–1945* (Harvard University Press, 1968) 189.

[40] See Elias Canetti, 'Hitler, According to Speer' in *The Conscience of Words* (Seabury, 1979) 145. Canetti's argument was put to use by Miguel Abensour in his compelling critique of Leon Krier's attempts to rehabilitate Speer. In order to argue that totalitarian architecture condenses the social, Abensour draws also on Arendt's claim that totalitarianism destroys the social space by pressing men against each other. See, Miguel Abensour, *De la compacité: Architectures et régimes totalitaires* (Sens, 1997). Totalitarian architecture destroys time as much as it destroys space by precipitating the communion of generations.

[41] See, eg, Frampton (n 37) 218.

precludes an individual approach, the inner, interminable process of interpretation which consists in the individual reconstruction of an edifice's abstract unity. Like Kitsch, totalitarian architecture repels any interpretation that is projected onto it. As we shall see, this was also the case with fascist law. It insisted on its immediate comprehensibility, denied its own historicity and rejected the multiplicity of perspectives that is opened up by an interpretative approach.

VI. Interpretation and Constituent Power

According to the view of the collective body that I have sketched above, ultimate authority – in legal terms, constituent power – vests in the collective body of all generations. While constituent power is present in constitutional moments, in the course of normal politics it is projected outside of society and assumes the form of an immortal corporate body, in which generations succeed each other along the axis of linear historical time. In constitutional moments, that is, in the presence of the collective body, interpretation is impossible. Interpretation assumes distance from the ultimate source of law, and a plurality of points of view that does not exist in *communitas*.[42] The theory of the collective body lays in my view the foundations of a theory of adjudication. As priests of law, judges are ultimately authorised not by the legislator but by the collective body of the generations, and it is primarily to the collective body that they are obligated. Once the collective body is projected outside of society, judges engage in interpretation in order to anchor man-made legal materials in the absent, ultimate source of their reason and authority. Thus, in Dworkin's account, laws and precedents have to be read in light of the everlasting principles that can be attributed to the legal system from its foundation.[43] The question what is truly everlasting arises again and again and is never settled.

Dworkin's insistence on the necessity of disagreement between judges is thought-provoking.[44] He makes an anthropological assumption that excessive

[42] This explains the difficulty of courts to intervene in processes that take place in constitutional moments and in the course of constitutional transitions. Judging historical events long after they took place can also unsettle the law, because it unsettles law's mythical foundations, and asks the law to perform the paradoxical function of re-evaluating and recasting mythical narrative while resting on it. When courts are expected to evaluate transitional and other dramatic historical moments they are drawn outside of their comfort zone, the bleak routine that stretches between such historical moments. The courts find themselves operating in the midst of mythical time, without stable mythical/historical foundations for their and the law's authority.

[43] On law and Law in Dworkin, see L Alexander, 'Was Dworkin an Originalist?' in Wil Waluchow and Stefan Sciaraffa (eds), *The Legacy of Ronald Dworkin* (Oxford University Press, 2016) 299.

[44] Dworkin writes: 'The dynamics of interpretation resist as well as promote convergence, and the centrifugal forces are particularly strong where the professional as well as the larger community is divided over justice. Different judges belong to different and rival political traditions, and the cutting edge of different judges' interpretations will be honed by different ideologies. Nor is this to be deplored. On the contrary, law gains in power when it is sensitive to the frictions and stresses of its intellectual sources. Law would founder if the various interpretive theories in play in court and classroom diverged too much in any one generation. Perhaps a shared sense of that danger provides yet another reason

agreement will lead to the collapse of law and to the stagnation of the social relations law regulates. His position suggests that interpretation not only brings the law down to earth, but also keeps it absent, elusive and ungraspable. This in fact may be the primary function of interpretation: interpretation guarantees that law remains transcendent and inaccessible, and that constituent power remains outside the social. By cultivating controversy over the everlasting commitments of the legal system, the interpretive approach guarantees that the ultimate meaning of the law remains beyond reach. Rather than bridging the gap between laws and the Law, interpretation celebrates the unbridgeability of this gap. By cultivating the differentiation between layers of meaning, practices of interpretation reproduce the corporate structure of society as a structure that is premised on the absence of the intergenerational, sovereign unity of the group. Without interpretive controversies, sovereignty would be re-appropriated by the living, and the law would become incapable of safeguarding individual rights in the face of a unified and unfettered community.

Pocock argued that the idea that law can be known directly is a radical idea that is alien to the traditionalist spirit of the common law.[45] According to Dworkin's account, we can gain access to the Law by working through legal materials that have been chaotically amassed in the course of the history of the legal system. Dworkin's point, however, seems to be that the law cannot be 'known' even indirectly due to the persistence of widespread controversy. Historical time is the dimension in which we pursue the Law without ever getting there. Interpretation not only assumes historical time: it consolidates historical time by deepening the gap between laws and the Law, and keeping the Law beyond reach and knowledge. Originalists believe that the law can be known directly. Solum formulates, as one of the theses of originalism, the 'textual constraint thesis' according to which, 'The original meaning of the text of the Constitution has legal force: the text is law and not a symbol of the law.'[46] It seems to me that Dworkin's model, in which the law has to be retrieved through complex negotiation between rules and principles and can never be ascertained, provides a better assurance against totalitarian and communitarian abuses of the law than originalism does.

VII. The End of Time and Interpretation

In his book on the rule of law, Franz Neumann pointed out a tension in the Nazi conception of the role of the judiciary: judges were expected to obey both

why they do not. But law would stagnate, and so founder in a different way, if it collapsed into the runic traditionalism I imagined as the final fate of courtesy.' See, Dworkin (n 1) 88–89. The 'centrifugal forces' to which Dworkin refers are those forces that seek to keep the collective body outside society. The danger of stagnation, in the terms I have been using, is the danger of return of the communal body into the group and the consequent collapse of law.

[45] Pocock (n 20) 264–68.

[46] Robert W Bennett and Lawrence B Solum, *Constitutional Originalism: A Debate* (Cornell University Press, 2011) 4.

the law and the tenets of National Socialism.[47] According to Neumann, the gap between the two expectations was largely bridged by a new conception of the law, which transformed the judiciary into a bureaucracy carrying out written executive decrees, including retroactive and individual ones, and which postulated an ultimate identity between the law and the will of the leader. The postwar literature on Nazi courts did not pursue Neumann's attempt to provide an account that synthesises formalism and teleology.

Immediately after the war, Radbruch famously painted a picture of Nazi courts as following formal methods of application of the law. Radbruch characterised Nazi law in terms of the judicial reluctance to examine law in light of considerations of principle. In contrast to Radbruch's account, some historians have argued that Nazi courts were not formalist enough and that they frequently deviated from the letter of the law. The judges identified with the goals of the regime and were willing to advance general policies by ignoring formal legal obstacles. The courts, it has been argued, engaged in full-blooded political, teleological reasoning in order to promote the interests of the people and the homeland.[48]

Contrary to these competing explanations, the distinctive characteristic of courts in totalitarian regimes is neither excess nor shortage of formalism, but the refusal to consider formal and substantive arguments in light of each other. The two competing accounts of Nazi courts are in fact similar. Nazi law emerges from both accounts as premised on the assumption that law is immediately present and directly accessible. Whether the law consists in formal commands or in general goals, it is immediately accessible to the judge. The law no longer consists of the historical accumulation of legal materials to be read in light of each other in judicial interpretations. Because the ultimate source of the law remains in theories of immanent sovereignty permanently active, the law applied by the court is always essentially contemporaneous with the judicial decision itself. The law is stripped of its historical dimension. The judicial decision derives its legitimacy not from its relation to past and future decisions and to everlasting principles, but directly from the communal body as it storms in the streets outside the courthouse.

The collapse of the rule of law under totalitarianism cannot be explained then in terms of the relative predominance of formal or substantive reasoning. It results from the impoverishment of the interpretive, controversy-generating mediation between different categories of considerations. Formalism can temper political abuses of the law only where it serves as a prelude to interpretation. When clear and formal rules are not applied in the light of underlying principles, their political abuse becomes inevitable. Similarly, the will of the people or the leader as the ever-present fount of the law that is not read in light of the hard rules of the legal

[47] F Neumann, *The Rule of Law: Political Theory and the Legal System in Modern Society* (Berg, 1986) 294–95.

[48] Gustav Radbruch, *Statutory Lawlessness and Supra-Statutory Law* [1946] (2006) 27(1) *Oxford Journal of Legal Studies* 1–11; Ingo Muller, *Hitler's Justice: The Courts of the Third Reich* (Harvard University Press, 1991).

tradition, implies an obscenity of meaning that leaves no room for interpretation. The rejection of interpretation is a general characteristic of a system that refuses to project constituent power and accept historical time. The obligation to interpret emerges as one of the essential commitments of political theologies that celebrate historical time and the absence of sovereignty, and respect individual liberties and the rule of law.

2

Time and the Law: The US Constitutional Experience

STEVEN G CALABRESI[1]

I. 'Time and the Law' and the US Constitution

Arguments about 'Time and the Law' and the US Constitution have been widespread since the Founding of the American Republic in the 1780s. I will begin with a famous exchange of letters on which I will offer my own observations.

A. Thomas Jefferson and James Madison

The exchange of letters was between Thomas Jefferson and James Madison on 'Time and the US Constitution' in the 1780s. Madison had sent to Jefferson, who was then in Paris, a copy of the new US Constitution, which had been ratified by the American people in 1787–1788. In response, on 6 September 1789, Thomas Jefferson wrote the following lengthy letter back to James Madison:

> I sit down to write to you without knowing by what occasion I shall send my letter. I do it because a subject comes into my head which I would wish to develop a little more than is practicable in the hurry of the moment of making up general dispatches.
>
> The question Whether one generation of men has a right to bind another, seems never to have been started either on this or our side of the water. Yet it is a question of such consequences as not only to merit decision, but place also, among the fundamental principles of every government. The course of reflection in which we are immersed here on the elementary principles of society has presented this question to my mind; and that no such obligation can be so transmitted I think very capable of proof. – I set out on this ground, which I suppose to be self evident, 'that the earth belongs in usufruct to

[1] I dedicate this chapter to Yale Law Professor Bruce Ackerman who has done more than anyone else in shaping my understanding of originalism and of its limits. While I disagree with him herein, I feel like I am only standing on the shoulders of a giant in Constitutional Theory. I would like to thank Gary Lawson for his helpful comments on this chapter.

the living': that the dead have neither powers nor rights over it. The portion occupied by an individual ceases to be his when himself ceases to be, and reverts to the society. If the society has formed no rules for the appropriation of it's lands in severality, it will be taken by the first occupants. These will generally be the wife and children of the decedent. If they have formed rules of appropriation, those rules may give it to the wife and children, or to some one of them, or to the legatee of the deceased. So they may give it to his creditor. But the child, the legatee, or creditor takes it, not by any natural right, but by a law of the society of which they are members, and to which they are subject. Then no man can, by natural right, oblige the lands he occupied, or the persons who succeed him in that occupation, to the paiment of debts contracted by him. For if he could, he might, during his own life, eat up the usufruct of the lands for several generations to come, and then the lands would belong to the dead, and not to the living, which would be the reverse of our principle.

What is true of every member of the society individually, is true of them all collectively, since the rights of the whole can be no more than the sum of the rights of the individuals. – To keep our ideas clear when applying them to a multitude, let us suppose a whole generation of men to be born on the same day, to attain mature age on the same day, and to die on the same day, leaving a succeeding generation in the moment of attaining their mature age all together. Let the ripe age be supposed of 21. years, and their period of life 34. years more, that being the average term given by the bills of mortality to persons who have already attained 21. years of age. Each successive generation would, in this way, come on, and go off the stage at a fixed moment, as individuals do now. Then I say the earth belongs to each of these generations, during it's course, fully, and in their own right. The 2d. generation receives it clear of the debts and incumberances of the 1st. the 3d of the 2d. and so on. For if the 1st. could charge it with a debt, then the earth would belong to the dead and not the living generation. Then no generation can contract debts greater than may be paid during the course of it's own existence. At 21. years of age they may bind themselves and their lands for 34. years to come: at 22. for 33: at 23. for 32. and at 54. for one year only; because these are the terms of life which remain to them at those respective epochs. – ***

What is true of a generation all arriving to self-government on the same day, and dying all on the same day, is true of those in a constant course of decay and renewal, with this only difference. A generation coming in and going out entire, as in the first case, would have a right in the 1st. year of their self-dominion to contract a debt for 33. years, in the 10th. for 24. in the 20th. for 14. in the 30th. for 4. whereas generations, changing daily by daily deaths and births, have one constant term, beginning at the date of their contract, and ending when a majority of those of full age at that date shall be dead. The length of that term may be estimated from the tables of mortality, corrected by the circumstances of climate, occupation &c. peculiar to the country of the contractors. Take, for instance, the table of M. de Buffon wherein he states 23,994 deaths, and the ages at which they happened. Suppose a society in which 23,994 persons are born every year, and live to the ages stated in this table. The conditions of that society will be as follows. 1st. It will consist constantly of 617,703. persons of all ages. 2ly. Of those living at any one instant of time, one half will be dead in 24. years 8. months. 3dly. 1[8],675 will arrive every year at the age of 21. years complete. 4ly. It will constantly have 348,417 persons of all ages above 21. years. 5ly. And the half of those of 21. years and upwards living at any one instant of time will be dead in 18. years 8. months, or say 19. years as the nearest integral number. Then 19. years is the term beyond which neither the representatives of a nation, nor even the whole nation itself assembled, can validly extend a debt. ***

I suppose that the received opinion, that the public debts of one generation devolve on the next, has been suggested by our seeing habitually in private life that he who succeeds to lands is required to pay the debts of his ancestor or testator: without considering that this requisition is municipal only, not moral; flowing from the will of the society, which has found it convenient to appropriate lands, become vacant by the death of their occupant, on the condition of a payment of his debts: but that between society and society, or generation and generation, there is no municipal obligation, no umpire but the law of nature. We seem not to have perceived that, by the law of nature, one generation is to another as one independent nation to another. ***

On similar ground it may be proved that no society can make a perpetual constitution, or even a perpetual law. The earth belongs always to the living generation. They may manage it then, and what proceeds from it, as they please, during their usufruct. They are masters too of their own persons, and consequently may govern them as they please. But persons and property make the sum of the objects of government. The constitution and the laws of their predecessors extinguished then in their natural course with those who gave them being. This could preserve that being till it ceased to be itself, and no longer. Every constitution then, and every law, naturally expires at the end of 19 years. If it be enforced longer, it is an act of force, and not of right. – It may be said that the succeeding generation exercising in fact the power of repeal, this leaves them as free as if the constitution or law has been expressly limited to 19 years only. In the first place, this objection admits the right, in proposing an equivalent. But the power of repeal is not an equivalent. It might be indeed if every form of government were so perfectly contrived that the will of the majority could always be obtained fairly and without impediment. But this is true of no form. The people cannot assemble themselves. Their representation is unequal and vicious. Various checks are opposed to every legislative proposition. Factions get possession of the public councils. Bribery corrupts them. Personal interests lead them astray from the general interests of their constituents: and other impediments arise so as to prove to every practical man that a law of limited duration is much more manageable than one which needs a repeal.

This principle that the earth belongs to the living, and not to the dead, is of very extensive application and consequences, in every country, and most especially in France. It enters into the resolution of the questions Whether the nation may change the descent of lands holden in tail? Whether they may change the appropriation of lands given antiently to the church, to hospitals, colleges, orders of chivalry, and otherwise in perpetuity? Whether they may abolish the charges and privileges attached on lands, including the whole catalogue ecclesiastical and feudal? It goes to hereditary offices, authorities and jurisdictions; to hereditary orders, distinctions and appellations; to perpetual monopolies in commerce, the arts and sciences; with a long train of et ceteras: and it renders the question of reimbursement a question of generosity and not of right. In all these cases, the legislature of the day could authorize such appropriations and establishments for their own time, but no longer; and the present holders, even where they, or their ancestors, have purchased, are in the case of bonâ fide purchasers of what the seller had no right to convey.

> Turn this subject in your mind, my dear Sir, and particularly as to the power of contracting debts; and develop it with that perspicuity and cogent logic so peculiarly yours. Your station in the councils of our country gives you an opportunity of producing it to public consideration, of forcing it into discussion. At first blush it may be rallied, as a theoretical speculation: but examination will prove it to be solid and salutary. It would furnish matter for a fine preamble to our first law for appropriating the public revenue; and it will exclude at the threshold of our new government the contagious and ruinous errors of this quarter of the globe, which have armed despots with means, not sanctioned by nature, for binding in chains their fellow men. We have already given in example one effectual check to the Dog of war by transferring the power of letting him loose from the Executive to the Legislative body, from those who are to spend to those who are to pay. I should be pleased to see this second obstacle held out by us also in the first instance. No nation can make a declaration against the validity of long-contracted debts so disinterestedly as we, since we do not owe a shilling which may not be paid with ease, principal and interest, within the time of our own lives. – Establish the principle also in the new law to be passed for protecting copyrights and new inventions, by securing the exclusive right for 19. instead of 14. years. Besides familiarising us to this term, it will be an instance the more of our taking reason for our guide, instead of English precedent, the habit of which fetters us with all the political heresies of a nation equally remarkeable for it's early excitement from some errors, and long slumbering under others.[2]

In Thomas Jefferson's view of 'Time and the Law', the US Constitution ought to have naturally expired after a mere 19 years, now commonly referred to as sunset clauses or provisions. When Jefferson died on 4 July 1826, the US Constitution was already 43 years old and today it is 230 years old. Jefferson never repeated his comment that the Constitution ought to expire in 1808, even though he was the President of the United States by then. Evidently, Jefferson changed his mind on this 'Time and the Law' issue. James Madison vehemently disagreed with his good friend that the US Constitution ought to carry a sunset clause after 19 years and expire. Madison's argument is carefully made and, since his view was the majority view at that time, it ought to be studied with great care. On 4 February 1970, James Madison wrote that:

> Your favor of the 9th. of Jany. inclosing one of Sepr. last did not get to hand till a few days ago. The idea which the latter evolves is a great one, and suggests many interesting reflections to legislators; particularly when contracting and providing for public debts. Whether it can be received in the extent your reasonings give it, is a question which I ought to turn more in my thoughts than I have yet been able to do, before I should be justified in making up a full opinion on it. My first thoughts though coinciding with many of yours, lead me to view the doctrine as not in all respects compatible with the course of human affairs. I will endeavor to sketch the grounds of my skepticism.
>
> As the earth belongs to the living, not to the dead, a living generation can bind itself only: In every society the will of the majority binds the whole: According to the laws of mortality, a majority of those ripe at any moment for the exercise of their will do not

[2] Jefferson's Papers: 27 March 1789 to 30 November 1789 (Princeton University Press, 1958) Volume 15, 392–97.

live beyond nineteen years: To that term then is limited the validity of every act of the Society: Nor within that limitation, can any declaration of the public will be valid which is not express.

This I understand to be the outline of the argument.

The Acts of a political Society may be divided into three classes.

1. The fundamental Constitution of the Government.

2. Laws involving stipulations which render them irrevocable at the will of the Legislature

3. Laws involving no such irrevocable quality.

However applicable in Theory the doctrine may be to a Constitution, it [sic] seems liable in practice to some very powerful objections. Would not a Government so often revised become too mutable to retain those prejudices in its favor which antiquity inspires, and which are perhaps a salutary aid to the most rational Government in the most enlightened age? Would not such a periodical revision engender pernicious factions that might not otherwise come into existence? Would not, in fine, a Government depending for its existence beyond a fixed date, on some positive and authentic intervention of the Society itself, be too subject to the casualty and consequences of an actual interregnum?

In the 2d. class, exceptions at least to the doctrine seem to be requisite both in Theory and practice.

If the earth be the gift of nature to the living their title can extend to the earth in its natural State only. The improvements made by the dead form a charge against the living who take the benefit of them. This charge can no otherwise be satisfyed than by executing the will of the dead accompanying the improvements.

Debts may be incurred for purposes which interest the unborn, as well as the living: such are debts for repelling a conquest, the evils of which descend through many generations. Debts may even be incurred principally for the benefit of posterity: such perhaps is the present debt of the U. States, which far exceeds any burdens which the present generation could well apprehend for itself. The term of 19 years might not be sufficient for discharging the debts in either of these cases.

There seems then to be a foundation in the nature of things, in the relation which one generation bears to another, for the descent of obligations from one to another. Equity requires it. Mutual good is promoted by it. All that is indispensable in adjusting the account between the dead & the living is to see that the debits against the latter do not exceed the advances made by the former. Few of the incumbrances entailed on Nations would bear a liquidation even on this principle.

The objections to the doctrine as applied to the 3d. class of acts may perhaps be merely practical. But in that view they appear to be of great force.

Unless such laws should be kept in force by new acts regularly anticipating the end of the term, all the rights depending on positive laws, that is, most of the rights of property would become absolutely defunct; and the most violent struggles be generated between those interested in reviving and those interested in new-modelling the former State of property. Nor would events of this kind be improbable. The obstacles to the passage of laws which render a power to repeal inferior to an opportunity of rejecting, as a security agst. oppression, would here render an opportunity of rejecting, an insecure provision agst. anarchy. Add, that the possibility of an event so hazardous to the rights of property could not fail to depreciate its value; that the approach of the crisis would increase this effect; that the frequent return of periods superseding all the obligations depending on antecedent laws & usages, must by weak[en]ing the reverence for those obligations, co-operate with motives to licentiousness already too powerful; and that the uncertainty incident to such a state of things would on one side discourage the steady exertions of industry produced by permanent laws, and on the other, give a disproportionate advantage to the more, over the less, sagacious and interprizing part of the Society.

I find no relief from these consequences, but in the received doctrine that a tacit assent may be given to established Constitutions and laws, and that this assent may be inferred, where no positive dissent appears. It seems less impracticable to remedy, by wise plans of Government, the dangerous operation of this doctrine, than to find a remedy for the difficulties inseparable from the other.

May it not be questioned whether it be possible to exclude wholly the idea of tacit assent, without subverting the foundation of civil Society? On what principle does the voice of the majority bind the minority? It does not result I conceive from the law of nature, but from compact founded on conveniency. A greater proportion might be required by the fundamental constitution of a Society, if it were judged eligible. Prior then to the establishment of this principle, unanimity was necessary; and strict Theory at all times presupposses the assent of every member to the establishment of the rule itself. If this assent can not be given tacitly, or be not implied where no positive evidence forbids, persons born in Society would not on attaining ripe age be bound by acts of the Majority; and either a unanimous repetition of every law would be necessary on the accession of new members, or an express assent must be obtained from these to the rule by which the voice of the Majority is made the voice of the whole.

If the observations I have hazarded be not misapplied, it follows that a limitation of the validity of national acts to the computed life of a nation, is in some instances not required by Theory, and in others cannot be accomodated to practice. The observations are not meant however to impeach either the utility of the principle in some particular cases; or the general importance of it in the eye of the philosophical Legislator. On the contrary it would give me singular pleasure to see it first announced in the proceedings of the U. States, and always kept in their view, as a salutary curb on the living generation from imposing unjust or unnecessary burdens on their successors. But this is a pleasure which I have little hope of enjoying. The spirit of philosophical legislation has never reached some parts of the Union, and is by no means the fashion here, either within or without Congress. The evils suffered & feared from weakness in Government, and licentiousness in the people, have turned the attention more towards the means of strengthening the former, than of narrowing its extent in the minds of the latter. Besides this, it is so much easier to espy the little difficulties immediately incident to every great plan, than to comprehend its general and remote benefits, that our hemisphere must be still more enlightened before many of the sublime truths which are seen thro' the medium of Philosophy, become visible to the naked eye of the ordinary Politician.[3]

[3] 'Madison's Rebuttal to "The Earth Belongs to the Living" Precept, 4 February 1790 (Editorial Note)', *Founders Online*, National Archives, accessed 11 April 2019, https://founders.archives.gov/documents/

James Madison responds to Thomas Jefferson's proposed 19-year sunset of the Constitution by politely saying that it is a terrible idea that would lead, if implemented, to chaos, to an increase in the risk factor for investment, to less economic growth, scientific invention, and writing which subsequently occurs with a Constitution that expires every 19 years. Moreover, Madison points out that new generations inherit from the old improvements that have been made, and so they should inherit not only the benefits of their ancestors but also the debts and laws made by their ancestors. Constitutional law is an exercise in inter-generational lawmaking. We agree to be bound by our great-grandfather's Constitution because we can then bind our great-grandchildren to some constitution norm such as the equality between the sexes.

As Madison shows, countries borrow money to fight wars like World War II and the Revolutionary War, even though those debts fall on succeeding generations of citizens because the benefits of winning those wars also accrue to succeeding generations of citizens. This is entirely just to new generations. They should have to take both the bitter and the sweet of what they inherit from their ancestors. There are some problems associated with this, such as paying for necessary wars, which are simply too big for one generation to handle. The solution is intergenerational debt and lawmaking. The 230-year-old US Constitution stands as the most sustained effort at intergenerational lawmaking and debt-sharing in modern history. It is a brilliant success that would not have been possible, but for Madison shooting down Jefferson's idea of a 19-year sunset rule for all constitutions.

The 230-year-old US Constitution makes it possible for me to write a book today, and be confident that I won't be prosecuted for something political I say in it in 20 years' time. Likewise, the 230-year-old Constitution makes it possible for someone to start a business today, and be confident it won't be taken away from him or her without just compensation being paid 20 years from now. The rigidity and entrenchment of the US Constitution enhances liberty because it promotes private ordering and planning and it reduces the costs of the risk factor inherent in all forms of investing. It is for this reason that the US economy is the largest in the world and that its GDP per capita ranks the US in 11th place worldwide out of more than 200 nations.[4] China, in contrast, has the world's second largest economy because it has five times as many people as the US, but its GDP per capita ranks it much lower in 79th place. The US Constitution is super-entrenched, providing low levels of risk to investors while the Chinese constitutional situation is highly fluid. It is for this reason that almost no one emigrates from the US to a foreign country while millions of Latin Americans, Asians, Africans, and Europeans seek to immigrate to the US. People vote for or against a constitutional regime with their feet

Madison/01-13-02-0019. [Original source: Charles F Hobson and Robert A Rutland (eds), *The Papers of James Madison*, vol 13, *20 January 1790–31 March 1791* (Charlottesville, University Press of Virginia, 1981, 18.]

[4] International Monetary Fund, 2018 Survey.

and millions of people seek to move to the US, with all its problems, while almost no one seeks to leave it.

Moreover, not only is the US Constitution super-entrenched because of our constitutions, our system of checks and balances, our system of the separation of powers, and our system of federalism. Ordinary laws made in the US by Congress and the President are also super-entrenched. No law can be repealed without the concurrence of the House of Representatives, the Senate and the President, which is a form of super-entrenchment. Many statutes like the Civil Rights Act of 1964 and the Voting Rights Act of 1965 are so well established that they cannot be repealed. This, too, creates certainty and reduces the risk factor for investment and for book writing. What Jefferson would have called the dead hand of the past is in fact very liberating for modern day Americans.

After 1790, Jefferson never again revisited his idea of a 19-year sunset clause for the US Constitution. The 50 state constitutions in the US do not carry sunset clauses either, although a few of them have clauses providing that every 50 years the voters of the state in question should vote on whether to call a new state constitutional convention.

The US is one of the youngest nations in the world, but its Constitution and Article I, Section 7 lawmaking process address the 'Time and the Law' question by providing for a huge amount of deference by the current generation of Americans to laws made by prior generations of Americans. This American system has come to be seen as very desirable by other democracies, and so almost all of the world's democracies today have super-entrenched constitutions. In Germany, India and Brazil, there are even constitutional provisions that are unamendable like the protection of human dignity. Among the countries with entrenched constitutions that we can include from the G-20 nations are: the US, Germany, France, Italy, Japan, India, Canada, Australia, South Korea, Brazil, South Africa, Indonesia and Mexico. Other entities like the European Union, Argentina and Turkey all have some constitutional entrenchment.[5] In non G-20 constitutional democracies, one also finds a lot of constitutional entrenchment. This is true for: Spain, Portugal, Hungary, Poland, Columbia, Chile, Nigeria, Kenya and Malaysia, as well as many others.

The global trend from 1945, and especially 1981, to the present day has been to favour entrenched constitutions backed up by judicial review of the constitutionality of legislative as well as executive actions. This global trend is one that privileges a point in time when a constitution is made over subsequent points in time, because doing so promotes liberty and investment. Admittedly, in many constitutional democracies Supreme Courts and Constitutional Courts update their constitutions by using what the Canadians call 'A Living Tree' approach to

[5] Steven Gow Calabresi et al, *The U.S. Constitution and Comparative Constitutional Law: Texts, Cases, and Materials* (Foundation Press, 2016); Steven Gow Calabresi and Jasmin Owens, *The Origins and Growth of Judicial Review in the G-20 Nations* (forthcoming in 2020) Volumes I & II.

constitutional interpretation. But, this phenomenon pales into insignificance with the extent to which all constitutions privilege a point A time in constitutional law when there is a Constitutional Moment, as compared to a point B in the life of a nation, which is a time of only ordinary lawmaking. Written constitutions backed up by judicial review always, at least to some extent, privilege Constitutional Moments over periods of ordinary lawmaking. This is a major 'Time and the Law' phenomenon evident all over the world today.[6]

II. Popular Sovereignty

Prior to the adoption of the US Constitution, the 13 original colonies all became sovereign states upon the adoption of the Declaration of Independence on 4 July 1776.[7] The US Constitution, however, identifies 'We the People of the United States' as being sovereign,[8] and in my opinion neither the US Federal Government nor the 50 states can properly be described as being sovereign. To alter and abolish the present Constitution would require a majority of the people in three quarters of the states since that was the way by which 'We the People of the United States' made the present Constitution. Twelve states sent delegates to the Philadelphia Constitutional Convention. The Convention prescribed that the Constitution would go into effect if ratified by state-ratifying conventions in nine states.[9] Hence, the majority in three quarters of the states rule. That is the only way under Article V by which the Constitution can be changed, altered or abolished.

The Framers of the Constitution took the dangerous concept of sovereignty by recognising the sovereignty of 'We the People of the United States' and burying it so that only a majority of the people in three quarters of the states was sovereign. It is so hard today to assemble such a majority that one can fairly say true 'sovereignty' plays only a small role in American law. Nutcases occasionally refer to the states as being sovereign or the Federal Government as being sovereign and both entities are given sovereign immunity by the federal courts. But, a closer analysis of the US Constitution quickly reveals that under Article V and the precedent of the Philadelphia Convention of 1787, it is only a majority of 'We the People' in three quarters of the states who have the sovereign power to alter, abolish or amend the Constitution of the United States.

This is a very good thing because 'sovereignty' is a very dangerous idea. Lord Acton once correctly wrote that 'Power tends to corrupt and absolute power corrupts absolutely.'[10] The power of the sovereign is an absolute power,

[6] *Ibid.*

[7] The Declaration of Independence (1776); Articles of Confederation Preamble & Article II (1777).

[8] US Constitution, Preamble.

[9] US Constitution, Article VII.

[10] Letter from Lord Acton to Bishop Mandel Creighton, 5 April 1887 in JN Figgis and RV Laurence (eds) *Historical Essays and Studies* (1907).

and so it is very dangerous, and it needs to be buried and encumbered as much as possible. The US Constitution accomplishes that while the Constitution of the United Kingdom does not. Under the UK Constitution, the Queen-in-Parliament with the House of Lords and the House of Commons is sovereign.[11] In practice, what this has generally meant in my lifetime is that a political party in the UK, which gets 43 per cent of the vote nationwide among three parties, elects a majority of the House of Commons, which is then sovereign and can change the UK's unwritten, un-entrenched Constitution.

Prime Minister Tony Blair did this by devolving power to Scotland, Wales, Northern Ireland and the City of London all with a majority of the House of Commons after receiving 43.2 per cent of the national vote. David Cameron then changed the UK Constitution to require a fixed term of parliament of five years with no power to hold a new popular election even if a vote of no confidence was passed. Cameron was elected in 2010 with 32 per cent of the vote while his liberal democratic coalition partner, Nick Clegg, received 22 per cent of the vote. Theresa May, who had succeeded Cameron as Britain's Prime Minister, then amended the UK Constitution again by a majority vote of the House of Commons to hold a general election sooner than five years after the previous general election. Theresa May returned as Prime Minister in 2017 after receiving 42.4 per cent of the vote. At the time of writing, May and her party in the House of Commons had the sovereign power in the UK to alter, abolish or reform the UK Constitution from 2017 to the present day. Today, Boris Johnson has that power. This is insane. No political party that receives less than 50.1 per cent of the vote should be able to change the Constitution, and I would set the threshold for constitutional change at more than one majority in one election. How did Britain get into such a horrible mess?

Seventeenth and eighteenth century Britain was a Mixed Regime that was partly monarchical, partly aristocratic, and partly democratic. This is explained in MJC Vile's *Constitutionalism and the Separation of Powers*.[12] As Professor Vile shows, Aristotle, Polybius, Cicero, St Thomas Aquinas and Machiavelli all praised such Mixed Regimes because they divided power among the three great Estates of the land: the Monarchy; the Aristocracy; and the People.[13] These political philosophers favoured Mixed Regimes over pure Monarchies; Aristocracies; or Democracies because they intuitively knew that 'Power tends to corrupt and absolute power corrupts absolutely.'[14] William Blackstone, writing in the eighteenth century, proclaimed that in the UK sovereignty rested in a mysterious entity called the King-in-Parliament, which represented the three great Estates of the Realm.[15] Since 1911, the Monarch and the House of Lords have been

[11] MJC Vile, *Constitutionalism and the Separation of Powers* (Liberty Fund Inc, republished in 1998).
[12] *Ibid.*
[13] Steven G Calabresi, et al, The Rise and Fall of the Separation of Powers (2012) 106 *Northwestern University Law Review* 527–49.
[14] *Ibid.*
[15] Vile (n 11).

reduced to being cyphers, because they have no electoral mandate from the people of the UK.[16] But, what that means in practice is that a party that wins a majority of the House of Commons with 43 per cent of the vote nationwide is *sovereign*. This is a really dangerous idea.

To put this in a 'Time and the Law' focus, it is very easy for the sovereign to act in the present day in the UK, while it is very hard for the sovereign to act in the present day in the US. The US system, as explained above, fosters certainty, it reduces the risk factor in investment and it fosters liberty. The UK system does the opposite. One day the UK is a part of the European Union, the next day it is withdrawing from it. One day Scotland is a part of the UK, the next day it is on the verge of declaring independence. This helps explain why the US has the 11th highest GDP per capita worldwide, while the UK comes in with the 25th highest GDP per capita according to the International Monetary Fund.[17]

Considerations of 'Time and the Law' suggest that sovereign power, which is absolute power, should be encumbered since it increases the risk factor for investment and for authors, inventors and artists, while also being corrupting.

III. The 27th Amendment

The text of the 27th Amendment reads as follows: 'No law, varying the compensation for the services of the Senators and Representatives, shall take effect, until an election of Representatives shall have intervened.' The Amendment was written by James Madison, as part of the original Federal Bill of Rights in 1789, and it passed Congress by the required two-thirds majority of both the House of Representatives and the Senate in 1789. A few states ratified it right away, but it was not until May 1992 that it was ratified by three-quarters of the states, 202 years, five months, and ten days after the Amendment had been proposed by Congress. Seven more states subsequently ratified the 27th Amendment, and only four states have not ratified it.

The question naturally arose whether the 27th Amendment was validly ratified, given that ratification had occurred over a period of time of more than two centuries, or whether instead Article V of the US Constitution, which sets forth the rules for constitutional amendments, implicitly includes some kind of requirement that constitutional amendments must be proposed by Congress and ratified by the states within say a period of seven years so one can be sure there is a contemporaneous constitutional consensus, which supports the constitutional change.

The question as to whether the 27th Amendment had been validly adopted was put before Congress on 20 May 1992, and in keeping with the precedent

[16] Chris Ballinger, *The House of Lords 1911–2011: A Century of Non-Reform* (Hart Publishing, 2012).
[17] International Monetary Fund, 2018 Survey.

first established regarding the ratification of the 14th Amendment, each house of the 102nd Congress passed its own version of a concurrent resolution agreeing that the amendment was validly ratified, despite the unorthodox period of more than 202 years for the completion of the task. The Senate's approval of the resolution was unanimous (99 to 0) and the House vote was 414 to 3.The 'Time and the Law' question on constitutional amendments under Article V of the US Constitution has thus been answered by Congress. It is constitutional for the states to ratify a constitutional amendment, which was proposed by Congress 202 years earlier. The United States was, in 1992, the same polity it was, in 1789, when the 27th Amendment was first proposed.

IV. 'Time and the Law' and the Scholarship of Professor Bruce Ackerman

Yale Law School Professor Bruce Ackerman is one of the leading constitutional theorists of American constitutional law in the US today. Professor Ackerman has proposed a positive account of American constitutional law in a series of books entitled *We the People* in which he argues that the US has had three major *Constitutional Moments*: the Founding in the 1780s; Reconstruction and the adoption of the 14th Amendment in the 1860s; and the New Deal constitutional moment of 1937 when New Deal Supreme Court justices took control of the Supreme Court and of constitutional law.[18] Professor Ackerman argues the US has had, in effect, at least three constitutions just as France's current constitution is that of the Fifth Republic.

Professor Ackerman's books and theory are too complex to be summarised fairly here, but Ackerman does, as a positive matter, accurately describe three very different periods in American constitutional law. Having described these three constitutional moments, Professor Ackerman then goes on to claim that as a normative matter we ought to defer to the most recent constitutional moment, 1937, rather than more distant constitutional moments like 1868 and the 1780s. Professor Ackerman insists that the Constitution 'changed' in 1937 even though no formal textual amendments were passed under Article V. He thinks present day Americans are bound by the constitutional moment of 1937.[19]

The easy and uncontroversial ratification by Congress of the amendment process followed over more than 202 years suggests that Professor Ackerman is wrong in describing the US today as a third republic. Most members of Congress,

[18] Bruce Ackerman, *We the People: Foundations* (Harvard University Press, 1991); Bruce Ackerman, *We the People: Transformations* (Harvard University Press, 1998); Bruce Ackerman, *The Failure of the Founding Fathers* (Harvard University Press, 2005).

[19] But see Steven G Calabresi and Gary Lawson, *The Depravity of the 1930's and the Modern Administrative State* (2018) 94 *Notre Dame Law Review* 821–66.

99 Senators and 414 Representatives thought in May 1992 that Americans today still live under the same Republic that they lived under in 1789. Only three members of the House of Representatives thought otherwise as to the 27th Amendment. Professor Ackerman insists that constitutional change happened in 1937, and he also insists that the 27th Amendment was not validly passed in 1992. But, if one buys a copy of the text of the Constitution or consults a copy of the Constitution used by the federal or state courts or by law professors in their casebooks, one will find that the constitutional text did NOT change in 1937 and that it did change to include the 27th Amendment in 1992.

Professor Ackerman may have his own copy of a document in his office drawer, which includes changes made in 1937, but not the 27th Amendment adopted in 1992. Such a document is not the one used by the American people, their elected representatives, or by other law professors. Professor Ackerman's theory is a theory as to 'Time and the Law', but it is one that has not been embraced by his fellow Americans.

V. 'Time and the Law' and Originalism

Originalism is a doctrine of American constitutional law, according to which constitutional and statutory texts ought to be given the original public meaning that they had when they were enacted into law. Former Supreme Court Justice Antonin Scalia[20] developed the idea out of important preliminary efforts by Ed Meese,[21] President Reagan's second Attorney General, and former Judge Robert H Bork,[22] both of whom advocated following original intent. There are four Scalia-style Originalists in the present-day US Supreme Court: Justices Clarence Thomas; Samuel Alito; Neil Gorsuch; and Brett Kavanaugh. One other Justice relies on constitutional Originalism to at least some extent: Chief Justice John Roberts. Justice Elena Kagan said after the death of Justice Scalia that 'we are all originalists now'.

Originalism is a theory of 'Time and the Law', which privileges the original public meaning of constitutional texts on the date they were enacted into law. For the original Constitution, this means privileging original public meanings when the Constitution was ratified in 1787 to 1788. For 14th Amendment cases, Originalists ought to privilege the original public meaning of the text in 1868 when the 14th Amendment was enacted.

Originalism is usually described as contrasting with 'Living Constitutionalism' or 'Evolving Constitutionalism' or 'Pragmatism'.[23] These latter theorists believe the

[20] Antonin Scalia, *A Matter of Interpretation* (Princeton University Press, 1998).
[21] Steven G Calabresi (ed), *Originalism: A Quarter Century of Debate* (Regnery Publishing, 2007).
[22] Robert H Bork, *The Tempting of America: The Political Seduction of the Law* (Free Press, 1997).
[23] See, eg, Stephen Breyer, *Active Liberty: Interpreting Our Democratic Constitution* (Clarendon Law Lectures, 2005).

Constitution evolves in Darwinian fashion over time. Originalists challenge their foes by saying that constitutional evolution can erode old rights as well as creating new ones. Sometimes, as Scalia once said, a society and its constitution rot over time instead of progressing. For these reasons, Scalia responded to a book by Yale Law School Professor Jack Balkin entitled *Living Originalism*[24] by arguing to the contrary that the Constitution was 'dead, dead, dead'. Scalia's passion on this subject was driven by his belief that Supreme Courts and Constitutional Courts in G-20 democracies were the Ayatollahs of the West: Powerful cultural elites who sought to impose their own questionable values on a majority of their countrymen in an undemocratic fashion.[25]

Originalism as so described is most definitely a theory about 'Time and the Law' just as Living Constitutionalism is also such a theory. Scalia valued Originalism for many of the same reasons I gave above in defending written constitutions backed up by courts with the power of judicial review. Originalism could, if applied, slow down change in the law thus reducing the risk in making investments, making private ordering easier, and enhancing liberty.

Nonetheless, there are some real limitations on Originalism's ability to eliminate change in constitutional law. First, even if constitutional provisions did not change in meaning over time, the world around them may change radically. As Judge Robert H Bork once put it, 'The world changes in which unchanging constitutional values find their application'.[26] For example, the First Amendment protects the freedom of 'the press'. Judge Bork argued quite correctly that this meant freedom of broadcasting, as well as the freedom of printing presses, and Justice Scalia signed many Supreme Court opinions applying the freedom of 'the press' to the Internet. This seems entirely correct to me, but it does suggest that the First Amendment is being read purposively and not literally and that doctrine is changing as the world in which the Constitution applies changes.

Second, there are some constitutional rights, which are worded at a very high level of generality. For example, the constitutional right, which I believe is most deeply rooted in history and tradition, a test devised by Justice Scalia to make it difficult for the Supreme Court to find new constitutional rights,[27] is the *Born Free and Equal* right, which appears as the following text in Article I of the Massachusetts Constitution of 1780. This text provides that:

> Art. I. – All men are born free and equal, and have certain natural, essential, and unalienable rights; among which may be reckoned the right of enjoying and defending their lives and liberties; that of acquiring, possessing, and protecting property; in fine, that of seeking and obtaining their safety and happiness.[28]

[24] Jack Balkin, *Living Originalism* (Belknap Press, 2014).

[25] Christopher J Scalia and Edward Whelan (eds), *Scalia Speaks: Reflections on Law, Faith, and Life Well Lived* (Crown Forum, 2017) 260–70.

[26] Concurring opinion in *Olman v Evans & Novak*, 750 F.2nd 970 (DC Cir *en banc* 1985).

[27] *Michael H v Gerald D* 491 US 110 (1989) as modified by *Washington v Glucksberg* 521 US 702 (1997).

[28] Massachusetts Constitution 1780, Art 1.

Born Free and Equal clauses, like the one quoted above, appeared in 24 out of 37 state constitutions in 1868, when the 14th Amendment was ratified, which is a solid majority of the state Bills of Rights at that time. Such clauses also appear in 37 out of 50 state constitutions today.[29] However, it is hard for me to see how LGTBQ people, for example, could 'enjoy liberty' today, or 'seek and obtain their safety and happiness' today if they cannot marry the way heterosexuals do, and yet this undeniably leads to a change in constitutional law. I agree with former Justice Scalia about the importance of originalism, and I consider myself to be an originalist. But, I think the Born Free and Equal clauses do amount to a constitutional right, which the federal courts must protect.[30] I construe the Born Free and Equal clauses as justifying Supreme Court decisions such as *Griswold v Connecticut*;[31] *Lawrence v Texas*;[32] and *Obergefell v Hodges*,[33] all cases which Scalia thought had been wrongly decided.

In sum, Originalism is a theory about 'Time and the Law', which privileges the original public meaning of legal texts when they were enacted into law as a way of effectuating and empowering democracy. The limits of Originalism are that: 1) the world changes in which unchanging values find their application; and 2) some legal texts are so broadly worded, like the Born Free and Equal clauses that they seem to invite constitutional update and change.

Ironically, the huge hurdles that Article V of the Constitution puts in place for enacting constitutional amendments causes the US Supreme Court to engage in more unilateral constitutional updating since the Court can change the Constitution by a five to four vote, while Article V requires two-thirds of both Houses of Congress to effectuate constitutional change. Both the Democratic and the Republican Parties in the US now seek to effectuate constitutional change through the appointment of Supreme Court justices with views similar to their own rather than by using the Article V process. What this shows is that if a constitutional designer makes formal processes of constitutional change too hard to use, the system will find some way around the formal barriers to change. This process also occurs with respect to statutes as to which it may be impossible for a present day majority to repeal perhaps because of a filibuster threat in the Senate even if a change in statutory law is desired by 51 per cent of the people. Change in this context can happen through a presidential executive order or proclamation or by administrative agency rulemaking, which circumvents the Article I, Section 7 requirements of bicameralism and presentment. The political system finds ways of working around efforts to tie the law to some past event or understanding when present

[29] Steven G Calabresi et al, *Individual Rights under State Constitutions in 2018: What Rights are Deeply Rooted in a Modern Day Consensus of the States?* (2018) 94 *Notre Dame Law Review* 49–153.

[30] Steven G Calabresi and Sophia M Vickery, *On Liberty and the Fourteenth Amendment: The Original Understanding of the Lockean Natural Rights Guarantees* (2015) 93 *Texas Law Review* 1299.

[31] 381 US 479 (1965).

[32] 539 US 558 (2003).

[33] 576 US ____ (2015).

day needs demand chance. This is just the way 'Time and the Law' interact in the US constitutional law and administrative law systems.

VI. Living Constitutionalism

Living Constitutionalists consider the original meaning of constitutional texts, but then ask what those texts mean in the present day. At least one living Constitutionalist, Jack Balkin, is also an Originalist as his book on *Living Originalism* reveals.[34] Living Constitutionalism is a theory of 'Time and the Law', which calls for applying the meaning of the Constitution, today, rather than when it was originally enacted. Four justices of the US Supreme Court are Living Constitutionalists: Justices Ruth Bader Ginsburg; Stephen Breyer; Sonia Sotomayor; and Elena Kagan. Justice Kennedy frequently joined this camp when he was on the Supreme Court, but I think he is best thought of as being a Living Originalist like myself.

The advantage to Living Constitutionalism is that it keeps the 230-year-old American Constitution up-to-date and relevant in the modern world by asking what do the words of the Constitution and the 14th Amendment mean to Americans living today. Professor Paul Brest, who is a Living Constitutionalist, argues that the relationship of present day Americans to the original Constitution is as distant as the relationship of a person living today who had one of 120 ancestors come over to the US.[35] Brest thinks the Living Constitution is more just and relevant to Americans today than the original Constitution is. Living Constitutionalists point to Chief Justice John Marshall's famous statement in *McCulloch v Maryland*, which appears below:

> A Constitution, to contain an accurate detail of all the subdivisions of which its great powers will admit, and of all the means by which they may be carried into execution, would partake of the prolixity of a legal code, and could scarcely be embraced by the human mind. It would probably never be understood by the public. Its nature, therefore, requires that only its great outlines should be marked, its important objects designated, and the minor ingredients which compose those objects be deduced from the nature of the objects themselves. That this idea was entertained by the framers of the American Constitution is not only to be inferred from the nature of the instrument, but from the language. Why else were some of the limitations found in the 9th section of the 1st article introduced? It is also in some degree warranted by their having omitted to use any restrictive term which might prevent its receiving a fair and just interpretation. In considering this question, then, *we must never forget that it is a Constitution we are expounding*.[36]

[34] Balkin (n 24).

[35] Paul Brest, 'The Misconceived Quest for the Original Understanding' (1980) 60 *Boston University Law Review* 204.

[36] 17 US 316 (1819).

The advantage, as a matter of policy, to construing the words of the Constitution according to what they mean to Americans today, rather than according to Americans in 1788 or 1868 is that the world has changed enormously since 1868 and those changes must be taken into account. The disadvantage to Living Constitutionalism is that as Justice Scalia once said 'sometimes countries rots rather than making progress'.[37] A Living Constitution allows for the erasing of constitutional rights that are out of fashion, as well as the creation of new rights. Professor Michael McConnell argues that this is what happened during the Jim Crow era between the end of Reconstruction in 1877 and *Brown v Board of Education* in 1954.[38] Living Constitutionalists are stuck believing that *Plessy v Ferguson*[39] upholding segregation in railway cars in 1896 was rightly decided because most Americans in 1896 were Segregationists.

Professor McConnell and I are both Living Originalists, and we believe the original public meaning of the 14th Amendment in 1868 was to ban all forms of racial discrimination as to civil rights, as well as all systems of caste such as feudalism and the Hindu Caste system.[40] We think the 14th Amendment constitutionalised the *Born Free and Equal* idea in 1868, and we therefore think *Plessy v Ferguson* was wrong on the day it was decided in 1896. McConnell has written about this in a law review article entitled 'The Forgotten Constitutional Moment'. McConnell argues that under Professor Ackerman's theory for recognising constitutional moments, the emergence of Jim Crow segregation on the basis of race qualifies as a constitutional moment under Ackerman's test for determining the appearance of constitutional moments.

I think giving the words of the Constitution anything other than their original public meaning is bizarre because, as law professor Stephen Smith of San Diego law school points out, words without their original public meaning are merely scratches on a piece of paper.[41] Why should we be governed by the meaning those scratches have to Americans today instead of by the runes in the desert caused by high winds? Both are equally arbitrary and undemocratic. I am a Living Originalist because I think Living Originalism adequately deals with change over the past two centuries in several ways: I believe along with Judge Bork that: 1) 'The world changes in which unchanging constitutional values find their application'; and 2) in which key constitutional rights like the *Born Free and Equal* right find their application. My belief in these two ideas is what makes me a Living Originalist. I think as a matter of 'Time and the Law' that my belief in these two

[37] Scalia and Whelan (eds) (n 25) 2017). The comment in quotes was made to the author personally in a conversation with Justice Scalia.

[38] Michael W McConnell, 'The Forgotten Constitutional Moment' (1994)11 *Constitutional Commentary* 115.

[39] 163 US 537 (1896).

[40] Steven G Calabresi and Michael Perl, 'Originalism and Brown v. Board of Education' (2014) *Michigan State Law Review* 429.

[41] Steven D Smith, 'Law Without Mind' (1989) 88 *Michigan Law Review* 104.

caveats to Originalism is more than enough to keep the Constitution and the 14th Amendment up-to-date.

VII. Precedent and 'Time and the Law'

Another theory of 'Time and the Law' in US Constitutionalism is that the Supreme Court ought to follow super precedents arrived at by the justices at an earlier point in time. Many Americans have thus argued that the Supreme Court ought to follow *Roe v Wade*,[42] the 1973 case in which the US Supreme Court found that pregnant women have a constitutional right to have an abortion almost on demand during the first two trimesters of pregnancy. In a 1992 case, the Court trimmed this right back to say that the right to an abortion ends once the foetus becomes viable outside the pregnant woman's body at say 24 weeks, but the Court wrote at great length about its obligation to follow the 1973 *Roe v Wade* precedent, even if the case was as, an initial matter, wrongly decided.

I do believe *Roe v Wade* was wrongly decided because I think the question of when life begins and the foetus becomes a rights' holder is a political question, which should be decided by the legislature and not the courts. As a personal matter, I believe life begins only after a fertilised egg successfully implants itself in the mother's womb. I therefore think that experimentation with foetal stem cells harvested from laboratory-fertilised eggs is fine and that use of intrauterine devices for contraception is fine. I personally would not allow abortions after implantation of the fertilised egg, except in cases of anencephaly, but I recognise that different people of various faiths have many different ideas as to where to draw the permissible abortion line. To me, abortion regulation during the first two trimesters of pregnancy is a classic political question, which ought to be resolved by the legislature and not by courts.

I do not believe that *Roe v Wade* is a super precedent, and I do not believe the Supreme Court in the present day should decide an abortion case a certain way because an earlier court decided such a case that way in 1973. This, then, is a question about 'Time and the Law'. I believe the Supreme Court ought not to follow precedent for four reasons.[43]

First, the text of the Constitution makes no mention of judicial precedents as being a source of law. The text says that only: 1) the Constitution; 2) laws enacted pursuant to the Constitution; and 3) all treaties made or which shall be made are the Supreme Law of the Land. Put quite simply, Supreme Court case law is not

[42] 410 US 113 (1973).

[43] Steven G Calabresi, 'Text, Precedent, and the Constitution: Some Originalist and Normative Arguments for Overruling Planned Parenthood of Southeastern Pennsylvania v. Casey' (2005) 22 *Constitutional Commentary* 311–48 (symposium issue on text and precedent); Steven G Calabresi, 'The Tradition of the Written Constitution: Text, Precedent, and Burke' (2006) 57 *Alabama Law Review* 635–87.

listed as being the Supreme Law of the Land. Such case law binds courts inferior to the Supreme Court, but it does not bind future sittings of the Supreme Court.

Second, there is no evidence in the original understanding of the Constitution that the Supreme Court would be bound by its precedents. The framers never said as much, and they would have been much more concerned about the Supreme Court's power if they had thought one session of the Supreme Court could bind future sessions of the Court. Given the enormous length at which the framers discussed and debated other government powers, the absence of discussion of one session of the Court having the power to bind future sessions is quite telling. In the framer's world, case law precedent evolved over time and precedents could be cut back or distinguished away. The framers also believed, along with William Blackstone, that one session of Parliament could not bind future sessions of Parliament. It seems most unlikely then that the framers would have thought the Supreme Court acting in 1973 could have bound the Supreme Court in the present day.

Third, it is contrary to precedent and to long-standing practice for the Supreme Court to follow its own precedents. This can be illustrated by giving ten examples of cases where the Supreme Court either evaded or over-ruled some prior precedent outright: 1) In the Legal Tender cases, *Knox v Lee*[44] overruled *Hepburn v Griswold*;[45] 2) *Plessy v Ferguson*[46] was effectively overruled by *Brown v Board of Education*;[47] 3) *Hammer v Dagenhart*[48] was overruled by *United States v Darby*;[49] 4) *Minersville School District v Gobitis*[50] was overruled by *West Virginia State Board of Education v Barnette*;[51] 5) *Mapp v Ohio*[52] overruled *Wolf v Colorado*;[53] 6) *Maryland v Wirtz*[54] was overruled by *National League of Cities v Usery*,[55] which was in turn overruled by *Garcia v San Antonio Metropolitan Transit System*;[56] 7) *Bowers v Hardwick*[57] was overruled by *Lawrence v Texas*;[58] 8) *Furman v Georgia*[59] was overruled by *Gregg v Georgia*;[60] 9) *Citizens United v FEC* overruled *Austin v Michigan Chamber of Commerce*[61] and partially overruled *McConnell v Federal Election Commission*;[62]

[44] 79 US 457 (1871).
[45] 75 US 603 (1870).
[46] See n 39.
[47] 347 US 483 (1954).
[48] 247 US 251 (1918).
[49] 312 US 100 (1941).
[50] 310 US 586 (1940).
[51] 319 US 624 (1943).
[52] 367 US 643 (1961).
[53] 338 US 25 (1949).
[54] 392 US 183 (1968).
[55] 426 US 833 (1976).
[56] 469 US 528 (1985).
[57] 478 US 186 (1986).
[58] 539 US 558 (2003).
[59] 408 US 238 (1972).
[60] 428 US 153 (1976).
[61] 494 US 652 (1990).
[62] 558 US 310 (2010).

and 10) *Employment Division v Smith*[63] overruled *Sherbert v Verner*[64] but was itself arguably overruled by *Hosana Tabor v EEOC*.[65]

Fourth, as a matter of policy, it would be a bad thing for the Supreme Court to follow wrongly decided precedent since it is almost impossible to amend the Constitution to correct such errors. In more than two centuries, only four Supreme Court decisions have been overturned by constitutional amendments, and it takes only one house in 13 out of 50 state legislatures to block a constitutional amendment. If the 13 least populous states voted to block a constitutional amendment five per cent of the population would be able to block a change that 95 per cent of the population desired. This is too preposterous to contemplate!

As a matter of 'Time and the Law', the Supreme Court at time A ought not to be able to block what the Supreme Court wants to do at a later time B. The Supreme Court should arguably follow precedents to promote certainty in the law when the precedents are not clearly wrong, but when precedents are clearly wrong they ought to be overruled. I believe this is the situation with *Roe v Wade*, a 1973 Supreme Court decision, which I think ought to be overruled.

Yale Law Professor Bruce Ackerman has argued that Americans ought to consider themselves bound today by the precedents on the scope of national power and economic liberties set by the New Dealers as a result of the events of 1937. At this time, President Franklin Roosevelt proposed a plan to pack the Supreme Court. One of the five conservative justices then resigned giving Roosevelt a majority in the Supreme Court. At that point, Congress voted down the Court-packing plan, which was no longer necessary to get Roosevelt's New Deal programmes approved by the Court. As I noted earlier, the text of the Constitution was not amended in 1937, but Professor Ackerman seems to think Americans should be guided by the *Zeitgeist of 1937* and not by the *Zeitgeist of the 1780s*; of the *Zeitgeist of 1868*, when the 14th Amendment was adopted; or even by the *Zeitgeist of 2019*. All of these alternatives are arguments about 'Time and the Law'. To what time period should American constitutional interpreters direct their attention? A brief consideration of the four time periods of: 1) the 1930s; 2) the 1780s; 3) 1868; and 4) the present day makes it clear that the 1930s is the worst time to look to for guidance on questions of constitutional meaning. I will draw here on Steven G Calabresi and Gary Lawson's, *The Depravity of the1930's and the Modern Administrative State*.[66]

A. The 1930s

The modern Administrative State had its origins in the 1930s both because President Franklin Roosevelt's Administration created a number of new regulatory

[63] 494 US 872 (1990).
[64] 374 US 398 (1963).
[65] 565 US 171 (2012).
[66] Steven G Calabresi and Gary Lawson, 'The Depravity of the1930's and the Modern Administrative State' (2018) 94 *Notre Dame Law Review* 821–66.

agencies including the Securities and Exchange Commission and the National Labor Relations Board and because the Supreme Court, quite wrongly, in *Humphrey's Executor v the United States*[67] held that Congress could create agencies that were independent of the President and the commissioners of which could be fired at will. President Franklin Roosevelt thought quite correctly that *Humphries Executor* was wrongly decided and that it was a politically made decision made by a conservative Supreme Court to clip his wings. I also believe *Humphrey's Executor* was wrongly decided as I explain at length in a book co-authored with Professor Christopher Yoo entitled *The Unitary Executive: Presidential Power from Washington to Bush*.[68]

More broadly, I would note that the Constitution was not amended in the 1930s to authorise the creation of an Administrative State. The Constitution provided for in the 1930s, as it does today, a separation of legislative, executive and judicial power. The Administrative State as it presently exists violates the separation of powers as we will prove later in the chapter. The major framework statute, which governs the Administrative State, was enacted in the 1940s by a Republican Congress and President Truman. It was an effort to roll back some but not all of the excesses and constitutional violations, which occurred when the Administrative State was created in the 1930s. There is thus no legal text, either constitutional or statutory, from the 1930s, which Professor Ackerman can point to as authorising the modern Administrative State. The modern Administrative State is in fact both unconstitutional and unwise for policy reasons. There is no text from the 1930s on this topic, which binds us today.

Lacking a text to work with, Professor Ackerman appeals to the *Zeitgeist of the 1930s*, which he believes was 'a constitutional moment'. The 1930s was clearly the worst decade of the last 100 years and is one of the worst decades in the history of mankind. It began with the Great Depression and it ended with the start of World War II. In 1933, Germany elected a Parliament, which chose Nazi Party leader Adolf Hitler to be the Chancellor of Germany. On 24 March 1933, the German Parliament delegated all of its legislative powers to Hitler and his Nazi Cabinet in the Enabling Act of 1933. When the President of Germany, Paul von Hindenburg, died in August 1934, Hitler used his delegated powers to declare himself to be the Fuhrer of the Third Reich, led by the Nazi Party, and democracy in Germany came to an end.

One of the first things Hitler did when he came to power in March 1933 was to pass a German style Eugenics Law, modelled on the many American state Eugenics Laws, which provided for the compulsory sterilisation of the 'feeble minded' so they would not pollute the gene pool of the Third Reich.[69] In the US

[67] 295 US 602 (1935).

[68] Steven G Calabresi and Christopher S Yoo, *The Unitary Executive: Presidential Power from Washington to Bush* (Yale University Press, 2008).

[69] James Q Whitman, *Hitler's American Model and the Making of Nazi Race Law* (Princeton University Press, 2017).

at this time, the Supreme Court had quite wrongly in *Buck v Bell*[70] upheld the constitutionality of compulsory sterilisation in an eight to one decision with the majority opinion written by Justice Oliver Wendell Holmes. Holmes infamously wrote that: 'The same principle which sustains compulsory vaccination sustains the compulsory cutting of the fallopian tubes. Three generations of imbeciles are enough.'[71] Approximately 60,000 Americans were compulsorily sterilised until *Skinner v Oklahoma*[72] put an end to the practice in the 1940s.

Hitler adopted his Eugenics Law to emulate the US and out of concern, as one Nazi official put it, that 'the Americans are beating us at our own game'. Perhaps not surprisingly, the German eugenics movement paved the way for the Holocaust. After all, if it is permissible to compulsorily sterilise inferior races of people, then why not just kill them in the here and now? Yale Law Professor James Q Whitman writes compellingly about this phenomenon in his book *Hitler's American Model: The United States and the Making of Nazi Race Law.*[73]

The 1930s was not just a decade which saw the triumph of Hitler, the Nazis, Eugenics Laws, and Anti-Semitism, it was also a decade of Fascism both in Italy and in the US. During the 1930s, Mussolini took over Italian industry in a programme of national socialism, which was then emulated by Franklin Roosevelt in the National Industrial Recovery Act (NIRA) of 1933, the legislative centrepiece of Roosevelt's first 100 days. Under the NIRA, all industries were to be governed by a criminally enforceable code of conduct drawn up by committee of corporate managers and labour union chiefs and then promulgated by Roosevelt himself.

The NIRA was challenged on constitutional grounds as being, among other things, an unconstitutional delegation of legislative power. The Supreme Court unanimously agreed in *Schecter Poultry v United States.*[74] We must emphasise that liberal Supreme Court justices Louis Brandeis, Benjamin Cardozo and Harlan Fiske Stone all agreed that Roosevelt's fascist law was an unconstitutional delegation of power. If we are to look at all at the 1930s for guidance, it is the unanimous opinion in *Schecter Poultry* that we should look at.

Aside from the grinding poverty of the 1930s in America, we must also remember that in the South it was yet another decade of Jim Crow segregation. Roosevelt's New Deal coalition was a union of ethnic groups in the Northeast and Midwest and of southern Segregationists in the South. Many members of Roosevelt's cabinet and administrative team were southern Segregationists. It was not until the end of the 1940s, when President Truman desegregated the armed services of the US that Jim Crow began to be questioned. The *Zeitgeist of the 1930s* was thus one of: 1) Nazism; 2) Fascism; 3) Eugenics Laws; 4) Anti-Semitism, and of Germany

[70] 274 US 200 (1927).
[71] *Ibid.*
[72] 316 US 535 (1942).
[73] Whitman (n 69).
[74] 295 US 495 (1935).

laying the groundwork for the Holocaust. The 1930s is an awful decade to look to for guidance in the present day. It was a time when Social Darwinists were abandoning the Enlightenment and Christianity to kill the Jews. I cannot begin to understand those who find something morally appealing about the 1930s. The fact that Franklin Roosevelt, in the 1930s, was not as extreme as Hitler, Stalin or Mussolini does not change the analysis. The 1930s was a bad time in the US as well as in Germany, the Soviet Union, Italy and Japan.

B. The 1780s

The 1780s was one of the greatest and best decades of the last 250 years. It started with Americans winning their independence at the battle of Yorktown, and the decade ended with a peaceful French Revolution, which aspired to bring American constitutional democracy to France. The 1780s were the culmination of the Age of the Enlightenment. The Enlightenment began in 1715 with the death of King Louis XIV, and it ended in 1789 with the French Revolution. It was preceded by a Scientific Revolution led by Galileo Galilei; Sir Francis Bacon; and many other individuals. Key Enlightenment intellectuals whose ideas influenced the *Zeitgeist of the 1780s* included: John Locke; Voltaire; Rousseau; Benjamin Franklin; Thomas Jefferson; and John Adams. They believed in the words of the Massachusetts Constitution of 1780 that:

> Art. I. – All men are born free and equal, and have certain natural, essential, and unalienable rights; among which may be reckoned the right of enjoying and defending their lives and liberties; that of acquiring, possessing, and protecting property; in fine, that of seeking and obtaining their safety and happiness.

The Social Darwinism of the 1930s had not yet been born and all intellectuals agreed that men were born free and equal, thus putting an end to feudalism under which aristocrats and commoners were unequal. The language quoted above from the Massachusetts Constitution of 1780 appears today in 40 of the 50 state constitutions. It is also the language that Thomas Jefferson artfully plagiarised in the Declaration of Independence:

> We hold these truths to be self-evident, that all men are created equal, that they are endowed by their Creator with certain unalienable Rights, that among these are Life, Liberty and the pursuit of Happiness.

The 'born free and equal' idea was overwhelmingly dominant in intellectual circles in the 1780s both in France and in the US. It is evident as well in the first articles of the French Revolutionary Declaration of the Rights of Man and of the Citizen of 1789, which stated that:

> **1.** Men are born and remain free and equal in rights. Social distinctions may be founded only upon the general good.

> **2.** The aim of all political association is the preservation of the natural and imprescriptible rights of man. These rights are liberty, property, security, and resistance to oppression. ***
>
> **4.** Liberty consists in the freedom to do everything which injures no one else; hence the exercise of the natural rights of each man has no limits except those which assure to the other members of the society the enjoyment of the same rights. These limits can only be determined by law.
>
> **5.** Law can only prohibit such actions as are hurtful to society. Nothing may be prevented which is not forbidden by law, and no one may be forced to do anything not provided for by law.

This Declaration of Rights was written and introduced in the French National Assembly by the Marquis de Lafayette who had served in the American Revolution under George Washington with whom he became a close friend, and he had read the many American state Bills of Rights, which had *Born Free and Equal* clauses in them.

The main political events of the 1780s were the writing and ratification of the US Constitution, the writing of the Federalist Papers to persuade Americans to ratify the Constitution, and the peaceful Revolution in France, which initially replaced absolute monarchy with a constitutional Monarch and a unicameral legislature called the National Assembly. While the French Revolution eventually turned into a Reign of Terror and a Napoleonic liberal dictatorship, these deplorable events occurred in the 1790s, not the 1780s. The Zeitgeist *of the 1780s* is captured by the *Born Free and Equal* clauses with their libertarian and egalitarian commitments to replace European Feudalism with something new and better. Recourse to the 1780s is thus desirable not only for Constitutional Originalists but also for Constitutional Pragmatists or Legal Realists. There are thus powerful normative as well as constitutional reasons for looking to the 1780s.

It is true that not all was bright in the 1780s because in the US African-American slavery was still in place and native Americans were being mistreated, and, in France, the kidnapping of King Louis XVI and his forced return from Versailles to Paris in October 1789 foreshadowed dangers to come. But on 31 December 1789, King Louis XVI's problems were only a storm cloud on the horizon, and, in the US, intellectuals all agreed that slavery was immoral and that it should be abolished, and the states of Massachusetts, Rhode Island and Vermont all abolished slavery. The Northwest Ordinance of 1787 was passed by the Continental Congress, and it abolished slavery in the land that would become the states of Ohio, Indiana, Illinois, Michigan, Wisconsin and Minnesota. Slavery seemed to be on its way out in the 1780s, and it was not until the invention of the cotton gin by Ely Whitney in 1793 that American slavery was given a new lease on life. The Vice President of the Confederacy, Alexander Stephens, described the cornerstone of the US Constitution as being the idea that 'all men were created equal' – an idea he explicitly disagreed with in his infamous Cornerstone Speech. Stephens acquitted Thomas Jefferson and the American framers of any charge that they thought, as he did, that 'science' showed African Americans to be inferior to white Americans.

Another of the core intellectual ideas of the 1780s was a belief in the importance of the separation of powers, as it was described by the French political philosopher Montesquieu in his *Spirit of the Laws*. Montesquieu argued for a separation of legislative, executive and judicial powers because he believed, as Lord Acton would later say, that 'power corrupts and absolute power corrupts absolutely'. All of the American state constitutions in the 1780s had a separation of powers and the federal constitution of 1787 does as well. American constitutionalism goes even further than the formal separation of powers and creates a system of checks and balances, as James Madison famously argued in *The Federalist Papers, Nos 47 to 51*.

Thus, the American Constitution gives the President a share of the legislative power by giving him the veto power; and it gives the Senate a share of the executive power by giving the Senate the power to confirm presidential nomination and to ratify presidentially negotiated treaties. Madison defended his system of checks and balances in the following excerpt from *The Federalist, No 51*:

> TO WHAT expedient, then, shall we finally resort, for maintaining in practice the necessary partition of power among the several departments, as laid down in the Constitution? The only answer that can be given is, that as all these exterior provisions are found to be inadequate, the defect must be supplied, by so contriving the interior structure of the government as that its several constituent parts may, by their mutual relations, be the means of keeping each other in their proper places. Without presuming to undertake a full development of this important idea, I will hazard a few general observations, which may perhaps place it in a clearer light, and enable us to form a more correct judgment of the principles and structure of the government planned by the convention. ***
>
> It is *** evident, that the members of each department should be as little dependent as possible on those of the others, for the emoluments annexed to their offices. Were the executive magistrate, or the judges, not independent of the legislature in this particular, their independence in every other would be merely nominal. But the great security against a gradual concentration of the several powers in the same department, consists in giving to those who administer each department the necessary constitutional means and personal motives to resist encroachments of the others. The provision for defense must in this, as in all other cases, be made commensurate to the danger of attack. *Ambition must be made to counteract ambition. The interest of the man must be connected with the constitutional rights of the place. It may be a reflection on human nature, that such devices should be necessary to control the abuses of government. But what is government itself, but the greatest of all reflections on human nature? If men were angels, no government would be necessary. If angels were to govern men, neither external nor internal controls on government would be necessary. In framing a government which is to be administered by men over men, the great difficulty lies in this: you must first enable the government to control the governed; and in the next place oblige it to control itself.*
>
> A dependence on the people is, no doubt, the primary control on the government; but experience has taught mankind the necessity of auxiliary precautions. This policy of supplying, by opposite and rival interests, the defect of better motives, might be traced through the whole system of human affairs, private as well as public. We see it particularly displayed in all the subordinate distributions of power, where the constant aim is to

> divide and arrange the several offices in such a manner as that each may be a check on the other that the private interest of every individual may be a sentinel over the public rights. These inventions of prudence cannot be less requisite in the distribution of the supreme powers of the State. But it is not possible to give to each department an equal power of self-defense. In republican government, the legislative authority necessarily predominates. The remedy for this inconveniency is to divide the legislature into different branches; and to render them, by different modes of election and different principles of action, as little connected with each other as the nature of their common functions and their common dependence on the society will admit. ***

As this excerpt from *The Federalist No 51* shows, intellectuals like Madison in the 1780s were influenced by Montesquieu, by Adam Smith, and by David Hume. The 1780s is one of the greatest decades in human history and certainly in the last 250 years. Unlike the 1930s, the 1780s is a very good place in which to find normatively appealing ideas. The *Zeitgeist* of *the 1780s* was *Born Free and Equal* clauses while the *Zeitgeist of the 1930s* was filled with Social Darwinism, master races of people, and rule by experts and not by democracies. For proof, the reader should consult Aldous Huxley, *Brave New World*.[75] These two decades of human history could not be more different.

C. The 1860s

The 1860s in the US were much more like the 1780s than the 1930s. Twenty four out of 37 states in 1868 when the 14th Amendment was ratified had a *Born Free and Equal* clause in their state constitutional Bills of Rights. The big event of the decade was the American Civil War and the abolition of African American slavery thanks to Abraham Lincoln's Emancipation Proclamation and the adoption in 1865 of the 13th Amendment. In his Gettysburg Address, President Lincoln described the United States as having been 'conceived in liberty and dedicated to the proposition that all men are created equal'. He added that the Civil War would determine 'whether a nation so conceived and so dedicated could long endure'.

The South tried in 1865 to legislate Black Codes imposing second class citizenship on African Americans, but Congress outlawed such Codes, and all other systems of caste, by adopting the 14th Amendment. The 14th Amendment thus forbade not only the Black Codes, but also European feudalism and the traditional Hindu Caste system. The 14th Amendment thus made the *Born Free and Equal* idea the Supreme Law of the Land. The Declaration of Independence and the Gettysburg Address were thus vindicated.

In Europe, the 1860s were ushered in with the publication of John Stuart Mill's, *On Liberty* in 1859. Mill's book had a huge impact in the US, in the British Empire, and in Continental Europe. Mill's book is the definitive book on classical liberalism

[75] Aldous Huxley, *Brave New World* (Chatto & Windus, 1932).

and it revived Lockean thinking after Edmund Burke and Jeremy Bentham had rejected it. Most Europeans in the 1860s were classical liberals and the UK, France, and the German states were all moving in a classical liberal direction. Feudalism and absolute monarchy were dead or were on the way out almost everywhere.

There were intellectual storm clouds on the horizon, however, because of the writings of the economist Thomas Mathus and the political philosopher Herbert Spencer. Malthus warned that people would continue breeding until everyone was impoverished and Spencer suggested in the 1860s that if that were so then we should expect 'the survival of the fittest' – a phrase Spencer coined. But, it was not until the publication in 1871 of Charles Darwin's, *The Descent of Man* that the foundations were really in place for the emergence of Social Darwinism as a creed. Darwin argued famously and correctly that human beings and apes were all descended from a common ancestor.

It was not until 1877 that the phrase 'Social Darwinism' appeared in print, and the Social Darwinist movement led by Francis Galton, Charles Darwin's half cousin. Social Darwinist ideas helped to pave the way for the introduction of Jim Crow segregation and of lynching in the South from 1877 to the 1930s. In Europe, Social Darwinism led Germany, France, and the UK to conquer and subdue most of Africa and major parts of Asia so that Europeans could 'take up the white man's burden', in Rudyard Kipling's infamous phrase to govern what were supposedly inferior races of people. The *Born Free and Equal* idea of the Enlightenment was explicitly rejected by the Social Darwinists whose ideas attained their ultimate expression in Adolf Hitler's Nazi Third Reich.

On balance, the 1860s in the US were a decade of classical liberalism since the 14th Amendment constitutionalised in 1868 the *Born Free and Equal* idea. Sadly, the 1860s were the last decade in which Americans were committed to the *Born Free and Equal* idea before the tide of Jim Crow and Social Darwinism came rushing in to bring an end to what was an Indian summer of the Enlightenment intellectual order. The *Zeitgeist of the 1860s* in the US was marked by the *Born Free and Equal* idea, which is part of the original meaning of the 14th Amendment. This explains why we as Originalists feel as confident as we do when we say that *Brown v Board of Education*[76] was right and *Plessy v Ferguson*[77] was wrong.

D. The Present Day

The present day begins for me on 10 December 1948 when the US Ambassador to the United Nations, Eleanor Roosevelt, appointed by President Truman, secured the ratification by the United Nations of the Universal Declaration of Human Rights. This Declaration was adopted by the revulsion and horror generated when

[76] *Brown v Board of Education* (n 47).
[77] *Plessy v Ferguson* (n 39).

the victorious allied armed forces crushed Hitler and the Nazis only to discover that Hitler had killed six million Jews in the Holocaust for Social Darwinist reasons. The Declaration begins as follows:

> Article 1.
>
> All human beings are born free and equal in dignity and rights. They are endowed with reason and conscience and should act towards one another in a spirit of brotherhood.
>
> Article 2.
>
> Everyone is entitled to all the rights and freedoms set forth in this Declaration, without distinction of any kind, such as race, colour, sex, language, religion, political or other opinion, national or social origin, property, birth or other status. Furthermore, no distinction shall be made on the basis of the political, jurisdictional or international status of the country or territory to which a person belongs, whether it be independent, trust, non-self-governing or under any other limitation of sovereignty.
>
> Article 3.
>
> Everyone has the right to life, liberty and security of person.
>
> Article 4.
>
> No one shall be held in slavery or servitude; slavery and the slave trade shall be prohibited in all their forms.

What is the legal effect of these articles of the Declaration of Human Rights? Quite obviously, it is a global affirmation of the continuing goodness of the *Born Free and Equal* idea of the Enlightenment and of the 1780s and a repudiation of the Social Darwinism that had led to the Holocaust.

In light of the modern day triumph of the *Born Free and Equal* idea of the 1780s and the spread of constitutional democracy and judicial review and bills of rights all over the globe, what are we to make of Professor Bruce Ackerman's idea that present day Americans ought to be governed by experts in administrative law and not by themselves? This idea of the Platonic Philosopher King as being an Administrative Agency is plainly unauthorised either by the text of the Constitution or by any statute. Professor Ackerman argues in favour of this anti-democratic idea by appeal to the *Zeitgeist of the 1930s*, but we have already seen that the *Zeitgeist of 1948* was a rejection of the 1930s and a return to the *Born Free and Equal* idea of the 1780s.

It seems to me that Professor Ackerman is hopelessly mistaken about the history of ideas over the last 250 years. The Enlightenment lives. The modern world is so committed to the *Born Free and Equal* idea that constitutional democracy, bills of rights, and judicial review to enforce those rights have spread all over the planet. Germany, Japan, Italy, India, Indonesia, South Africa, Brazil, Argentina, Mexico and even France are constitutional democracies where independent courts decide cases and controversies. Only very few countries like Russia, China and Saudi Arabia continue to challenge the *Born Free and Equal* idea.

Unfortunately, the American Administrative state with its underpinnings in the idea of the Platonic Philosopher King as an expert autocrat is a backwater of

American law, which has resisted freedom and democracy and has instead tried to strangle them. We are now told by Professor Ackerman that we should keep Administrative Law as it is because of the *Zeitgeist of the 1930s*. I disagree both on originalist grounds and for policy reasons. I believe in liberty, equality and democracy and not in government by either Experts or a Master Aryan Race.

This then is a theory about 'Time and the Law' in American Constitutionalism. I think we should value decisions made in the 1780s, in 1868, and in the present day over decisions made in 1937. The 1930s were just a very bad time in the law.

E. The Magna Carta, the Ancient Constitution and 'Time and the Law'

John Philip Reid[78] puts forth a theory about 'Time and the Law', which was propounded by Sir Edward Coke in the years between 1600 and his death in 1634. This theory was accepted by all the Massachusetts Bay and Connecticut colonists who were followers of Coke and Oliver Cromwell during the English Civil War and opponents of King Charles I. The theory of the supremacy over the King and Parliament was revived by James Otis and the American colonists in the 1760s and most Americans before the Revolutionary War believed they had the rights of Englishmen under the Ancient Constitution and that King George III and Parliament were trying to deprive them of those rights unlawfully in the years between 1760 and 1776.

Sir Edward Coke wrote about the Ancient Constitution between 1600 and his death in 1634. Coke believed King James I, King Charles I, and Parliament were unconstitutionally denying Englishmen both in England and in the North American colonies of constitutional rights which they had as Englishmen in violation of the Ancient Constitution and of the Magna Carta, which Coke thought was decided under the Ancient Constitution. The Ancient Constitution was a theory of 'Time and the Law', which is why I will discuss it here.

According to Sir Edward Coke, the Ancient Constitution was codified by King Edward the Confessor prior to the conquest of England by the Normans in 1066. It was sometimes referred to as the *Leges Edwardii* or the laws of King Edward the Confessor. Under the Ancient Constitution, the King was a constitutional and not an absolute monarch who had to consult the Witan – the King's lords and the common people – before legislating. Englishmen enjoyed the same liberty that the Anglo-Saxons had enjoyed when they were members of the German tribes in the forests of Germany as described by the Roman historian Tacitus in the first century AD in his book *Germania*.

[78] John Philip Reid, *The Ancient Constitution and the Origins of Anglo-American Liberty* (Northern Illinois University Press, 2005).

When William the Conqueror conquered England in 1066 with an army of 10,000 Normans, he had to govern a country of between two and three million Anglo-Saxons and Celts with a very small army. Accordingly, William the Conqueror had himself crowned King in Westminster Abby, a church, which Edward the Confessor had built, and he swore in his coronation oath that he would follow King Edward's laws – the *Leges Edwardii*. In 1100 AD, William the Conqueror's son, King Henry I, also swore in his coronation oath to follow King Edward's laws. When King John in the early 1200s violated the *Leges Edwardii*, his barons forced him in 1215 AD to sign the Magna Carta by which King John in effect pledged to follow the *Leges Edwardii*. Sir Edward Coke believed that the Magna Carta was no new thing, but that it instead constituted a pledge by King John that he would follow the Ancient pre-Norman conquest constitution, ie the *Leges Edwardii*.

Sir Edward Coke thus argues that from the Norman conquest up to 1603, England was a constitutional and not an absolute monarchy because its Kings and Queens were bound by the *Leges Edwardii*. It should be noted in this regard that between 1300 and 1487, five English monarchs were overthrown by Parliament and were executed. These five deposed Kings were: Edward II; Richard II; Henry VI; Edward VI; and Richard III. Coke rejected outright the claims of the Divine Right of Kings made by James I and Charles I between 1603 and 1641 to rule absolutely. As a Member of Parliament, Coke forced King Charles I to sign the Petition of Right in which he pledged to rule as a constitutional monarch. Charles I signed the Petition of Right, dismissed Parliament, and ruled as an absolute monarch. The English Civil War of 1641 to 1649 and the Glorious Revolution of 1688 finally established as a matter of English constitutional law that the King was only a constitutional monarch who was under the law and not above it.

The English Puritans, who lived in New England, all favoured Parliament over Charles I in the English Civil War. They believed in the Ancient Constitution propounded by Sir Edward Coke, and they thought they had rights under it. They claimed that they were living under the Yoke of the Norman Oppression. They longed to enjoy the liberties that Englishmen had enjoyed under the Anglo-Saxon Kings and in the forests of Germany. When trouble began between England and her 13 North American colonies between 1760 and 1776, Americans longed for their rights under the Ancient Constitution and complained that under George III, they were suffering from the Yoke of the Norman Oppression. Thomas Jefferson believed in the Ancient Constitution, and he frequently urged guests to read Tacitus, *Germania* to see what the rights of the American colonists were.

The historian JGA Pocock wrote a book entitled *The Ancient Constitution and the Feudal Law: A Study of English Historical Thought in the Seventeenth Century*[79] criticising Sir Edward Coke's account of English history. But, for my purposes

[79] JGA Pocock, *The Ancient Constitution and the Feudal Law: A Study of English Historical Thought in the Seventeenth Century* (Cambridge University Press, 1987).

it does not matter whether Coke's history of England is accurate because ALL New Englanders and the American Revolutionaries believed that Coke's historical account was correct, and they acted on this belief in starting and winning the American Revolution. Thus, even if Pocock is right that Englishmen do not enjoy the protection of the *Leges Edwardii*, colonial Americans thought and acted successfully on the belief that they did. Thus, once peace with England was signed in 1783, Americans believed they had the ancient rights of Englishmen that date back to before 1066 to the reign of King Edward the Confessor. This belief informs the meaning of the *Born Free and Equal* clauses and of the unenumerated rights enjoyed by Americans under the Ninth Amendment to the US Constitution.

Belief in the present day US constitutional law status of the rights described in the Ancient Constitution and in the Magna Carta is a radical theory of 'Time and the Law', which constitutionally privileges rights enjoyed by Englishmen prior to the Norman Conquest of 1066. Thomas Jefferson wrote of the need of Americans to throw off the Yoke of the Norman Oppression imposed by King George III. Many American conservatives and libertarians believe they are groaning under the Yoke of the New Deal oppression. This belief has ancient roots in American and English constitutional law, and it is an account of 'time and the law'.

VIII. 'Time and the Law' in the US Congress

I now turn to some special problems of 'Time and the Law' in the rules that govern the US Constitution. Article I, Section 5 provides that: 'Each House may determine the rules of its proceedings, punish its members for disorderly behavior, and, with the concurrence of two thirds, expel a member.' Both the Senate and the House of Representatives have adopted extensive rules to govern procedure in their own house, and these rules have their own distinctive effects, which are massive. I will start in Part A by discussing the unique status of the rules adopted by the Senate, and I will then in Part B discuss rules common to both the Senate and the House of Representative. This will be done from a 'Time and the Law' perspective.

A. 'Time and the Law' and the US Senate

The US Senate, unlike the US House of Representatives, is a continuing body. This is the case because in each new session of the Senate only one third of the Senators are newly elected or re-elected while two thirds of the Senators who sit in the new session of the Senate also sat in the prior session of the Senate. In contrast, all members of the House of Representatives are newly elected or re-elected and so each session of the House must start anew. One consequence of this difference is that in the Senate the rules of the Senate carry over from the previous session whereas in the House of Representatives new rules are adopted from scratch at the beginning of each session.

This is especially consequential because Senate Rule 22 requires a two-thirds majority of the Senate to end debate on a rules change. Moreover, one of the rules of the Senate that carries over is the rule requiring 60 votes in the Senate to end debate on a bill thus making the filibuster of legislation possible. The US Senate is the only legislative body in the world that does not operate by majority vote but which instead requires a three-fifths vote to get anything major done. This is especially undemocratic since the Senate itself is very undemocratic on account of the fact that each state gets two and only two seats in the Senate. California, which has a population bigger than that of Canada with more than 39 million people, has two Senate seats along with Wyoming, which also gets two Senate seats with a population of around 400,000 people.

During the Administration of President George W Bush, Senate Democrats began filibustering Bush's nominees to the lower federal courts. I testified before the Senate Judiciary Committee arguing that a simple majority of the Senate could and should change the Senate rules by a simple majority vote to end the filibuster of judicial nominees. Here is what I said on that occasion:

> The U.S. Constitution was written to establish majority rule. The historical reasons for this are clear. A major defect with the Constitution's precursor, the Articles of Confederation, was that it required super majorities for making many important decisions. The Framers deliberately set out to remedy this defect by empowering Congress to make most decisions by a simple majority. The only exceptions to this principle are in seven express situations where a two thirds vote is required. Each House of Congress does have the power by majority vote to establish the rules of its proceedings but there is no evidence this clause was meant to authorize filibusters. From 1789 to 1806, the Senate's rules allowed for cutting off debate by moving the previous question, a motion which required only a simple majority to pass.
>
> The filibuster of legislation did not originate until 1841, when it was employed by Senator John C. Calhoun to defend slavery and an extreme version of minority rights. Calhoun was called a filibusterer from a Dutch word for pirate, or as we would say today terrorist, because he was subverting majority rule. From 1841 to the present, the principle use of the filibuster has been to defend Jim Crow laws oppressing African Americans. Now for the first time in 214 years, a minority of senators are seeking to extend filibustering from legislation to the whole new area of judicial nominees – nominees who they know enjoy the support of a majority of the Senate.
>
> This is a bad idea for three reasons. First, such filibusters weaken the power of the President, who is one of only two officers of government elected to represent all of the American people. Second, filibusters of judges undermine judicial independence by giving a minority of senators led by special interest groups a veto over who can become a judge. It's already hard enough for talented and capable individuals to be appointed judges without a minority of senators imposing a litmus test. Third, the filibuster of legislation can at least be defended on the ground the federal legislation ought to be considered with extraordinary care. In contrast, the nomination of one out of one hundred and seventy-five appellate judges is a much less momentous matter. ***
>
> The Senate can always change its rules by majority vote. To the extent that Senate Rule Twenty-Two purports to require a two-thirds majority for rules changes,

> Rule Twenty-Two is unconstitutional. It is an ancient principle of Anglo-American law that one session of a legislature cannot bind a succeeding session of the same legislature. This principle goes back to the great William Blackstone, who said in his commentaries 'Acts of Parliament derogatory from the power of subsequent parliaments be naught.' Three vice presidents of the United States presiding over the Senate, Richard Nixon, Hubert Humphrey, and Nelson Rockefeller, have all ruled that the Senate rules can be changed by majority vote. Lloyd Cutler, White House Counsel to Presidents Jimmy Carter and Bill Clinton, has written in the Washington Post that Senate Rule Twenty-Two is plainly unconstitutional. The Senate can and should now amend Rule Twenty-Two by simple majority vote to ban filibusters of judicial nominations.[80]

My effort to get the Senate by majority vote to abolish the filibuster of judicial nominations failed in 2003, but Senate Democrats under President Obama ended up abolishing the filibuster of lower court judicial nominees and executive branch nominees in 2013. Senate Republicans then abolished the filibuster of Supreme Court nominees in 2017 so that they could confirm Neil Gorsuch's nomination to the US Supreme Court.

The 'Time and the Law' rule in the Senate today is that nominations can no longer be filibustered but that legislation, except for Reconciliation Tax and Spending Bills, can be filibustered. The Senate filibuster weakens the power of newly elected senators to change the status quo by requiring 60 votes to get anything important done.

B. The House of Representatives

There is no filibuster in the House of Representatives and the old rules of the previous session of the House do not carry over to govern new sessions of the House in the way that such rules do carry over in the Senate. Newly elected Representatives are more equal to longstanding Representatives than is the case in the Senate. In both the House and the Senate, legislation, nominations, and treaties proposed in an old session die when that session ends. Such legislation, nominations, and treaties must be reintroduced in a new session of the Senate and of the House of Representatives. In this respect, newly elected Members of Congress are fully equal to Members of Congress who have served in the previous session. This is a good thing since it promotes democracy.

C. The Seniority System and the Congressional Committees

Both the Senate and the House of Representatives have elaborate Committee structures to facilitate their oversight of the Executive Branch. These Committee

[80] *Testimony of Professor Steven G Calabresi*, Senate Judiciary Committee Hearing on Judicial Nominations and Filibusters (6 May 2003).

structures almost always remain the same when a new Congress is seated and members usually remain assigned to the same Committees in a new session of Congress. Both Houses use a seniority system that rewards long serving Members of Congress who move slowly up the ladder on their Committees in seniority. Committee Chairs are picked by seniority in the Senate, but in the House seniority counts a lot but is not dispositive in who becomes a Committee Chair.

The bottom line as a matter of 'Time and the Law' is that newly elected senators and representatives have less power than do long-serving senators and representatives, which is undemocratic. The seniority and committee systems do, however, effectuate a lot more meaningful oversight of the executive branch than otherwise would be possible. This benefit strengthens Congress as an institution relative to the President and the Supreme Court, which indirectly benefits newly elected Members of Congress as well.

IX. Conclusion

This chapter has offered a positive account of how problems of 'Time and the Law' are addressed both by the US Constitution and by the US Congress in the rules it adopts to govern itself. I hope this description of current US practice will be of interest both to Americans and to people from other constitutional democracies.

3

Time and Change in Constitutional Amendment

RICHARD ALBERT†

I. Introduction

We cannot understand constitutional change without understanding its relation to time. Yet the temporal dimension of constitutional amendment remains understudied and undertheorised, despite the prevalence of democratic constitutions that require political actors to adhere to certain specifications as to the timing of various steps in the amendment process, whether at the initiation, proposal or ratification stages, or indeed all of these steps.[1] For example, amendment rules sometimes establish deliberation floors or ceilings to compel political actors to consider an amendment proposal during a defined period of time, establishing either a minimum or maximum period of deliberation.[2] Amendment rules sometimes also create safe harbour provisions that altogether prohibit political actors from proposing amendments for a defined period of time, either after a new constitution has come into force or after an amendment has failed or succeeded.[3] In this chapter, I explore and evaluate the use of temporal limitations in constitutional amendment from a comparative perspective in order both to demonstrate the possibilities for the design of temporal limitations and to expose the trade-offs between political brinkmanship and constitutional contemporaneity in constitutional amendment.

Time is only one dimension along which formal amendment rules may fragment power. They may also fragment power among political parties, as does the Japanese Constitution, by requiring supermajority agreement in the national legislature.[4] They may in addition fragment power among institutional branches

† I am grateful to the editors for inviting me to present an earlier draft at their conference on 'Time and the Law', and to conference participants for their criticisms and suggestions. I examine all aspects of constitutional amendment, not only in relation to time, in Richard Albert, *Constitutional Amendments: Making, Breaking, and Changing Constitutions* (Oxford University Press, 2019).

[1] See part II.

[2] See section II.A.

[3] See section II.B.

[4] Japan Constitution, Art 96.

of government, as does the French Constitution, authorising the executive and legislature each to initiate a constitutional amendment.[5] Amendment rules may also fragment power along ethnic or cultural identities, as does the Constitution of Bosnia and Herzegovina, which authorises amendment only by a Parliamentary Assembly whose members must include Croats, Bosniacs and Serbs,[6] or the Constitution of Kiribati, which does not permit an amendment to the rights of Banabans unless the amendment is supported by the nominated or elected representative of the Banaban community.[7] Geographic boundaries are another way to fragment the amendment power, as in Iraq, where the Constitution protects the country's regions by disallowing certain amendments unless the regional legislative authority has approved it and so have the people of the region.[8] Federalist structures are another way of fragmenting amendment power: the constitutions of Australia and South Africa each sometimes require a subnational entity affected by a given amendment to consent to the change.[9]

Fragmenting power is the core design strategy of formal amendment rules in constitutional democracies: it serves the purpose of rallying a broad, representative and sustainable base of support behind a ratified amendment.[10] Fragmenting the amendment power almost always exacerbates amendment difficulty but this is better understood as its consequence not its purpose. The fragmentation of power in constitutional amendment is a screen through which may pass only those amendments reinforced by a breadth and depth of political and popular agreement that may potentially reflect multiple layers of legitimacy. Not only the legal legitimacy that comes from successfully navigating the codified rules of amendment and the sociological legitimacy reflected in the approval of political actors representing disparate groups, but also the moral legitimacy associated with modern forms of collaborative governance that privilege consent and cooperation over conquest and the consolidation of power.[11]

The fragmentation of the amendment power across time has unique properties and consequences whose study can fill a void in the field. The study of the temporal dimension of constitutional amendment moreover responds to Paul Pierson's call for greater attention to the *structure* as opposed to the *number* of veto points in institutional design, particularly as to how and why political institutions are structured to be resistant to change.[12] I therefore take this as an invitation, both to fill this void and to advance our learning and interest in the relationship between time and change. Although I focus primarily on Canada and the United States, the analysis may be applicable elsewhere and indeed is intended to invite further study.

[5] France Constitution, Art 89.

[6] Bosnia & Herzegovina Constitution, Art X.

[7] Kiribati Constitution, Art 124.

[8] Iraq Constitution, Art 126.

[9] Australia Constitution, Art 128; South Africa Constitution, Art 74.

[10] In this chapter, I use 'amendment' to mean 'formal amendment' unless otherwise noted.

[11] For a discussion of these three forms of legitimacy, see Richard H Fallon, Jr, 'Legitimacy and the Constitution' (2005) 118 *Harvard Legal Review* 1787 at 1794–97.

[12] See Paul Pierson, *Politics in Time* (Princeton University Press, 2004) 144–46.

I begin from the proposition that democratic constitutions constrain conduct and choice.[13] They structure the institutions of government, they shape the interactions of political actors and they determine whether and how political actors may operationalise their political preferences into law. The choice of constitutional design itself also reflects objectives. For instance, whether constitutional designers adopt weak- or strong-form judicial review, the choice reflects an effort to manage governance and its outcomes according to prior values about how political actors should make decisions and more importantly why, perhaps for political insurance as Tom Ginsburg theorises,[14] or for hegemonic preservation as Ran Hirschl maintains.[15] The point is that constitutions create and enable institutions and then constrain them; they confer powers and then make rules for their use; they define a collective purpose and then qualify it when necessary.[16]

Just as democratic constitutions structure the conduct of their subjects and objects – the who and what of constitutional law – constitutions sometimes also structure the timing – the when – of decisions political actors make on the authority of the constitutional text. For example, the Constitution Act 1982 required the Prime Minister of Canada to convene a first ministers' constitutional conference within 15 years of its coming-into-force.[17] The United States Constitution established a temporal rule of its own: the slave trade was protected from abolition for the first 20 years of the Constitution.[18] These and other uses of time in constitutional design to structure conduct and choice have, with good reason, drawn recent attention from scholars of comparative public law.[19] But there remains much to learn about the temporal dimension of constitutional change, specifically about the architecture of the rules of formal amendment.

My purpose in this chapter is to illuminate the options available to constitutional designers as they consider whether and why to codify temporal limitations

[13] See Jon Elster, *Ulysses Unbound: Studies in Rationality, Precommitment, and Constraints* (Cambridge University Press, 2000) 89–90. Unless otherwise noted in this chapter, 'democratic constitutions' or 'constitutions' mean constitutions that are both democratic and codified. I do not focus on either uncodified constitutions, which do not codify formal amendment rules in the conventional form, or sham constitutions, which have no constraining effect on political actors.

[14] Tom Ginsburg, *Judicial Review in New Democracies: Constitutional Courts in Asian Cases* (Cambridge University Press, 2003).

[15] Ran Hirschl, *Towards Juristocracy: The Origins and Consequences of the New Constitutionalism* (Harvard University Press, 2004).

[16] See Richard Albert, 'The Cult of Constitutionalism' (2012) 39 *Florida State University Law Review* 373 at 390.

[17] See Constitution Act 1982, s 49, being Schedule B to the Canada Act 1982 (UK), 1982, c 11 [Constitution Act 1982]. The conference was ultimately held in 1996 but there is some question whether the gathering fulfilled the spirit of the constitutional requirement. See John D Whyte, '"A Constitutional Conference … Shall be Convened …" Living with Constitutional Promises' (1996) 8 *Constitutional Forum* 15.

[18] US Constitution, Art V.

[19] See, eg, Sofia Ranchordás, *Constitutional Sunsets and Experimental Legislation* (Edward Elgar, 2014); Ozan O Varol, 'Temporary Constitutions' (2014) 102 *California Law Review* 409 (2014); Rosalind Dixon and Tom Ginsburg, 'Deciding Not to Decide: Deferral in Constitutional Design' (2011) 9 *International Journal of Constitutional Law* 636.

on how political actors deploy amendment rules.[20] This study of amendment has real implications for the present day, as we have seen and will likely continue to see around the democratic world efforts by political actors to circumvent the onerous rules of constitutional amendment.[21] Constitutional designers have at their disposal resources to help them understand the relationship between formal amendment difficulty and informal constitutional change,[22] as well as how to identify when political actors deploy the democratic procedures of amendment and ordinary lawmaking to achieve non-democratic ends.[23] But they have few resources to understand and evaluate the use of temporal limitations in the design of formal amendment rules. I seek here to fill that void.

II. Time and Change in Constitutional States

There are two major forms of temporal limitations in constitutional amendment: deliberation requirements and safe harbours.[24] I focus in this chapter only on deliberation requirements, though it is useful here to distinguish them from safe harbours.[25] A deliberation requirement compels political actors to evaluate an amendment proposal during a defined period of time. This period of time may be either a floor or a ceiling, the former referring to a minimum amount of time for which an amendment proposal must remain open to deliberation by political actors and the public prior to its ratification, and the latter to the maximum amount of time during which political actors and the public may deliberate on an amendment before a ratification vote must be held. A safe harbour creates an outright prohibition on constitutional amendment during a specified period of time. Both kinds of limitations are peremptory and variable. They are peremptory in that political actors are bound by these codified temporal limitations,[26] and they

[20] This is an updated version of a paper on 'Temporal Limitations on Constitutional Amendment', appearing in Volume 21 of the *Review of Constitutional Studies*.

[21] See Richard Albert, 'Amending Constitutional Amendment Rules' (2015) 13 *International Journal of Constitutional Law* 655 at 658–59.

[22] See, eg, Arend Lijphart, *Patterns of Democracy: Government Forms and Performance in Thirty-Six Countries* (Yale University Press, 1999) 225–30; Edward Schneier, *Crafting Constitutional Democracies: The Politics of Institutional Design* (Rowman & Littlefield Publishers, Inc, 2006) 223; Heather K Gerken, 'The Hydraulics of Constitutional Reform: A Skeptical Response to Our Undemocratic Constitution' (2007) 55 *Drake Law Review* 925 at 927.

[23] See, eg, Richard Albert, 'Constitutional Amendment by Stealth' (2015) 60 *McGill Law Journal* 673; Ozan O Varol, 'Stealth Authoritarianism' (2015) 100 *Iowa Law Review* 1673; David Landau, 'Abusive Constitutionalism' (2013) 47 *University of California Davis Law Review* 189.

[24] See Richard Albert, 'The Structure of Constitutional Amendment Rules' (2014) 49 *Wake Forest Law Review* 913 at 952–54.

[25] Yaniv Roznai is currently developing a project on safe habours, what he calls 'circumstantial unamendability'.

[26] Political actors may of course always amend a deliberation requirement if it is not absolutely entrenched against amendment, and they may also rewrite a constitution that makes a deliberation requirement or safe harbour formally unamendable. The political costs of both, however, are high

are variable in that designers may choose to codify deliberation requirements or safe harbours of varying durations of time, either separately or in combination. Safe harbours are occasionally subject to override, as in Portugal, where political actors may by an extraordinary supermajority vote to initiate the amendment process for exceptional reasons.[27]

A. Safe Harbours

Constitutional designers codify different forms of safe harbours in connection with constitutional amendment. We can identify at least five general periods of time during which designers will impose safe harbours: (1) states of emergency; (2) periods of succession or regency; (3) the interval following a failed amendment; (4) the interval following a successful amendment; and (5) the period immediately following the adoption of a new constitution. Each of these forms of safe habour disables the amendment process during specified periods of time.

Consider a safe harbour during a state of emergency. According to the Spanish Constitution, 'the process of constitutional amendment may not be initiated in time of war or under any of the states contemplated in section 116',[28] a reference to states of 'alarm, emergency and siege (martial law)'.[29] Constitutions also codify safe habours in connection with succession or where a ruler is unable to lead. In Tonga, for example, 'it shall be lawful for the Legislative Assembly to discuss amendments to the Constitution provided that such amendments shall not affect the law of liberty, the succession to the Throne, and the titles and hereditary estates of the nobles'.[30] Likewise, in Luxembourg, 'during a regency, no change can be made to the Constitution the constitutional prerogatives of the Grand Duke, his status as well as the order of succession'.[31]

Safe harbours sometimes also prohibit amendment in the immediate aftermath of a failed or successful amendment, and in the period following the adoption of a new constitution. In Estonia, 'an amendment to the Constitution regarding the same issue shall not be initiated within one year after the rejection of a corresponding bill by a referendum or by the Riigikogu',[32] which is the unicameral legislature authorised to amend the Constitution in collaboration with other institutions in Estonia.[33] In contrast, under the Greek Constitution, 'revision of the

in democratic states. See Yaniv Roznai, *Unconstitutional Constitutional Amendments: The Limits of Constitutional Amendment Powers* (Oxford University Press, 2017).

[27] Portugal Constitution, Art 284.

[28] Spain Constitution, s 169.

[29] *Ibid*, s 116.

[30] Tongo Constitution, Art 79.

[31] Luxembourgh Constitution, Art 115.

[32] Estonia Constitution, Art 168.

[33] *Ibid*, Arts 161–68.

Constitution is not permitted before the lapse of five years from the completion of a previous revision'.[34] The Cape Verdean Constitution illustrates the fifth form of safe harbour, which disabled the amendment power for five years after the adoption of the 1980 Constitution: 'This Constitution may be revised, in whole or in part, by the National Assembly after five years from the date of its promulgation'.[35] The Constitution did, however, create an escape-hatch authorising an extraordinary supermajority of the National Assembly to bypass this safe harbour.[36] One of the earliest safe harbours, if not the first, appeared in the first French Constitution, which disallowed amendments to the new constitution for the first two terms of the national legislature.[37]

B. Deliberation Floors and Ceilings

This chapter is concerned principally with deliberation requirements. The distinction between a deliberation floor and ceiling is important to what follows, so let us review some examples of each to concretise the difference. A deliberation floor establishes the minimum period of time to deliberate on an amendment proposal prior to a binding vote or action to ratify it, or to move the proposal forward to the next steps in the amendment process. In contrast, a deliberation ceiling establishes the maximum period of time for this kind of deliberation on an amendment.

Consider the Italian Constitution. It creates a deliberation floor requiring at least three months between legislative debates on an amendment proposal:

> Laws amending the Constitution and other constitutional laws shall be adopted by each House after two successive debates at intervals of not less than three months, and shall be approved by an absolute majority of the members of each House in the second voting.[38]

Similarly, although its deliberation floor is directed to the public not the legislature, the South Korean Constitution requires the President to give the public a minimum amount of time to evaluate an amendment: 'Proposed amendments to the Constitution shall be put before the public by the President for twenty days or more'.[39]

In contrast, the Costa Rican Constitution codifies a deliberation ceiling. The Legislative Assembly must review the amendment proposal 'three times at intervals of six days, to decide if it is admitted or not for discussion'.[40] Interestingly,

[34] Greece Constitution, Art 110.
[35] Cape Verde Constitution, Art 309.
[36] *Ibid.*
[37] French Constitution, Art 3 (1791) (superseded).
[38] Italy Constitution, Art 138.
[39] South Korea Constitution, Art 129.
[40] Costa Rica Constitution, Art 195.

the Australian Constitution merges both a deliberation floor and ceiling into its conditions for ratifying an amendment:

> The proposed law for the alteration [of the Constitution] must be passed by an absolute majority of each House of the Parliament, and not less than two nor more than six months after its passage through both Houses the proposed law shall be submitted in each State and Territory to the electors qualified to vote for the election of members of the House of Representatives.[41]

There is a third variety of deliberation requirement: the model of intervening elections. This model of constitutional change combines time with the design of representative institutions, for instance by requiring successive parliaments to consent to an amendment. The same parliament is prohibited from both proposing and ratifying a formal amendment without an intervening national election to reconstitute the parliament between each of these steps. This model is prominent in Scandinavia, where Denmark, Norway and Sweden each structure their formal amendment in this way.[42] In this chapter, I focus only on deliberation floors and ceilings.

III. Two Models of Constitutional Consensus: Canada and the United States

Both deliberation floors and ceilings structure how political actors and the public arrive at the consensus required to legitimate a constitutional amendment. But each is anchored in a different perspective on the nature and form of the political agreement that legitimises a constitutional amendment, and each privileges different values in the formation of constitutional consensus. In this part, I compare two competing approaches to the codification of deliberation floors and ceilings. The American model imposes neither a deliberation floor nor a ceiling and authorises the inter-generational ratification of a constitutional amendment. In contrast, the Canadian model imposes both a deliberation floor and a ceiling, and therefore makes constitutional amendment conditional on intra-generational ratification. Both constitutional traditions have lived experiences suggesting caution in codifying deliberation requirements.

A. Inter-Generational Ratification

The text of the original United States Constitution is silent on when amendment proposals must be ratified. As I discuss below, however, Congress has

[41] Australia Constitution, Art 128.

[42] See Denmark Constitution, s 88; Norway Constitution, Art 121; Sweden Inst of Gov, Art 16.

sometimes imposed a ratification deadline on amendment proposals, an option the Constitution leaves open by its very silence.[43] The text of the Constitution says only that an amendment will be valid where two-thirds of Congress votes to propose one and thereafter three-quarters of the states vote to ratify it either in state legislatures or conventions.[44] By law, though not required by the constitutional text, the Archivist of the United States issues a certification when the requisite number of states have ratified an amendment.[45] Historically, the average time span from proposal to ratification has been under two years and three months for 24 of the 27 amendments to the Constitution; two of them took under four years; and one took nearly 203 years to ratify.[46] The absence of deliberation requirements, however, has raised questions about the timeliness of state ratification.

On 2 March 1861, one month before the first major battle in the Civil War, the United States Congress adopted an amendment proposal protecting slavery in the states. Known as the 'Corwin Amendment' for Representative Thomas Corwin,[47] this amendment proposed that

> no amendment shall be made to the Constitution which will authorize or give to Congress the power to abolish or interfere, within any State, with the domestic institutions thereof, including that of persons held to labor or service by the laws of said State.[48]

Outgoing President James Buchanan signed the proposal,[49] newly-elected President Abraham Lincoln endorsed it,[50] and three states ratified it.[51] But the onset of the Civil War interrupted the ratification process.[52] The Corwin Amendment would have become the 13th Amendment had it been ratified,[53] but instead the United States ultimately codified a different 13th Amendment abolishing slavery.[54]

[43] I leave open for now the possibility that these congressionally-imposed time limits are unconstitutional constraints on the amendment power.

[44] US Constitution, Art V. The Constitution also authorises amendment by constitutional convention but this convention process has never been successfully used. See William B Fisch, 'Constitutional Referendum in the United States of America' (2006) 54 *American Journal of Comparative Law* 485 at 490.

[45] See 1 USC § 106b (1988).

[46] See The US National Archives and Records Administration: www.archives.gov/exhibits/charters/constitution_amendments_11-27.html (last accessed 10 September 2019).

[47] Ewen Cameron Mac Veagh, 'The Other Rejected Amendments' (1925) 222 *The North American Review* 274 at 281.

[48] HR 80, 36th Cong, 12 Stat 251 (1861).

[49] Rogers M Smith, 'Legitimating Reconstruction: The Limits of Legalism' (1999) 108 *Yale Law Journal* 2039 at 2059, fn 89.

[50] Abraham Lincoln, *First Inaugural Address* (4 March 1861).

[51] Douglas Linder, 'What in the Constitution Cannot be Amended? (1981) 23 *Arizona Law Review* 717 at 728.

[52] Gary Jeffrey Jacobsohn, *Constitutional Identity* (Harvard University Press, 2010) 36.

[53] Alexander Tsesis, *The Thirteenth Amendment and American Freedom: A Legal History* (New York University Press, 2004) 2–3.

[54] US Constitution, Amendment XIII.

But there remains an open question: is the Corwin Amendment still today ratifiable?[55] The ratification of the 27th Amendment in 1992 – well over 200 years after Congress passed it and transmitted it to the states – suggests the answer could well be yes. The amendment states that 'no law, varying the compensation for the services of the Senators and Representatives, shall take effect, until an election of Representatives shall have intervened'.[56] James Madison initially proposed the amendment in the First Congress on 8 June 1789.[57] Congress adopted a resolution proposing the amendment in the same year, six states had ratified it by 1792, and a seventh state ratified it in 1873.[58] But it was not until 1978 that another state ratified the amendment, and this subsequently led another 30 states to jump aboard in the intervening 14 years.[59] In 1992, Michigan became the 38th state to ratify the amendment proposal, in so doing reaching the three-fourths threshold for satisfying the ratification requirement.[60] Despite having taken over 200 years to ratify, Congress saw no constitutional infirmity with the amendment,[61] the Department of Justice issued a memorandum defending its constitutional soundness,[62] and a federal court refused to hear a challenge to it.[63]

The amendment rules in Article V do not prohibit Congress from imposing a time limit for states to ratify an amendment proposal.[64] Congress first imposed a ratification deadline on the 18th Amendment,[65] insisting that

> this article shall become inoperative unless it shall have been ratified as an amendment to the Constitution by the legislatures of the several States as provided in the Constitution, within seven years from the date of the submission hereof to the State by the Congress.[66]

Similar language has appeared in all amendment proposals or authorising resolutions since the 20th Amendment.[67] The Corwin Amendment appears, then, to lay

[55] Michael Stokes Paulsen, 'A General Theory of Article V: The Constitutional Lessons of the Twenty-Seventh Amendment' (1993) 103 *Yale Law Journal* 677 at 701–704.

[56] US Constitution, Amendment XXVII.

[57] Louise Weinberg, '*Political Questions and the Guarantee Clause*' (1994) 65 *University* of Colorado Law Review 887 at 937, fn 179.

[58] Gideon M Hart, 'The "Original" Thirteenth Amendment: The Misunderstood Titles of Nobility Amendment' (2010) 94 *Marquette Law Review* 311 at 327, fn 88.

[59] Richard A Primus, 'When Should Original Meanings Matter?' (2008) 107 *Michigan Law Review* 165 at 209, fn 157.

[60] David P Currie, 'The Constitution in Congress: Substantive Issues in the First Congress' (1994) 61 *University of Chicago Law Review* 775 at 851, fn 449.

[61] Paul E McGreal, 'There is no Such Thing as Textualism: A Case Study in Constitutional Method' (2001) 69 *Fordham Law Review* 2393 at 2431.

[62] Memorandum Opinion for Counsel to the President, 16 Op Off Legal Counsel 87 (2 November 1992).

[63] See *Boehner v Anderson* 809 F Supp 138 (DDC 1992).

[64] Adam M Samaha, 'Dead Hand Arguments and Constitutional Interpretation' (2008) 108 *Columbia Law Review* 606 at 649.

[65] Peter Suber, 'Population Changes and Constitutional Amendments: Federalism Versus Democracy' (1987) 20 *University of Michigan Journal of Law Reform* 409 at 423–24.

[66] US Constitution, Amendment XVIII, § 3.

[67] Michael J Lynch, 'The Other Amendments: Constitutional Amendment that Failed' (2001) 93 *Law Library Journal* 303 at 305.

dormant, ready to be ratified by the requisite number of states today. The same is true of three other amendments proposed by the Congress years ago but not yet ratified by the states.

Each of these unratified amendments has been passed by both houses of Congress, transmitted to the states, and contains no expiration date. Each apparently remains viable pending ratification by the required three-quarters of states.[68] The first proposes to change the size and number of congressional districts.[69] Proposed in 1789, it has thus far been ratified by roughly ten states.[70] The second would strip American citizenship from anyone who accepts a foreign title of nobility, honour or dispensation without congressional permission.[71] It was successfully proposed in 1810 by a wide margin in the Senate and the House.[72] The third proposes to grant Congress the power to regulate child labour.[73] Proposed in 1924, it has been ratified by 28 states.[74] The fourth outstanding amendment is the Corwin Amendment.

These four outstanding amendment proposals were transmitted to the states in 1789, 1810, 1861 and 1926. The long interval between proposal and ratification raises the question whether an amendment without a ratification deadline nonetheless expires after a significant period of time. The answer from political practice is no: the 27th Amendment was ratified over 200 years after its proposal. The answer from the case law of the United States Supreme Court appears also to be no: lapse of time does not by itself negate the ratifiability of an amendment passed by Congress and transmitted to the states.[75] Whether an amendment has been ratified with sufficient contemporaneity to its proposal is a judgement for Congress to make,[76] and Congress' judgement is moreover a political question unreviewable by courts.[77] The congressional role is collateral to the larger point here, however, which is that the United States Constitution authorises inter-generational ratification: an amendment proposal may be validly ratified by a future generation whose ratifiers may not even have been alive when it was first proposed.

[68] Congress has also transmitted to the states amendment proposals that were defeated in part due to expiration dates. Those amendments include the Equal Rights Amendment and the District of Columbia Congressional Representation Amendment. See HRJ Res 554, 95th Cong, 92 Stat 3795 (1978) (proposing District of Columbia Representation Amendment); HRJ Res 208, 92nd Cong, 86 Stat 1523 (1972) (proposing Equal Rights Amendment).

[69] 1 Pub Res 3, 1st Cong, 1 Stat 97 (1789).

[70] Gabriel J Chin and Anjali Abraham, 'Beyond the Supermajority: Post-Adoption Ratification of the Equality Amendments' (1008) 50 *Arizona Law Review* 25 at 29.

[71] 11 Pub Res 2, 11th Cong, 2 Stat 613 (1810).

[72] Curt E Conklin, 'The Case of the Phantom Thirteenth Amendment: A Historical and Bibliographic Nightmare' (1996) 88 *Law Library Journal* 121 at 123.

[73] HRJ Res 184, 68th Cong, 43 Stat 670 (1924).

[74] Jol A Silversmith, 'The "Missing Thirteenth Amendment": Constitutional Nonsense and Titles of Nobility' (1999) 8 *Southern California Interdisciplinary Law Journal* 577 at 580, fn 20.

[75] *Coleman v Miller* 307 US 433 (1939).

[76] *Ibid* at 454.

[77] *Ibid*. *Coleman* refined the earlier holding in *Dillon v Gloss* 256 US 368 (1921), which held that ratification 'must be within some reasonable time after the proposal'. *Ibid* at 375.

B. Intra-Generational Ratification

In contrast, the general amendment procedure in the Constitution of Canada requires intra-generational ratification and indeed denies the possibility of inter-generational ratification. Here I stress the *general* amendment procedure because the Constitution Act 1982 codifies five separate amendment procedures,[78] each one designated for amendments to specific provisions and principles, and each increasing in difficulty according to the importance of the codified provision or principle to which it is assigned.[79] It is beyond the scope of this chapter to explain and evaluate all five amendment procedures but a short word on each is appropriate.[80]

The first, the unilateral provincial amendment procedure, authorises provinces to amend their own constitution by simple legislative majority.[81] The second, the unilateral federal amendment procedure, authorises a majority in both houses of Parliament to amend Parliament's own internal constitution and matters of federal executive government.[82] The regional amendment procedure requires both houses of Parliament and the legislatures of one or more but not all provinces affected by a given amendment to agree by majority vote to the amendment.[83] The most onerous amendment rule, the unanimity procedure, requires approval resolutions from both houses of Parliament and from each provincial legislature, and it applies for amendments to the provisions and principles thought to be most important in Canada, including the monarchy, the composition of the Supreme Court and the rules of formal amendment themselves.[84] None of these four amendment procedures codifies a temporal limitation on proposal or ratification.

But the general amendment procedure codifies both a deliberation floor and ceiling, in contrast to the United States Constitution, which codifies neither. This procedure requires approval from both houses of Parliament and from at least two-thirds of the provinces whose aggregate population represents at least half of Canada's total provincial population.[85] This 'general' amendment procedure serves as both the default amendment procedure and a more targeted one: it must be used to amend all provisions and principles not otherwise assigned to another amendment procedure, and it also applies to certain designated provisions and principles,

[78] Parliament also possesses a narrow power of amendment outside of the Constitution Act 1982. See Constitution Act, 1867, 30 & 31 Victoria, c 3 (UK), pt VII, s 101.

[79] See Constitution Act 1982, Procedure for Amending Constitution of Canada, Part V. For a theoretical perspective on the purpose of these escalating amendment thresholds in Canada, see Richard Albert, 'The Theory and Doctrine of Unconstitutional Constitutional Amendment in Canada' (2016) 41 *Queen's Law Journal* 153.

[80] For a detailed analysis of Canada's formal amendment rules, see Richard Albert, 'The Difficult of Constitutional Amendment in Canada' (2015) 53 *Alberta Law Review* 85.

[81] Constitution Act 1982, s 45.

[82] *Ibid* at s 44.

[83] *Ibid* at s 43.

[84] *Ibid* at s 41.

[85] *Ibid* at s 38(1).

for instance senatorial selection, power and representation.[86] For our purposes, the key parts of the general amendment procedure are the temporal limitations it puts on ratifying an amendment:

> A proclamation shall not be issued [] before the expiration of one year from the adoption of the resolution initiating the amendment procedure thereunder, unless the legislative assembly of each province has previously adopted a resolution of assent or dissent.
>
> A proclamation shall not be issued [] after the expiration of three years from the adoption of the resolution initiating the amendment procedure thereunder.[87]

The first of two parts reflects a deliberation floor and the second a deliberation ceiling. Together, they generate the rule that no amendment may become official without giving political actors at least one year from the date of its proposal to consider it, nor may an amendment pass after three years from the same date. This is a small window of time within which to authorise a material change to the Constitution of Canada. But there are both theoretical and actual reasons why this rule makes sense in the Canadian context. First, as a matter of theory that can be applied elsewhere, the rationale for the three-year limit was threefold: first, 'to bring closure to an amendment process that was dragging on without ever capturing the necessary support'; second, to 'ensure that a forgotten resolution supporting an amendment would not later catch a government by surprise if the requisite support was gained'; and third, 'to ensure a proposal was debated at a time when the circumstances surrounding its initiation were still current'.[88] As a Canada-specific matter, however, the one-year rule must be read alongside the cluster of rules allowing provinces to opt-out of amendments that affect provincial powers, rights or privileges.[89] In order to invoke this protection, a province 'needs a reasonable time to decide whether or not to exercise this option, and one year does not seem unduly long to consider a change that is likely to last for generations'. And so the Constitution is designed to give provinces one year to evaluate whether to proceed with ratifying the amendment or to opt-out from its application.[90]

The Constitution of Canada codifies other amendment rules in connection with time. For one, the House or Senate or indeed any legislative assembly may rescind an earlier-passed resolution of assent to a proposed amendment at any time before the amendment is proclaimed.[91] Indeed, Newfoundland exercised this power of rescission when a change of government occurred while the Meech Lake Accord was pending before the legislative assemblies.[92] Another temporal

[86] *Ibid* at s 42(1).
[87] *Ibid* at s 39.
[88] Katherine Swinton, 'Amending the Canadian Constitution: Lessons from Meech Lake' (1992) 42 *University of Toronto Law Journal* 139 at 146.
[89] Constitution Act 1982, s 38(2)-(4), s 40.
[90] Swinton (n 88).
[91] Constitution Act 1982, s 46(2).
[92] James Ross Hurley, *Amending Canada's Constitution: History, Processes, Problems and Prospects* (Minister of Supply and Services Canada, 1996) 112.

amendment rule in Canada allows the House to overcome Senate inaction: an amendment made using the regional, general or unanimity procedure may be made without an authorising Senate resolution if the Senate has not adopted one within 180 days of the House of Commons adopting its own authorising resolution and again adopting it after 180 days.[93] The rules of amendment specify that this 180-day period does not run while Parliament is prorogued or dissolved.[94]

Some have attributed the failure of the Meech Lake Accord to the three-year deliberation ceiling, which required provinces to ratify it within the specified time or the entire amendment package would expire.[95] The Meech Lake Accord sought to fulfill Quebec's requests for more powers in the aftermath of the process that led to the Constitution Act 1982 – a process in which Quebec had been arginalised.[96] Negotiated by the heads of government in Canada, the Meech Lake Accord was both perceived as and indeed was in fact the result of executive federalism,[97] a term with negative connotations for excluding the people from meaningful participation in its design and negotiation. The Accord proposed amendments to recognise Quebec as 'a distinct society',[98] to give all provincial governments the formal power to suggest senatorial nominees for appointment,[99] to grant all provinces some control over immigration, to constitutionalise the Supreme Court,[100] to mandate constitutional conferences,[101] and to grant all provinces a veto in constitutional amendments on major items concerning proportional representation, the Supreme Court and the Senate.[102] Most of what Quebec had demanded was later offered to all provinces.

There was some doubt, however, whether the Meech Lake Accord was indeed subject to the three-year time limit in the general amendment procedure.[103] The uncertainty arose from the Meech Lake amendment package itself, parts of

[93] Constitution Act 1982, s 47(1).

[94] *Ibid* at s 47(2).

[95] See, eg, Government of Canada, *Special Joint Committee on the Process for Amending the Constitution of Canada, The Process of Amending the Constitution of Canada: The Report of the Special Joint Committee* (Ottawa, Supply and Services, 1991) 31; Gordon Robertson, *Memoirs of a Very Civil Servant: Mackenzie King to Pierre Trudeau* (University of Toronto Press, 2000) 342–48; Patrick J Monahan, 'After Meech Lake: An Insider's View' (The Inaugural Thomas G. Feeney Memorial Lecture, 13 October 1990) 9–10, available at: https://qweri.lexum.com/w/canlii/1990CanLIIDocs13.pdf.

[96] See Peter W Hogg, *Meech Lake Constitutional Accord Annotated* (Carswell, 1988) 3–4.

[97] David Cameron and Richard Simeon, 'Intergovernmental Relations in Canada: The Emergency of Collaborative Federalism' (2002) 32 *Publius* 49 at 52.

[98] The 1987 Constitutional Accord, Ottawa, Ontario, 3 June 1987, at Schedule s 1 ('Meech Lake Accord').

[99] *Ibid* at Schedule s 2.

[100] *Ibid* at Schedule s 3.

[101] *Ibid* at Schedule s 13.

[102] *Ibid* at Schedule s 9.

[103] Compare Gordon Robertson, 'Meech Lake – The Myth of the Time Limit' Choices (Supplement to *Institute for Research on Public Policy Newsletter*) Vol 11, No 3, May/June 1989 (arguing that the time limit should not apply), with RE Hawkins, 'Meech Lake – The Reality of the Time Limit' (1989) 35 *McGill Law Journal* 196 (arguing that the time limit should apply) and FL Morton, 'How Not to Amend the Constitution' (1989) 12 *Canadian Parliamentary Review* 9 (arguing that the entire debate was flawed).

which on their own would trigger the general amendment procedure while others would fall under the unanimity procedure. Only the general amendment procedure requires that an amendment be ratified within three years of its initiation; the unanimity procedure does not. Yet political actors proposed the Meech Lake Accord as an omnibus bill of amendments, and subjected it to the most exacting requirements of both the general and unanimity procedures, requiring Parliament and each of the provinces to approve the proposal within three years. This may have been the nub of the problem in the ratification of the Meech Lake Accord because it was not constitutionally necessary to subject the entire Meech Lake Accord to the three-year requirement.[104] Political actors could have split the package into two parts: one with amendments in relation to matters under the unanimity rule in Section 41, which does not impose a deliberation requirement; and another with amendments in relation to matters under Section 38, which does impose a deliberation requirement. Nonetheless, the lead advisor to the Government of Canada on constitutional matters at the time recently explained the choice made at the time:

> The Meech Lake Accord included some amendments that called for the general procedure and others that required unanimous approval. The draft amendments were part of one interrelated package. I advised that both the three-year limitation period and the need for unanimity would apply simultaneously.[105]

Political actors therefore chose, correctly or not, a ratification strategy reflecting the concept of *cumul*, which refers to the informal combination of requirements in two or more amendment procedures.[106]

Soon after its negotiation in 1987, the Meech Lake Accord seemed on its way toward ratification, with the Parliament and over two-thirds of the provinces having ratified it.[107] But the Accord began to show signs of distress in the face of opposition from political actors across the country.[108] As the deadline approached with three provinces yet to ratify the amendment package, the first ministers gathered to negotiate a way toward ratification. They arrived at an agreement: in exchange for the three holdout premiers putting the Accord to a vote before the expiration of the deadline, all premiers in turn agreed to place before their legislatures a separate

[104] See Warren J Newman, 'Living with the Amending Procedures: Prospects for Future Constitutional Reform in Canada' (2007) 37 *South Carolina Law Review* (2d) 384 at 400.

[105] See Mary Dawson, 'From the Backroom to the Front Line: Making Constitutional History or Encounters with the Constitution: Patriation, Meech Lake, and Charlottetown' (2012) 57 *McGill Law Journal* 955 at 983.

[106] For a discussion of *cumul*, see Jacques-Yvan Morin et José Woehrling, '*Les constitutions du Canada et du Québec – du régime français à nos jours* (Les Éditions Thémis, 2004) 531.

[107] See Bruce P Ellman and A Anne McClellan, 'Canada After Meech' (1990) 2 *Constitutional Forum* 63 at 64.

[108] See Michael B Stein, 'Improving the Process of Constitutional Reform in Canada: Lessons from the Meech Lake and Charlottetown Constitutional Rounds' (1997) 30 *Canadian Journal of Political Science* 307 at 320.

resolution that would address the concerns of the three holdouts.[109] Despite these 11th hour efforts, two provincial legislatures failed to ratify by the deadline, leading to the defeat of the entire package. In the final analysis, the outcome may understandably be perplexing for some observers because the Accord had nonetheless been approved by all parties in the Parliament of Canada, as well as eight of ten provinces representing almost 95 per cent of the entire population of Canada.[110] The unravelling of the Meech Lake Accord cannot of course be explained by one factor alone but, as Peter Oliver observes, correctly in my view,

> as the last days of that three-year period elapsed and as two small provinces succeeded in blocking the way forward for the others, the amending formula came to be seen as more than just a procedure, but in fact part of the problem.[111]

IV. Designing Temporal Limitations

These contrasting Canadian and American experiences with constitutional amendment expose the trade-offs involved between political brinkmanship and constitutional contemporaneity when political actors choose or not to associate temporal limitations to the ratification of an amendment proposal. The risk of political brinkmanship rises as a ratification deadline approaches, but the absence of a ratification deadline makes possible inter-generational ratification, which might undermine the political and moral value of contemporaneity between proposal and ratification. The question whether constitutional designers should codify deliberation requirements does not yield a definitive answer as to the better practice in constitutional design. The best answer can come only from deep reflection about the purpose of constitutional amendment and about the values most important to the formation of constitutional consensus. In either case, the choice to codify or reject temporal limitations is not one that would be wise to recommend for universal application. The choice must instead fit the unique cultural, historical, legal, political and social specificities of a given jurisdiction, as with all matters of constitutional design. The choice need not always be a trade-off between two; one can imagine a middle path that strikes a constructive balance between both ends. But exploring the trade-offs between brinkmanship and contemporaneity can nonetheless help inform the choice.

[109] Ronald L Watts, 'Canadian Federalism in the 1990s: Once More in Question' (1991) 21 *Publius* 169 at 178.

[110] See CES Franks, 'The Myths and Symbols of the Constitutional Debate in Canada' in Reflections Paper No 11 (Kingston, Institute of Intergovernmental Relations, 1993) available at: www.queensu.ca/iigr/sites/webpublish.queensu.ca.iigrwww/files/files/pub/archive/reflectionpapers/Reflection11CDFranksMythsandSymbols.pdf.

[111] Peter Oliver, 'Canada, Quebec, and Constitutional Amendment' (1999) 49 *University of Toronto Law Journal* 519 at 592.

A. Time and Brinkmanship

The United States has lived its own Meech Lake moment. The failure of the Equal Rights Amendment in the United States likewise demonstrates the risk of political brinkmanship when a ratification deadline approaches. In 1972, Congress adopted an amendment proposal to formally codify sex equality. The text of the proposal transmitted to the states read as follows:

> *Resolved by the Senate and the House of Representatives of the United States of America in Congress assembled (two-thirds of each House concurring therein)*, That the following article is proposed as an amendment to the Constitution of the United States, which shall be valid to all intents and purposes as part of the Constitution when ratified by the legislatures of three-fourths of the several States within seven years from the date of its submission by the Congress:
>
> Article—
>
> Section 1. Equality of rights under law shall not be denied or abridged by the United States or by any State on account of sex.
>
> Section 2. The Congress shall have the power to enforce, by appropriate legislation, the provisions of this article.
>
> Section 3. This amendment shall take effect two years after the date of ratification.[112]

Congress attached a seven-year ratification deadline to its proposal to the states.[113] Just like the Meech Lake Accord, early days proved promising for the Equal Rights Amendment: within one week, seven states had ratified it; within one month, 14 states; and within one year, 30 states – just eight states fewer than the 38 required for ratification – had ratified the proposal.[114] But in subsequent years, only five additional states ratified the proposal, bringing the number to 35, three short of the required supermajority.[115] As the seven-year ratification deadline approached and it seemed unlikely that three more states would ratify the amendment,[116] Congress passed a resolution extending the ratification period for just over three more years. The new resolution acknowledged the earlier ratification deadline but expressly replaced it:

> *Resolved by the Senate and House of Representatives of the United States of America in Congress assembled*, That notwithstanding any provision of House Joint Resolution 208

[112] HRJ Res 208, 92d Cong, 2d Sess, 86 Stat 1523 (1972).

[113] *Ibid.*

[114] Orrin G Hatch, 'The Equal Rights Amendment: A Critical Analysis' (1979) 2 *Harvard Journal of Law and Public Policy* 19 at 19–20.

[115] *Ibid* at 21. Four states, however, passed resolutions rescinding their earlier ratifications, though the legal validity of those rescissions was unclear. For a careful discussion of the competing considerations on whether a state can and should have the capacity to legally rescind an earlier ratification, see Walter Dellinger, 'The Legitimacy of Constitutional Change: Rethinking the Amendment Process' (1983) 97 *Harvard Law Review* 386 at 421–24.

[116] See Leo Kanowitz and Marilyn Klinger, 'Can a State Rescind its Equal Rights Amendment Ratification: Who Decides and How?' (1977) 28 *Hastings Law Journal* 979 at 981.

> of the Ninety-second Congress, second session, to the contrary, the article of amendment proposed to the States in such joint resolution shall be valid to all intents and purposes as part of the Constitution when ratified by the legislatures of three-fourths of the several States not later than June 30, 1982.[117]

This congressional extension attracted significant attention at the time. Scholars debated whether Congress had the authority to extend the period of ratification and if so by what margin, whether the rule of presentment required the President to sign the measure, and whether it was proper for Congress to change a deadline after it had already been set.[118] The Equal Rights Amendment ultimately failed, even with the ratification extension, although some later relied on the 200-year ratification of the 27th Amendment to argue that the time limit had been unconstitutional all along and that the Equal Rights Amendment remained open indefinitely for states to ratify until they achieved the three-quarters mark for ratification.[119] In the end, however, the deliberation requirements complicated the task of ratifying the amendment proposal.

There may nonetheless be good reason for constitutional designers to codify deliberation requirements. In Canada, the one-year deliberation floor was a complement to the right of provincial legislatures to opt-out of certain amendments from whose effect the Constitution of Canada authorises provinces to withdraw even if the requisite initiation and ratification thresholds are otherwise met.[120] The right to opt-out is available for amendments that are made using the general amendment procedure and that derogate from provincial legislative powers, proprietary rights or any other provincial rights or privileges.[121] Where a province chooses to exercise this opt-out right, the provincial legislature must properly register a timely dissent,[122] in which case it will be eligible for reasonable compensation if the amendment concerns the transfer of educational or cultural matters from provincial to federal jurisdiction.[123] The choice to opt-out is a serious one. A province requires a reasonable amount of time to evaluate whether to opt-out of amendments in this category, and less than one year might not be

[117] HRJ Res 638, 95th Cong, 2d Sess, 92 Stat 3799 (1978).

[118] See Ruth Bader Ginsburg, 'Ratification of the Equal Rights Amendment: A Question of Time' (1979) 57 *Texas Law Review* 919; J William Heckman, Jr, 'Ratification of a Constitutional Amendment: Can a State Change its Mind?' (1973) 6 *Connecticut Law Review* 35; Grover Rees III, 'Throwing Away the Key: The Unconstitutionality of the Equal Rights Amendment Extension' (1980) 58 *Texas Law Review* 875; Comment, 'The Equal Rights Amendment and Article V: A Framework for Analysis of the Extension and Rescission Issues' (1978) 127 *University of Pennsylvania Law Review* 494.

[119] See Allison L Held et al, 'The Equal Rights Amendment: Why the ERA Remains Legally Viable and Properly Before the States' (1997) 3 *William and Mary Journal of Women and the Law* 113. However, the more persuasive view is that the Equal Rights Amendment proposal expired when the deadline – either one – passed without ratification. See Brannon P Denning and John R Vile, 'Necromancing the Equal Rights Amendment' (2000) 17 *Constitutional Comment* 593.

[120] Constitution Act 1982, ss 38(2)-(4), 40.

[121] *Ibid* at s 38(2)-(3).

[122] *Ibid* at s 38(3).

[123] *Ibid* at s 40.

long enough for a province to decide whether to exercise its constitutional right to opt-out.[124]

The three-year deliberation ceiling is not as closely connected to another amendment rule but it nonetheless derives from a theory of how to manage the formation of constitutional consensus. The rule was anchored in three rationales: first, to ensure a definitive end, whether rejection or codification, of an amendment proposal; second, to foreclose the possibility of ghost amendments that are proposed, then languish for some time only to be revived much later to the surprise of political actors; and third, to focus public awareness and political decision-making on an amendment proposal in the time it is initiated.[125] But although the theory seems soundly rooted in good reason, there was little thought given to how these temporal limitations would apply in practice at the time of the design of the Constitution Act 1982.[126]

It is worth asking whether three years is too little time to ratify an important amendment.[127] The late Richard Simeon observed that the failure of the Meech Lake Accord 'was more likely a result of the brinkmanship tactics employed than of the rule itself', and therefore that three years is not necessarily too short.[128] But perhaps the nature of the relationship between time limits and brinkmanship is altogether different, not correlative but rather causative. Perhaps rather than understanding brinkmanship as something to which political actors have recourse independently of and without instigation by the time limits, we should consider that time limits may cause political actors to engage in brinkmanship when their objective is either to defeat the amendment subject to the time limit or to extract concessions on the amendment itself or on other issues, related or not. On this understanding, the fragmentation of power across time gives political actors an important weapon to fight an amendment proposal they oppose or to improve their bargaining position as the deadline approaches and their vote increases in value. This would better explain the fate of the Meech Lake Accord and the Equal Rights Amendment. Each gave political actors a roadmap to a winning strategy where their own interests where concerned: to either seek concessions on the amendment or on some other matter of consequence to them, or alternatively to hold out until time expires should their demands go unfulfilled.

This risk of political brinkmanship need not dissuade constitutional designers from codifying deliberation ceilings. Although deliberation ceilings may aggravate

[124] Swinton (n 88) at 146.

[125] Swinton (n 88) at 146.

[126] See Richard Simeon, 'Why did the Meech Lake Accord Fail?' in Ronald L Watts and Douglas M Brown (eds), *Canada: The State of the Federation 1990* (Kingston, Institute of Intergovernmental Relations, 1990) 15 at 28.

[127] One might also ask whether three years is too long. Following the defeat of the Meech Lake Accord, a special parliamentary committee recommended shortening the time limit to two years. See Government of Canada, *Special Joint Committee on the Process for Amending the Constitution of Canada, The Process of Amending the Constitution of Canada: The Report of the Special Joint Committee* (Ottawa, Supply and Services, 1991) 30–31.

[128] Simeon (n 126).

the possibility of amendment failure, they nonetheless offer important advantages, though whether reward outweighs risk is a judgement for constitutional designers to make with due regard to local norms. In addition to the three advantages above – ensuring a definitive end, foreclosing ghost amendments, and focusing decision-making – deliberation ceilings concentrate the formation of constitutional consensus within a defined period of time. Where amendment rules fragment power across political institutions and actors by dispersing the initiation and ratification powers, deliberation ceilings promote both contemporary and representative consensus. Ratification on these terms fosters representative consensus in so far as the ratifying actors differ in form and interest from the initiating actors. Ratification on these terms also reflects intra-generational contemporaneity in their independent judgements of the amendment, provided the deliberation ceiling directs political actors to act within some narrow period of time. Contemporaneity and representativeness both reinforce the sociological legitimacy of the amendment.

B. Time and Contemporaneity

But inter-generational ratification may also encourage sociological legitimacy. Where an amendment is ratified across generations, its codification may be said to reflect the considered intertemporal judgement of the constitutional community. Inter-generational ratification is consistent with Jed Rubenfeld's thesis that 'written self-government does not demand that new constitutional principles be adopted whenever a majority so wills' but rather 'only when a people is prepared to make a significant *temporal* commitment to them'.[129] Rubenfeld argues that our understanding of self-government should require something more than the support of 'actual people of the here and now'[130] and be anchored in a less presentist notion of sovereignty. We must instead reimagine, he suggests, the formation and sustainability of constitutional consensus, that it takes shape over time and that in order to reflect the sociological legitimacy that only the people's popular will can confer, it cannot privilege the consent of the governed today over the consent of the governed over time. On this theory, political actors should seek legitimacy for their decisions made in the name of the people

> not in governance by the present will of the governed, or in governance by the a-temporal truths posited by one or another moral philosopher, but rather in a people's living out its own self-given political and legal commitments over time – apart from or even contrary to popular will at any given moment.[131]

[129] Jed Rubenfeld, *Freedom and Time: A Theory of Constitutional Self-Government* (Yale University Press, 2001) 12.
[130] *Ibid* at 11.
[131] *Ibid.*

This view counsels pause in answering the question whether an amendment proposal should remain ratifiable for generations. Intra-generational ratification may not necessarily reflect the considered judgement of the constitutional community where the ratifying supermajority is fleeting and unsustainable, and also where the people and their representatives are pressed to action by special circumstances, like a national emergency or crisis. In these circumstances, the supermajority approval of an important constitutional amendment may not in fact reflect stable and representative support. This is precisely why many national constitutions expressly prohibit political actors from amending the constitution during periods of great insecurity, like war or siege or succession, where passions may move the people to make decisions they would not otherwise make in normal times.[132] Not even extraordinary supermajorities may withstand this critique if they are temporary and susceptible to collapsing quickly after their formation.

Inter-generational ratification may make it possible to respond to this concern, though it would not necessarily solve it.[133] Assume an amendment rule is silent on whether an amendment must be ratified within a defined period of time, as is the case with the United States Constitution. This permissive amendment rule would permit an extended ratification period like the two centuries it took to ratify the 27th Amendment. But it would also authorise instantaneous ratification, a speedy ratification that would not test the durability of the supermajorities that had expressed their support for the amendment. Constitutional designers must therefore be explicit in their design of amendment rules if they wish to force inter-generational ratification. They may, for instance, prohibit ratification prior to the expiration of a certain period of time, like an extended deliberation floor. An inter-generational deliberation floor would be unusual: imagine a deliberation floor prohibiting political actors from ratifying an amendment within one generation, or 20 years after its initiation. By the time the ratification deadline had expired, the people may have adopted an entirely new constitution altogether.[134]

Perhaps instead of *requiring* inter-generational ratification, constitutional designers could adopt the United States Constitution's model of *allowing* it.[135] But other rules should be codified alongside the open-ended amendment ratification

[132] See Part II.A.

[133] It has been suggested that an effective design to combat the problem of fleeting supermajorities in constitutional democracies is the Scandinavian model of intervening election, which requires multiple ratification by successive legislatures. See Albert (n 21).

[134] The average lifespan of a constitution is 19 years. See Zachary Elkins et al, *The Endurance of National Constitutions* (Cambridge University Press, 2009) 2.

[135] Indeed, this was suggested by Clyde Wells, Premier of Newfoundland and Labrador, one of the leading political actors involved in the failed Meech Lake and Charlottetown Accords. See Clyde Wells, 'Constitutional Amendment and Constituent Assemblies' (1991) 14 *Canadian Parliamentary Review* 8 at 9–10.

rule – additional rules that would make it possible for political actors and the people to verify that the constitutional consensus behind an amendment has indeed remained stable and representative over time. We can verify the durability of the constitutional consensus behind a constitutional amendment by designing rules requiring political actors to confirm a prior rescission or ratification where the ratification process extends across more than one generation or some other significant period. Without the power to confirm a prior rescission or ratification, we cannot *really* speak of inter-generational ratification because the successful ratification of an amendment across generations would reflect separate generations acting in isolation rather than in conversation.

The United States Constitution exposes a design flaw on this point because it does not state clear rules on whether a state has the power, while an amendment is pending, to rescind a prior ratification, or to ratify an amendment that it has in the past rejected. As a consequence, an amendment ratified across generations in a regime where the constitutional text imposes no ratification deadline creates a serious risk of creating the artificial appearance of considered supermajority approval for the amendment. This constitutional design conceals the reality that there had never existed, in any single period, a durable supermajority to ratify an amendment. Sanford Levinson calls the silence of the Constitution on this point the 'easiest example' of something to change in the design of Article V.[136] Indeed, the Constitution's silence on the power to rescind a prior ratification has generated significant scholarly interest in exploring whether states should have the power to change their mind on a pending amendment.[137]

To avoid uncertainty, constitutional designers should be explicit about whether ratifying bodies – state legislatures, state conventions or indeed others – possess the right to rescind a prior amendment ratification or to ratify an amendment previously rejected before political actors arrive at the ratification threshold required to codify a constitutional amendment. To illustrate, where the ratification threshold in a federalist constitution requires two-thirds of subnational states to consent to the amendment proposed by the national government, it should be clear from the text of the constitution whether and how a state may negate its prior ratification of an amendment, or do the opposite, as long as the two-thirds ratification threshold has not yet been met. If the objective of designing amendment rules in this way is to foster the kind of non-presentist sociological legitimacy that comes from inter-generational ratification, then it would not be enough simply to authorise

[136] Sanford Levinson, 'Designing an Amendment Process' in John Ferejohn et al (eds), *Constitutional Culture and Democratic Rule* (Cambridge University Press, 2001) 271 at 281.

[137] See, eg Charles L Black, Jr, 'On Article I, Section 7, Clause 3 – and the Amendment of the Constitution' (1978) 87 *Yale Law Journal* 896; Brenda Feigen Fasteau and Marc Feigen Fasteau, 'May a State Legislature Rescind its Ratification of a Pending Constitutional Amendment' (1978) 1 *Harvard Women's Law Journal* 27; Philip L Martin, 'State Legislative Ratification of Federal Constitutional Amendments: An Overview' (1974) 9 *University of Richmond Law Review* 271; Robert M Rhodes and Michael P Mabile, 'Ratification of Proposed Federal Constitutional Amendments – The States May Rescind' (1978) 45 *Tennessee Law Review* 703.

subnational states to rescind or ratify a prior decision. Here it would be advisable for constitutional designers either to require subnational states to confirm or reject their prior decision if significant time had passed between the original amendment proposal and the final ratification satisfying the three-quarters threshold, or to state a presumption that the prior decision remains valid unless the subnational state chooses to reverse it.

The first option would be more difficult to design and to oversee. It would require constitutional designers to designate a specific period of time after which final ratification of a pending amendment would require confirmations of prior ratifications or rejections. Identifying the right period of time may prove difficult, but political actors regularly draw lines in their work, and there is no evident reason why they should not be trusted to make this choice too. The second option would be less difficult both in terms of constitutional design and political enforcement. It would require no specific designation of the period of time for which a prior ratification or rejection remains valid and until it must be confirmed, nor would it pose challenges as to its application because there would be an understanding that the original decision on ratification remained valid until the relevant political actors made an intervening decision reflecting the contrary intent, specifically to reverse a ratification or rescission.

Three other considerations in the design of temporal limitations merit some mention. First, constitutional designers may vary the duration of time for which a pending amendment remains valid according to the importance of the subject-matter of the amendment. For matters of heightened importance, political actors and the people could be required by constitutional design to deliberate for a longer period of time than they must devote to less important matters, where the importance of the matter is determined by the designers themselves.[138] The variability of temporal limitations within the larger structure of amendment rules is not unusual, as many constitutional democracies vary the amendment thresholds according to the amendable subject-matter.[139] Second, temporal limitations like deliberation ceilings and floors should not be associated exclusively with federal states like Canada and the United States. To the extent that constitutional designers see value in them, they may be used as well in unitary states, parliamentary and presidential forms of government, republics and constitutional monarchies, and indeed all democratic states where amendment rules are taken seriously, as they should be. Third, deliberation ceilings and floors may not in fact be *deliberative*. Establishing minimum or maximum periods of time for ratifying an amendment does not, without more, ensure that the choice will be

[138] Note that constitutional designers may also vary the method of amendment, quite apart from the deliberation requirements, according to the importance of the amendable matter. See Richard Albert, 'The Expressive Function of Constitutional Amendment Rules' (2013) 59 *McGill Law Journal* 225 at 247–57.

[139] See Albert (n 24) at 942–46.

informed or even debated, nor does it encourage deliberative decision-making. This is another important feature of constitutional amendment rules that constitutional designers should consider incorporating in how they structure the requirements for constitutional change.

V. Conclusion

I have sought in this chapter to explain and evaluate some of the options available to constitutional designers in the design of deliberation requirements in constitutional amendment, part of a larger category of temporal limitations, which also includes safe harbours.[140] My purpose has been to diagnose not to prescribe, in an effort to highlight some of the considerations that constitutional designers must confront in fragmenting the constitutional amendment power. Amendment rules commonly fragment powers along institutional branches of government, political parties, geographic boundaries, federalism, ethnic and cultural differences, as well as linguistic divisions. Time, however, as I have shown, is an additional dimension along which to fragment the amendment power in constitutional design. My larger purpose has been to suggest a research agenda for further inquiry into some of the questions associated with time and change.

Many new lines of inquiry present themselves. As to Canada and the United States, perhaps the distinction between inter- and intra-generational ratification maps onto the deeper divide between originalism and living constitutionalism. More broadly, the degree of the unity or diffusiveness of the amending authors is relevant in designing deliberation requirements: the more unified are the political actors, the more time constitutional designers can afford to build into the amendment process; the more diffuse they are, the higher the risk of brinkmanship as the ratification deadline approaches. The process of constitution-making may raise an instructive parallel inasmuch as it may be profitable to compare the comparative risk and reward of imposing temporal limitations in designing and amending constitutions. Separately, legal systems and their cultural contexts may affect how time is perceived and structured, and this may explain variations in the use of time in common and civil law jurisdictions. Analogous questions in other fields may also provide useful insights, namely in connection with the ratification of treaties. Interdisciplinary perspectives from political science, history and philosophy would both complicate and perhaps clarify the analysis. The point is simple but powerful: there remains much to learn about the options and implications of manipulating time in the design of amendment rules.

[140] See part II.

4

Night Laws: How Nightfall Shapes Regulation

GUY I SEIDMAN*

I. Introduction: Night and Law – What is the Connection?

This volume studies 'time, law and change'. This chapter offers a variation on this theme, examining what the law has to say about one particular time period: *the night*. *The night* has long ignited the imagination and piqued the interest of many disciplines of art, science and faith,[1] but *the law*? What does this practical, dry-eyed field have to say about *the night*, which is, after all, a natural phenomenon, not a legal construct?

At first blush, the idea that law would have much to say about the *night* seems unlikely. There is, indeed, limited legal writing on this topic.[2] What I suggest in this chapter is that the law – in its past, in the present, and likely in the future – has much to say about the night. This is the first offspring of my 'night law' project.

Each of the three parts that follow this introduction discusses one essential aspect of 'night law,' as I see it. Section II seeks to establish a factual and normative link between *night* (and *time*) and *the law*. It deals with what are likely the most conventional aspects: law's role in defining and standardising time and in regulating human conduct, especially at the time defined as night. The typical issues raised are nightly noise nuisance and working hours for individuals and businesses. The two following parts raise grander issues. Titled 'Regulating the Dark', section III reflects on the rich legal history of night-time regulation. The main tool used was criminal law, but involved some measures we would now

* My thanks to Jill Gross, Daniel Klein, Gary Lawson, Hillel Sommer, Ofer Vallach and especially Yael Nesher of the IDC Law Library and Sofia Ranchordás and Yaniv Roznai, the editors of this volume.

[1] For a recent example of interdisciplinary interest in the night, see: Jacques Galinier et al, 'Anthropology of the Night Cross-Disciplinary' (2010) 51(6) *Current Anthropology* 819–47.

[2] In my research, I came across one legal scholar who wrote on the wider theme of *law and time*, striking some of the themes of this paper, especially those addressed in section IV below. See: TD Rakoff, *A Time for Every Purpose: Law and the Balance of Life* (Harvard University Press, 2002).

categorise as civil (like tort liability) and administrative. Their aim was mostly that, at a prescribed time, the population would curb outdoor fires, go indoors and stay the night. The law accorded them increased protection if they stayed at home and limited their freedom to wander outside. Most of these measures are long gone, but may yet prove relevant. Looking into the future, section IV, titled 'Re-regulating the Night?' investigates the price that society has paid for effectively banishing the night. It suggests that major policy concerns – notably public health, socio-economic equality and ecology – yet unsettled, may require us, in the very near future, to consider re-introducing some regulation of night-time activities and practices.

II. Regulating Time, Regulating *the Night*

A good starting point to explain the relevance of *the law* to *the night* is to observe that, while night and day are physical occurrences, law – likely in all jurisdiction, for some time and diverse purposes, in an effort to establish civic order – all jurisdictions and for some time, in an effort to establish civic order and unity – *defines*, when one ends and the other begins, and *regulates* human behaviour which is permissible during each period.

It is actually quite often the case that definitions and rules prescribed by law are more relevant to people's quotidian conduct than physical facts. Britain's AV Dicey made famous the saying about Parliamentary sovereignty that 'Parliament can do everything but make a woman a man and a man a woman.'[3] As a physical fact, Parliament cannot assign or reassign gender. However, it is within its vast powers to *define* who is legally to be considered man or woman[4] and prescribe the substantive rights and duties of each, ie, what it *means* to be a man or woman.

All this and more applies to the terms *day* and *night*. Their physical meaning is relatively clear. *Black's*, the preeminent American Legal dictionary defines *night*, inter-alia, as '[t]he time from sunset to sunrise' and *day* as '[t]he period between the rising and the setting of the sun'.[5] These are undisputable facts of nature which no Parliament can alter. However, they are also almost irrelevant for most of us for the following reasons. First, in the modern world, few people outside rural communities regulate their daily activities in accordance with sunrise and sunset,

[3] AV Dicey, *Law of the Constitution*, 1st edn (Macmillan and Co, 1885) 39, quoting Swiss political theorist Jean-Louis De Lolme, *The Constitution of England* (1771) (himself likely quoting England's 2nd Earl of Pembroke's Statement before the House of Lords (1648)) calling it a 'grotesque expression which has become almost proverbial'.

[4] Such legal definitions have extensive social, psychological and economic ramifications. The issue of transgender people raises further complex questions about crossing genders, even challenging the binary male-female divide.

[5] 'Also termed *natural day*.' See: BA Garner (ed), *Black's Law Dictionary*, 10th edn (Thomson Reuters, 2014).

and the terms *day* and *night* no longer designate times of *activity* and *rest*. Expanding human activities at night and reduced sleeping times are some of the troubling modern afflictions, which may require future regulation, as noted in section IV. Second, as technology advanced mankind has become much better at measuring and telling time,[6] no longer needing to rely on the sun. Third, while *the day* – meaning the 'period of the earth's rotation on its axis' – has long served as 'the most fundamental unit of time-measurement in most societies',[7] we are now concerned with a variety of time frames, both longer and shorter. Indeed, we now conceptualise *time* itself in much more detailed and nuanced frames.[8] Thinkers of all disciplines have made enormous, sophisticated inroads in analysing the complex concept of *time*,[9] their huge analytic advances often surpassing the popular understanding of usage.[10]

Indeed, it is likely that we have *law* – with its stabilising effects of its regulation – to thank for this gap between science and practice. To explain this, we must understand lawyers' role in *defining* and *applying* time.

First, the law is particularly useful with definitions. The process of establishing legal definitions may be laborious, but the outcome is likely to clarify a term and ensure its uniform application within the jurisdiction. Standardisation is often beneficial.[11] Once time became standardised nationally and internationally, enshrined by domestic laws and international agreements, change, even for good reasons, is almost impossible.[12] Unbelievable as it now seems, well into the nineteenth century, many towns in England and the United States set their clocks

[6] Roman philosopher Seneca wrote – '[w]hat hour it was I cannot certainly tell; philosophers will agree more often than clocks', Seneca, *Apocolocyntosis* (translated by WHD Rouse) (Loeb Classical Library No 15, London, 1913) 373. On the history of timekeeping and measurement, see: P Janich, *Protophysics of Time – Constructive Foundation and History of Time Measurement* (D Reidel Publishing Co, 1985); P Glennie and N Thrift, *A History of Timekeeping in England and Wales 1300–1800* (Oxford University Press, 2009); A McCrossen, *Marking Modern Time: A History of Clocks, Watches and other Timekeepers in American Life* (University of Chicago Press, 2013).

[7] See: L Holford-Strevens, *The History of Time – A Very Short Introduction* (Oxford University Press, 2005) 1.

[8] *Ibid* xi; C Hammond, *Time Warped: Unlocking the Mysteries of the Time Perception* (Canongate, 2012).

[9] For a particularly complex treatment, see: M Heidegger, *History of the Concept of Time – Prolegomena*, 1925, 1979 (translated by Theodore Kisiel) (Indiana University Press, 1985). For a psychological and anthropological perspective, see: KK Birth, *Objects of Time – How Things Shape Temporality* (Palgrave Macmillan, 2012); on time and science, see: T Jones, *Splitting the Second – the Story of Atomic Time* (Institute of Physics Publishing, 2000); F Weinert *The March of Time – Evolving Conceptions of Time in Light of Scientific Discoveries* (Springer, 2013); see also, comprehensively: HJ Birx (ed), *Encyclopedia of Time – Science, Philosophy, Theology & Culture* (Sage, 2009).

[10] See: John Laverty, 'Introduction' in Jones (n 9) vii.

[11] *Cf.* Andrew Russel and Lee Vinsel, 'The Joy of Standards' *New York Times*, 16 February 2019.

[12] We can see the long 'tail' of custom in Black's Law Dictionary's (n 5) definition of *day*, which in 2014 still included specifically named and defined days ranging from *love day* (a day for amicable dispute resolution of neighbourly help without pay, 13th century), *quarter day* (one of four days per year when money owed was payable, 15th century), *return day* (for specific legal action, 17th century) to *business day* (when institutions are open for business, 1826).

to noon when the sun passed overhead, and each had a slightly different local time. All this changed with train services, which needed a uniform time for timetables and the telegraph that made long distance communication possible.[13] In England, the Great Western Railway first adopted *Greenwich Mean Time* (GMT) in 1840, and by 1848 almost all railways followed suit. Yet, this would become the mandatory, uniform time for England, Scotland and Wales only after the Statutes (Definition of Time) Act 1880 (43 & 44 Vict.) was passed.[14] The difficulty was not limited to England, or to trains.[15] As regards international navigation, indeed all nations of the world were looking for a uniform time system: '[I]n the 1870s scientific pursuits requiring simultaneous observations from scattered points became important, and those needs led to proposals for federal action in the early 1880s.'[16] The final step took place in 1884, when the International Meridian Conference accepted Greenwich as the prime meridian for longitude and GMT as the world's time standard.[17] Standard time zones are now defined in the United States by federal law – 15 USC §260. The time zone system was a matter of compromise and its nature, as a matter of convention rather than nature, is further reflected by the artifice of establishing daylight saving time by law[18] and by frequent public discussion in various countries regarding the change of their time zone.[19]

Second, the law is also realistic in that *rules must allow for exceptions (and interpretations)*. Already in 238 CE, Latin author *Censorinus* noted that the term *day* means **both** the entire period of the earth rotation and *daylight* time,[20] a distinction that remains in modern legal definitions.[21] Similarly, while the time in England is, by the 1880 statute, *GMT*, some norms still rely on *local times*. A famous 1898 case saw a man busted by Bristol police for riding his bicycle

[13] There were perhaps as many as 300 'local sun times' in the United States alone. See: https://en.wikipedia.org/wiki/Time_in_the_United_States#cite_note-Why_Do_We_Have_Time_Zones?-1; https://en.wikipedia.org/wiki/Time_in_the_United_Kingdom.

[14] See: https://en.wikipedia.org/wiki/Time_in_the_United_Kingdom; for an account of the process, see: C Blaise, *Time Lord – Sir Sandford Fleming and the Creation of Standard Time* (Vintage Books, 2002).

[15] On the impacts of the railway industry in the United States, see: N Allen, 'The Times They Are A-Changing: The Influence of Railroad Technology on the Adoption of Standard Time Zones' (2000) 33(2) *History Teacher* 241–56.

[16] See: IR Bartky, 'The Adoption of Standard Time' (1989) 30(1) *Technology and Culture* 25 (cite) (Bartky is dismissive of the role of railroads and telegraph in the process).

[17] The Conference was held in Washington DC with participants from 26 nations. See: https://en.wikipedia.org/wiki/International_Meridian_Conference; www.timeanddate.com/time/time-zones-history.html.

[18] Introduced in England by the Summer Time Act 1916 (6 & 7 Geo. V), and now in force under the Summer Act of 1972 and established by the Uniform Time Act of 1966, Pub. L. 89–387, 80 Stat. 107 in the United States.

[19] See, comprehensively: www.timeanddate.com/news/time/. In March 2019, the European Parliament approved by a vote of 410 to 192 a draft directive that would cancel Daylight Saving Time in the EU, each Member State having to decide whether it wanted to remain on 'permanent summer' or 'permanent winter' time. See: Daniel Boffey, 'European Parliament Votes to Scrap Daylight Saving Time from 2021', *The Guardian*, 26 March 2019.

[20] See: Holford-Strevens (n 7) 1.

[21] *Black's Law Dictionary* (n 5) makes the same observation.

at 8:15 pm without a lamp. Local law stated that riders must carry a lamp 'so lighted as to afford adequate means of signaling the approach of the bicycle, *during the period between one hour after sunset and one hour before sunrise*'. Sunset was at 7:13 pm under GMT but at 7:23 pm in Bristol, resulting in his acquittal on appeal.[22] The law also has a knack for allowing substantive objections to formal rules. This is why ancient Judaic law and seventeenth century English law, which both allowed the killing of a thief operating at night, made exceptions for well illuminated nights.[23]

You might think such laws are no longer common today. You would be wrong. They are actually quite abundant,[24] and their 'classic' role remains the same: defining when night begins and ends, for specific legal purposes (enacting social judgement calls) and regulating conduct at night. There are many regulations restricting night activity, both commercial and private; this can be part of a wider social policy as is the case with restrictions on alcohol sales,[25] but for the individual Israeli reveller it clearly 'defines' the night.[26] Another example is the regulation of noise nuisances at night. If we look at the Israeli ordinances,[27] we find, in one set of regulations, a series of different times that define 'night' for the purposes of maintaining the quiet in residential areas. One regulation forbids singing, shouting and operating a musical instrument, a radio receiver, a television, etc between 11 pm and 7 am, while[28] a second forbids the beating of carpets, mattresses and the like from 7 pm to 7 am.[29] Another prohibits carrying out repair and renovation work between 10 pm and 7 am, and[30] a fourth forbids operating machinery for the purposes of digging or construction between 7 pm and 6 am.[31] A fifth regulation forbids the moving of garbage bins, gas cylinders or similar containers between 10 pm and 6 am.[32] So when exactly is it quiet night-time in an Israeli

[22] See: www.markpack.org.uk/130873/statutes-definition-of-time-act-1880/ (emphasis in the original). J Gleick, *Time Travel: A History* (Vintage Books, 2016) 250.

[23] See more in section III. This aspect too is reflected in *Black's Law Dictionary* (n 5) 2014 definition of 'night'.

[24] See, eg, www.legislation.gov.uk/uksi/2009/3015/article/23/made#text%3Dsunset. *Black's Law Dictionary* (n 5) makes the same observation in its definition of *night*.

[25] Restricting alcohol sales after a certain hour is one of a wide array of alcohol laws, and is likely to be effective. See: https://en.wikipedia.org/wiki/Alcohol_law; RA Hahn et al, 'Effectiveness of Policies Restricting Hours of Alcohol Sales in Preventing Excessive Alcohol Consumption and Related Harms' (2010) 39(6) *American Journal of Preventive Medicine* 590–604.

[26] Israeli law prohibits the sale of alcohol in stores (but not in restaurants or pubs) between 11 pm and 6 am and prohibits possession or consumption of alcohol in public between 9 pm and 6 am. See: www.haaretz.com/.premium-cops-can-seize-booze-consumed-in-public-1.5240277.

[27] Promulgated under section 5 of the Abatement of Nuisances Law, 5721–1961 to implement section 2 which states that '[a] person shall not cause any considerable or unreasonable noise, from any source whatsoever, if the same disturbs or is liable to disturb a person in the vicinity or a passerby'.

[28] Regulation 3 of the Abatement of Nuisances Regulations (Prevention of Noise), 5752–1992. Exceptions are provided for a party or celebration held on holidays.

[29] Regulation 8 *ibid.*

[30] Regulation 4 *ibid.*

[31] Except where urgently required; Regulation 5, *ibid.*

[32] And not at all on rest day; Regulation 6 *ibid.*

residential area? A more general set of regulations prohibits 'unreasonable' noise at night, which it defines as a period from 10.01 pm to 05.59 am.[33] I consider the link between law and night now established. But there is much more.

III. Regulating the Dark

A. Introduction

It was only in the twentieth century, as electric lighting became common indoors and outdoors with the expanding installation of power lines,[34] that humanity finally vanquished two of its oldest foes: the darkness of the night and the need to use open fire for all artificial illumination.[35] Until that time, nights were very dark, especially by modern standards.[36] This alone likely made outdoor activities dangerous, as the dark provided cover for criminals. On the other hand, the only available antidote, an open fire source for heating, cooking and illumination, was also dangerous; as many human settlements were cramped and built of wood, fires frequently broke out, often with horrific consequences to life and property.

What were people expected to do at night? What did social conventions, religious teachings and legal norms have people do? How did it protect them from the hazards of the night? For most of history, it would seem that, for the sake of individuals and rulers alike, people were expected to remain indoors at night. They did not necessarily go to sleep right after sunset,[37] and it is possible that they split their night into two sleep segments, with an active part in-between.[38]

What role did *the law* play in all of this? This section looks at four legal mechanisms. Two of them supported the requirement that people stay at home at

[33] Abatement of Nuisances Regulations (Unreasonable Noise), 5750–1990.

[34] By 1920, 35% of urban and suburban homes in America had electricity; by World War II the number had reached 90%. See: P Bogard, *The End of Night – Searching for Natural Darkness in an Age of Artificial Light* (Little, Brown & Co, 2013) 22; also: TP Hughes, *Networks of Power – Electrification in Western Society, 1880–1930* (The Johns Hopkins University Press, 1983, 1993). Enormous efforts have seen the global electrification rate rise from 76% in 1990 to 85% in 2012 but an estimated 1.1 billion people remain without access to electricity and an additional one billion people lack access to advanced energy services. See: www.ruralelec.org/rural-electrification.

[35] On the electrification process, see: WJ Hausman and P Hertner, *Global Electrification – Multinational Enterprise and International Finance in the History of Light and Power, 1878–2007* (Cambridge University Press, 2008); C Otter, *Victorian Eye: A Political History of Light and Vision in Britain, 1800–1910* (University of Chicago Press, 2008) 10.

[36] Especially in nights without star- and moonlight, which were very common in cloudy, smoggy cities like London.

[37] Diaries suggest that adults typically retired to bed between nine and ten at night, and slept for about six to eight hours. See: AR Ekirch, 'Sleep We Have Lost: Pre-Industrial Slumber in the British Isles' (2001) 106(2) *American historical Review* 343, 349–50.

[38] The idea is primarily advocated by Ekirch, who notes that '[u]ntil the close of the early modern era, Western Europeans on most evenings experience *two major intervals of sleep* bridged by up to an hour or more of quiet wakefulness'. For a more comprehensive treatment see: AR Ekirch, *At Day's Close: Night in Times Past* (WW Norton & Co, 2005) 354.

night – the general rule mandating people to stay at home ('*curfew*') and the laws punishing people for venturing out unauthorised (literally, for 'nightwalking'). The other two concern the legal regimes that protected people at home from crime (especially from the 'night thief') and prescribed rules for fire prevention and liability for fire (including at night).

B. Your Home is Your Castle: Stay Home!

For most of human history, it was very clear what people were to do after dark: stay indoors. To upgrade this rule from a social convention to a binding legal norm, two mechanisms were used. One, that likely applied for a limited time, was *the curfew*; at a certain hour of the evening, the public would be notified – typically by the ringing of church bells – that it should curb all outdoor fires (hence, the origin of the term, the French '*couvre-feu*') and retreat home. English historiography suggests that curfew was a repressive measure imposed by William the Conqueror on the Anglo-Saxons in 1068, likely in an effort to prevent a rebellion.[39] Even if this account is true,[40] there are records showing the curfew bell was established by Anglo-Saxon king Alfred the Great in the late ninth century as a fire safety measure[41] and that while the curfew law established by William was likely repealed by King Henry I in 1103, curfew bells kept ringing in England many centuries later.[42]

While it is uncertain whether the ringing of the curfew bell was a legally binding order to go indoors for the night or merely a cultural tradition marking the end of day, there is clear evidence that venturing outdoors at night was dangerous, socially frowned upon *and* legally restricted: '[a]nyone on the streets at night with no good reason was automatically liable to arrest'.[43] In sixteenth century London,

> [n]ight-time was more dangerous, people were more on edge and more sensitive to the slightest tap or rattle in night's quiet. Law-abiding Londoners liked to be locked in their houses in an envelope of security as they slept in bed, leaving nothing to chance.[44]

[39] Which is much the way the general curfew is imposed today: as a short-term emergency measure. See: n 118, below; https://en.wikipedia.org/wiki/Curfew_bell. Also: 'The Laws of William the Conqueror by; The Statutes of the Realm by; The Reports of Savile, Rolle, Levinz, and Others by' (1840) 51(109) *The North American Review* 285, 287, 293; 'The Vesper Bell' (1853) 1(6) *The Illustrated Magazine of Art* 368.

[40] A revisionist account suggests that this is all but a myth: see: L Cresswell, 'The Curfew: Its Origin and History' (1895) 278(1974) *The Gentleman's Magazine* 599, 615.

[41] See: JM Miller and RA Wright (eds), *Encyclopedia of Criminology, Vol. 1* (Routledge, 2005) 360 ('*curfews*').

[42] The Ringing continued well into the 19th century. See: J Timbs, *Curiosities of London* (Longmans, Green, Reader & Dyer, 1868); www.londonancestor.com/misc/bermondsey-curfew-bell.htm; https://en.wikipedia.org/wiki/Curfew_bell#cite_note-brand222-10.

[43] 'In the medieval and early modern periods, strangers in the night were feared like evil spirits.' See M Beaumont, *Nightwalking – A Nocturnal History of London* (Verso, 2015) 19.

[44] P Griffiths, *Lost Londons: Change, Crime, and Control in the Capital City, 1550–1660* (Cambridge University Press, 2008) 332.

Fires 'seemed more terrible against night's black' and night crime, especially theft and robbery, was 'more dangerous because of night's thick cover'. People walking in the 'dark risked arrest after the curfew hour. Anyone outside at night without reason or permission was potentially criminal and suspicious.'[45]

Those who ventured outdoors at night – and there is always some nightlife in a major city –[46] were typically of either side of the law: *criminals*, like thieves, prostitutes ('women of the night') and louts and people who were needed to secure the night ('night watchmen') or provide vital services such as doctors, midwives and *night soil men*.[47]

The legal way to regulate outdoor activity seems to have been to create specific offences. For example, in England and her colonies, '[n]ight-walking has been a crime for a millennium or more'.[48] Nightwalking was a label that communicated the various anxieties of early modern society. These included 'the ordering of time, and specific disorders that were associated with darkness and the fall of night' and the regulation of sexuality and the policing of women.[49] Other forms of social regulation included *eavesdropping*[50] and *scolding*.[51] With gaming, these were the offences typically used in fifteenth century England to counter various forms of social misconduct, including malicious gossip, sexual misdeeds or the operation of a rowdy alehouse.[52] In practice, nightwalking was seldom presented as an offence

[45] *Ibid* 332–33 (cites) 286, 355.

[46] On 16th and 17th century London, see: P Ackroyd, *London – The Biography* (Anchor Books, 2003) 444.

[47] C Koslofsky, *Evening's Empire – A History of the Night in Early Modern Europe* (Cambridge University Press, 2011) 132. In the absence of a sewage system, Londoners relied, well into the 19th century, on 'night soil men' to dispose of human waste. The term was used since the work was legally restricted to the hours of darkness. See: L Jackson, *Dirty Old London – The Victorian Fight Against Filth* (Yale University Press, 2014) 46–47. On the sewage system, see also: LD Schwartz, *London in the Age of Industrialization* (Cambridge University Press, 1992) 151 and fn 75, 235–236; www.hamptonwickhistory.org.uk/ewExternalFiles/Down%20The%20Drain.pdf. On night soil men in Baltimore and Philadelphia, see: DG Roberts and D Barrett, 'Nightsoil Disposal Practices of the 19th Century and the Origin of Artifacts in Plowzone Proveniences' (1984) 18(1) *Historical Archaeology* 108–15.

[48] See: Beaumont (n 43) 15.

[49] See: P Griffiths, 'Meanings of Nightwalking in Early Modern England' (1998) 13(2) *The Seventeenth Century* 212.

[50] 'Eavesdroppers stood outside other people's houses, often at night, listening to their conversations or sometimes observing their private acts, while nightwalkers, found wandering around after sunset with no reason that seemed legitimate to respectable people, were suspected of having more nefarious intent.' (M Keniston McIntosh, *Controlling Misbehavior in England, 1370–1600* (Cambridge University Press, 1998) 65.) Other offences included sexual misconduct, disorderly alehouses, 'evil reputation' and poverty. See: JJ Chriss, *Social Control: An Introduction* (Polity Press, 2007) 88. On the nature and origins of eavesdropping in England and in other nations, see: JL Locke, *Eavesdropping – An Intimate History* (Oxford University Press, 2010) 17.

[51] One was 'a listener at windows and sower of discord between the neighbors'; the other 'a common listener at night who followed the said listening by increasing disputes'. (See: M Keniston McIntosh, 'Finding Language for Misconduct – Jurors in Fifteenth-Century Local Courts' in BA Hanawalt and D Wallace (eds), *Bodies and Disciplines* (University of Minnesota Press, 1996) 93 and fn 15).

[52] 'The four types of offenses reported most often during the first half of the century (scolding, nightwalking, eavesdropping, and gaming) were generally described by local people as threatening amicable interactions and the social tranquility of the community." See: Keniston McIntosh (n 50) 87 and 90 (cite).

unless it was linked with another form of wrongdoing, such as eavesdropping.[53] As court records suggest, these offences 'were usually portrayed as types of behavior suspect largely through association. Roaming around at night was connected sometimes with staying too late at a tavern, sometimes with eavesdropping'.[54] In terms of gender, available data suggests that most eavesdroppers and nearly all nightwalkers were men,[55] but it was predominantly women who were reported for the offence of scolding.[56]

The system of social control through public courts started around 1300 CE and was beginning to collapse with calls for law reform in England beginning in 1620.[57] But progress was slow. The concern over civic unrest increased as London grew, and nightwalking remained a relevant concern, among many in the first half of the seventeenth century.[58] In fact, in the early seventeenth century, we also see the change in the nature of nightwalking – from a 'gender neutral' offence to an offence mostly linked with female sex workers.[59] In earlier centuries, the term 'nightwalker' was just as likely to describe male clients of prostitutes since it 'belonged to a regulatory vocabulary describing questionable nocturnal activity'.[60] From the mid-sixteenth century on, nightwalkers were gradually imagined as sexual offenders and, in the seventeenth century, as women.[61] In fact, while nightwalking had already long been an offence, by the seventeenth century, 'many more nightwalkers were taken to court after 1600, and nearly all of them were women after 1620'.[62] But that did not mean 'that men could drift though London with the sort of freedom that daylight might bring'. Men would also get into trouble in the seventeenth century 'for walking in the streets after the curfew hour', but now the offence would not be called nightwalking.[63]

As we near the modern era, we see two more contradictory trends. On the one hand, more people ventured for nocturnal activities (legitimate or not). Curfew

[53] Wandering at night was normally said to be merely against the King's peace or to the harm of the neighbours. See: Keniston McIntosh (n 50) 66–67.

[54] See: Keniston McIntosh (n 50) 96.

[55] See: Keniston McIntosh (n 50) 65 (data is through the end of the 16th century). Also: Beaumont (n 43) 197.

[56] See: Keniston McIntosh (n 50) 95 (speaking of the period up to the 17th century). Scolding was 'a broad category of offenses, often charged against women, which included malicious gossiping, backbiting, and spreading rumors'. See: Chriss (n 50) 88.

[57] See: Chriss (n 50) 88.

[58] See: Griffiths (n 44) 39–40. Court records of 1604–1658 suggest a significant number of offenders were charged with nightwalking: Griffiths (n 44) 82–87.

[59] 'The feminisation of nightwalking', notes Griffiths (n 49) 213, 'was never inevitable; its near symmetry with female sexuality only fully emerged in the minds of seventeenth-century elites'. Griffiths dates the beginning of this change to around 1560 (*ibid* 216–17).

[60] '[W]hich also included "night-hunter", "night-rover", and reports of people "hanging", "straggling", "loitering", "rouging and ranging in the streets in the nightseason."' In a 1435 case, a brothel's clients were described as 'night-walking men'. See: Griffiths (n 49) 217.

[61] See: Griffiths (n 49) 217–20; Griffiths (n 44) 154–55; Koslofsky (n 47) 218 and fn 90.

[62] See: Griffiths (n 44) 334.

[63] They might be said to be 'taken in the watch' or 'wandering late'. *See*: Griffiths (n 44) 334.

regulations gradually lost effect as nightly activities grew enormously (at least for the wealthy):[64]

> [T]he hours of legitimate business were extended beyond the nine or ten o'clock curfews that might have been expected in the early seventeenth century [and the] theatres, the opera, and other entertainments ensured that increasing numbers of people would be on the City streets well past the time when the City might once have closed down in response to a curfew that was now too difficult to enforce.[65]

This seems to have happened all across Europe and was likely related to the installation of lighting in public spaces across the continent. London, to take an eminent example, was not only the most populous city in the world but also, thanks to gas street lighting, could boast that 'no other place on earth was as extensively lit, or as bright'.[66] On the other hand, this enthusiasm should be qualified; this brightness is judged by pre-electric lights standards,[67] and illumination in London was further limited by weather and environmental factors,[68] and so it could only partly reduce fear of falling prey to crime on the streets. The city remained dangerous. Data from 1888 suggests that London's Metropolitan Police, formed to curb the city's crime, would more than *triple* its beat patrol at night.[69] Looking at social attitude, we find evidence suggesting, first, that use of the term 'street' increased towards the nineteenth century, but negatively so,[70] and second, that nocturnal activity outdoors was not fully socially acceptable even well into the nineteenth century.[71]

[64] See: Koslofsky (n 47) 8 and fn 28 quoting French historian Alain Cabantous as concluding that 'one way or another, the vast majority of the population of Western Europe slowly began to see the night as a period of free time' (citing A Cabantous, *Histoire de la nuit: XVIIe–XVIIIe siècle* (Fayard, 2009) 305).

[65] See: JM Beattie, *Policing and Punishment in London, 1660–1750* (Oxford University Press, 2001) 83 (cite), 119, 172.

[66] See: Bogard (n 34) 40; also see: Koslofsky (n 47) 133 et seq.

[67] London still maintains almost 1,500 gas powered lights, some almost 200 years old. See: Maev Kennedy, 'Light Brigade: Carrying the Torch for London's Last Gas Street Lamps' *The Guardian*, 11 May 2018. On gaslights see generally: L Tomory, *Progressive Enlightenment: The Origins of the Gaslight Industry, 1780–1820* (The MIT Press, 2012).

[68] 'Fog and smoke threatened clear perception across the [19th] century'. See: Otter (n 35) 190; also Bogard (n 34) 14 et seq. Koslofsky (n 47) 128–56 (on the rise of street lighting) and 173 (on the London fog).

[69] From 921 to 3,126 the patrolling beats. See: PK Andersson, *Streetlife in Late Victorian London – The Constable and the Crowd* (Palgrave Macmillan, 2013) 34.

[70] See: Andersson, *ibid* 22; Griffiths (n 49) 218; D Epstein Nord, *Walking the Victorian Streets – Women, Representation, and the City* (Cornell University Press, 1995) 1–2; R Gilfillan, *Crime and Punishment in Victorian London – A Street Level of the City's Underworld* (Pen and Sword Books, 2014) 40 (cite), 41. Matters were no better in America's cities, see: PC Baldwin, *In the Watches of the Night: Life in the Nocturnal City, 1820–1930* (The University of Chicago Press, 2012) 6.

[71] This was the case in *England* (see: J Ewing Ritchie, *The Night Sight of London* (William Tweedie, 1858) 149, but also in the United States (where '[t]he laws were vague about whether people could legally walk abroad at night, but the custom was clear that law-abiding people were to stay indoors. Those found in the streets at night were strongly suspected of being up to no good.')); Baldwin (n 70) 9.

C. On Fires and Thieves: Defending the Castle against Burglars and Arsonists

From ancient times, law was acutely aware that crime and fire threatened social order, especially when they occur at night. This section looks at three of the legal responses: laws regarding criminal break-in or arson, legally sanctioned extra-judicial killing of night-thieves, and legal measures – civil, administrative, as well as criminal – taken to counter the threat of fire.

Over history, the death penalty was meted out for many offences, including theft and arson. This is easier to explain in ancient times, when rudimentary legal systems had to use the threat of a death penalty to deter and when human life perhaps held less value; but, as late as the eighteenth century, England made a large number of offences (mostly property crimes) punishable by death. This 'bloody code' was established after Parliament gained control in the Glorious Revolution of 1688 and is explained as an effort to keep order through the fearsome suppression of the population in the absence of a professional police force. The list of offences punishable by death included arson, larceny, theft and burglary.[72] The common law had an additional argument for the severity of punishment in cases of arson and burglary; these were not just crimes *against property* but crimes against *the security of habitation*.[73]

Arson was categorised as a capital offence in common law until the nineteenth century because of aggravating factors such as the threat fire poses to human life and the view that arson was evidence of the perpetrator's moral recklessness. Substantively, by the mid-eighteenth century, English common law defined arson as the malicious and willful burning of the house of another *by day or night*. However, statutes enacted in the nineteenth century sometimes imposed a harsher penalty for arson at night.[74]

Burglary in common law was defined by Lord Coke as the breaking and entering at night into a house of another with the *intent* to kill or commit some other felony.[75] *Night* – functionally defined as when 'darkeness comes' and 'you cannot discerne the countenance of a man'[76] – aggravates the offence, 'for the night is the time wherein man is to rest, and wherein beasts runne about seeking their prey'.[77] In the late eighteenth century, Blackstone follows suit, speaking of 'nocturnal

[72] See, generally: L Seal, 'Criminalisation and the Eighteenth-century's "Bloody Code"' (2008) 74(1) *Criminal Justice Matters* 16; and, in detail: J Walliss, *The Bloody Code in England and Wales, 1760–1830* (Palgrave Macmillan, 2018).

[73] See: J Dressler (ed), *Encyclopedia of Crime & Justice,* 2nd edn (Macmillan, 2001) 73. Also: W Blackstone, *Commentaries on the Laws of England,* 1st edn, Vol. 4 (Clarendon Press, 1765–1769) 220.

[74] See: Dressler, *ibid* 73–75.

[75] Whether the felonious intent was carried out or not. See: E Coke, *The Third Part of the Institutes of the Laws of England* (W Clarke & Sons, 1809) (1644) 63.

[76] *Ibid* 63.

[77] *Ibid* 63.

housebreaking … as a very heinous offense' and stressing that burglary 'must be by night, and not by day; for in the day time there is no burglary'.[78]

Over time, many of the elements of Coke and Blackstone's definitions of burglary have changed; burglary evolved into '[t]he modern statutory offense of breaking and entering any building – not just a dwelling, and not only at night – with the intent to commit a felony',[79] although thievery at night can be an aggravating factor.[80]

D. The Night-Thief

Burglary had an even harsher and more unique feature in legal history: the extra-judicial element. English common law may seem harsh in its treatment of crime. But any sentence carried out was, at least ostensibly, the result of a legitimate, rules based, official judicial system, and followed a legal procedure. But what of citizens' reactions to crime, as it happens? Legal systems have long wrestled with the issue of self-defence; when is it legitimate for one person to kill off a person, perceived as an assailant, acting as judge and jury? The question is more forceful when a person feels threatened at home, her castle. The law still struggles to answer these questions.[81] In history, we find a somewhat surprising law; a series of legal systems, well aware of requirements of justification and proportionality, waive these preconditions lawfully permitting virtually any killing of a night-thief: 'this was not a form of self-defense, but a right to kill with impunity …. [T]he slain man's kin did not treat the killing as an occasion for revenge, but rather as a rightful execution'.[82]

This rule has a long pedigree. In the Old Testament's book of *Exodus*, there is a section that can only be described as a law code.[83] This code regulates, among other matters, both thievery and arson, setting a general sensible rule that a thief or an arsonist must pay compensatory damages. The punishment for the night-thief, however, is dramatically more severe:

> If while tunneling, a thief should be found and is struck down and dies, there is no bloodguilt for him. If the sun rises upon him, there is bloodguilt for him ….[84]

[78] Blackstone (n 73) 223–24.

[79] As noted as a second meaning of the term 'burglary' in *Black's Law Dictionary* (n 5). See, similarly, the UK's Theft Act 1968 c. 60, section 9.

[80] That is the case in the definition of 'burglary' in the American Model Penal Code, § 221.1.

[81] See: CE Light, *A History of America's Love Affair with Lethal Self-Defense* (Beacon Press, 2017); C Light, 'From a Duty to Retreat to Stand Your Ground: The Race and Gender Politics of Do-It-Yourself-Defense' (2015) 15(4) *Cultural Studies ↔ Critical Methodologies* 292–98; SP Green, 'Castles and Carjackers: Proportionality and the use of Deadly Force in Defense of Dwellings and Vehicles' (1999) *University of Illinois Law Review* 1.

[82] See: GP Fletcher, 'The Metamorphosis of Larceny' (1976) 89 *Harvard Law Review* 469, 477.

[83] Starting at *Exodus* 21: 1.

[84] *Exodus* 22: 1–2 (see: R Alter, *The Hebrew Bible – A Translation* (WW Norton & Co, 2018)). *Exodus* 21:16 metes out death on he who steals a man and sells him to slavery.

'These few enigmatic words have spawned much controversy'[85] and much analysis. Late nineteenth century English theologian Charles Ellicott explains that the section deals with a 'thief who is making forcible entry into a dwelling-house'[86] and that '[m]ost codes agree with the Mosaic in allowing the inmates of the house to resist such an attempt if made at night, and to shed the blood of the burglar, if necessary'.[87] The explanation is that the residents may suspect a murderous intent on the burglar's part (if resisted in his thieving efforts).[88] In daytime, 'no violence is to be feared …. Moreover, assistance is readily obtainable, and thus there is no need of resorting to extreme measures. The English law makes exactly the same distinction as the Mosaic'.[89] Another commentator suggested that killing a night-thief is justifiable homicide because '[i]n the dark the householder would probably not be able to recognize the burglar, so as to bring him to justice, nor would he know whether he might not intend murder'.[90] The same rule applied under other ancient legal systems such as Greece[91] and Rome. The Roman Law of the Twelve Tables imposed a death *sentence* for intentional arson[92] and made clear that a night-thief may be killed with impunity, not in self-defence, and that no revenge or compensation can be extracted for it.[93]

These are very harsh rules and were very likely never strictly implemented, even in antiquity.[94] That said, the idea that *arson* and *night thievery* were aggravated offences was forceful for a very long time. With the advent of electricity and stricter building codes, surely, night is no longer the criminal haven it once was.[95]

[85] See, for analysis of Jewish glossators: JD Jacobs, 'Privileges for the Use of Deadly Force Against a Resident-Intruders: A Comparison of the Jewish Law and he United States Common Law' (1990) 63 *Temple Law Review* 31, 34 (cite), 34–42 (analysis); see also: GP Fletcher, 'Self-Defense as a Justification for Punishment' (1991) 12 *Cardozo Law Review* 559.

[86] See: CJ Ellicott (ed), *An Old Testament Commentary for English Readers, Vol. 1* (Cassel, Petter, Galpin & Co, 1882) 269.

[87] *Ibid.*

[88] Ellicott also offers a more modern rationale, that in breaking in at night, the thief 'may be considered as having dissolved the "social compact," and converted himself from a fellow-citizen into a public enemy'. *Ibid.*

[89] *Ibid.* On 18th century English law, see: Blackstone (n 73) 220–29.

[90] SR Driver, *Cambridge Bible for Schools and Colleges – The Book of Exodus* (Cambridge University Press, 1911) 223.

[91] M Gagarin, 'Self-defense in Athenian Homicide Law' (1978) 19(2) *Greek Roman and Byzantine Studies* 111, 1.

[92] Table VII, Law VI; see: www.constitution.org/sps/sps01_1.htm.

[93] The Tables say so not once but twice, and their purview is not limited to house breaking. See: Table II, Laws IV (*night-thief*) and V (*day-thief*); and Table VII, Law IV (*destruction of another's crop at night*).

[94] On this see: W Smith, *Dictionary of Greek and Roman Antiquities*, 2nd edn (Little, Brown & Co, 1848) 563–64 (defining *Furtum*, theft), 623–33 (defining *Incendium*, the crime of setting any object on fire, by which the property of a man is endangered).

[95] At least when electricity is up and running. There are many examples of how crime and fire levels often (but not always) rise during power outages and blackouts. See: DE Nye, *When the Lights Went Out – A History of Blackouts in America* (The MIT Press, 2010) 52, 89–90, 105–106, 133, 185, 230. On the devastation brought to Puerto-Rico by Hurricane Maria and its aftermath, see: www.vox.com/energy-and-environment/2018/6/13/17413828/puerto-rico-blackout-power-grid-hurricane-maria.

But was it ever the case? Was the night really so dangerous until the twentieth century? French historian *Cabantous* intriguingly argues that in both England and France, homicides and thefts were actually *not more numerous at night*, yet both crimes were classified differently and punished more severely when committed at night.[96] Whatever the facts, it is clear the public perception and the law historically saw night-time as a time of heightened risk, and reacted accordingly.

E. Containing and Regulating Fire

In domesticating fire,[97] humans overcame 'a strong and potentially destructive force of nature' and met 'a basic condition for the subsequent emergence of agriculture and industry'.[98] Fire also allowed humanity to remain active and social after dark. But during most of history, lighting up and maintaining fire was difficult and expensive,[99] and the illumination level achieved was not very bright by modern standards:[100]

> Five hundred years ago … cities, towns, and villages would have appeared nearly as dark as the oak forests … no streetlights would have shone. Within, candles and lamps no brighter than those of Roman times would have lit only a bowl of porridge, a book, a shirtsleeve in need of mending, another … Such small light was precious and meted out sparingly. For much of the evening, people lay in their houses after dousing their cooking fires, sleeping and dreaming away the hours.[101]

Change occurred gradually; technology allowed for better illumination and social pressures brought about the installation of lighting in public spaces. Yet, as long as *all artificial lighting* was based on an open flame, people had little option but to use fire at night, if they could afford to. Fire – intentional or accidental – was a constant danger throughout history, until the introduction of electric lighting presented a huge, if recent, relief.[102] It was a common occurrence, a scourge that ravaged many

'Old school' crime is not the only threat; see, eg, Z Lukszo, G Deconinck and MPC Weijn (eds), *Securing Electricity Supply in the Cyber Age* (Springer, 2010).

[96] See: Koslofsky (n 47) 8–9 and fn 29 (citing Cabantous (n 64) 140–84).

[97] For discussion of the ca. 10,000 year period when the technology of the uses of fire was slowly developed, see: JE Rehder, *The Mastery and Uses of Fire in Antiquity* (McGill-Queen's University Press, 2000).

[98] B De Vries and J Goudsblom (eds), *Mappae Mundi – Humans and their Habitats in a Long-Term Socio-Ecological Perspective*, 2nd updated printing (Amsterdam University Press, 2003) 28, 30; J Brox, *Brilliant – The Evolution of Artificial Light* (Houghton Mifflin Harcourt, 2010) 7–19.

[99] In the worst time, notes Brox, 'many saw only by the light of their cooking fires …. The poorest people might have no light at all'. (*Ibid* 15–17; cite: 17).

[100] *How feeble was pre-electric light*? 'A single 75-watt incandescent bulb burns one hundred times brighter than a candle.' (Bogard (n 34) 8).

[101] Brox (n 98) 1.

[102] Has the threat of fire increased, due to global warming? For a relatively reassuring view, see: Stefan H Doerr and Cristina Santín, 'Global Trends in Wildfire and its Impacts: Perceptions versus Realities

of the world's major urban and cultural centres,[103] including the infamous Great Fire of London (1666).[104] As we draw closer to our time, fire's danger actually *increased*: as cities grew larger, entire districts of tightly packed wooden houses were at the mercy of an overturned lamp, a stray cinder, a child careless with a candle'.[105] In the United States, fires 'grew in destructive power in parallel with the coming of the industrial revolution' as cities rapidly grew larger, denser, mixing residential, commercial and industrial use.[106]

Given the centrality of the fire experience, it is not surprising that we find throughout history varied legal regulation – from liability for arson, negligence in keeping fire,[107] or looting during one,[108] to fire prevention measures like building codes to setting up of fire brigades. Some of these measures bear witness to the heightened risk of night fires in densely populated urban areas.

To link the two issues, it should be noted that while fires often caused major loss in life and property, they also often afforded an opportunity to regroup and rebuild for the future, and part of the adaptation process included new regulation concerning future fire hazard. A good example is the Great Fire of London, which was followed by dynamic steps to rebuild, adapt, and better prepare for fire events.[109]

In early 1667, Parliament passed an Act for the rebuilding of the City of London. It was, in essence, a building code, with ideas on how to rebuild a less dangerous London. Wood was to be replaced with brick or stone as building materials; design restrictions were aimed at eliminating resident overcrowding and widening streets; the city of London was given the power to prohibit trades that they deemed perilous in respect to fire.[110] Direction from the same year

in a Changing World' (2016) 371(1696) *Philosophical Transactions of The Royal Society B Biological* 20150345.

[103] Often, more than once. See:https://en.wikipedia.org/wiki/List_of_town_and_city_fires; www.ddwei.info/pdf/subsequent/0.pdf.

[104] See: WG Bell, *The Great Fire of London in 1666* (Endeavour Press Ltd, 1920, 2016); N Hanson, *The Great Fire of London in that Apocalyptic Year, 1666* (John Wiley & Sons, Inc, 2001).

[105] 'Such danger might be reason enough to send the children to bed in the dark.' See: Brox (n 98) 17.

[106] See:https://pdf4pro.com/view/urban-conflagrations-in-the-united-states-by-154f.html. Also: C Meisner Rosen, *The Limits of Power: Great Fires and the Process of City Growth in America* (Cambridge University Press, 1986) 4–11.

[107] For which the Old Testament book of *Exodus* (22:6) sets compensation.

[108] For which §25 of the Hammurabi Code would impose death. See: RF Harper, *The Code of Hammurabi King of Babylon About 2250 B.C.*, 2nd edn (The University of Chicago Press, 1904) 19.

[109] Fire, however, remained a frequent, destructive occurrence: ca. 120 fires, destroying ten houses or more, occurred between 1666 and 1833. See: C van Lieshout, 'The Most Valuable Means of Extinguishing the Destroying Fires: Fire-fighting and the London Water Companies in the Long Eighteenth Century' (2017) 42(1) *London Journal* 53 and fn 1. On the rebuilding of London, see: JF Field, *London, Londoners and the Great Fire of 1666: Disaster and Recovery* (Routledge, 2018); also: Ackroyd (n 46) 220–36; Hanson (n 104) 175 et seq.

[110] See: www.british-history.ac.uk/statutes-realm/vol5/pp603-612#h2-0001; www.parliament.uk/about/living-heritage/transformingsociety/towncountry/towns/collections/collections-great-fire-

established easier access to water and prescribed a form of a fire-fighting bucket brigade, especially at night:[111]

> The twelve city companies and aldermen were required to provide buckets and squirts, and the bellmen had to walk up and down each ward from 10 P. M. to 5 A. M. Every householder at the cry of 'fire!' had to place a well armed man at his door and hang a light out, under a penalty of 20s. He was also to have ready at his door water in a vessel, and pumps were ordered to be fixed in every well in the streets.[112]

An Act of 1707 made churchwardens responsible for fixing fire plugs in water mains and instructed the bellmen to 'cry at night: – "Take care of your fire and candle," "Be charitable to the poor and pray for the dead," all of which was very wholesome advice.'[113]

The Great fire of London led not only to the evolution of fire protection through private and public efforts,[114] but also to the establishment and rapid expansion of fire insurance,[115] and to change in fire liability law; the traditional common law rule was strict liability for an escape for fire, essentially *imposing a special duty* on house holders to keep their fires safe and not looking for negligence of fault. The precise background for the change is unclear, but it likely the result of the transitioning of English society from feudal economy with closely knit domestic housing arrangement to dense urban communities and the harshness of the traditional common law rule that now became clear.[116] The Act of 1707, and then the Fires Prevention (Metropolis) Act 1774 changed the rule, so that there would be no liability for the escape of fire, liability in the absence of fault, or some blameworthy conduct.[117]

1666/1666-act-to-rebuild-the-city-of-london/; https://en.wikipedia.org/wiki/Rebuilding_of_London_Act_1666.

[111] See: GC Gillespie, 'Early Fire Protection and the Use of Fire Marks' (1922) 46(3) *The Pennsylvania Magazine of History and Biography* 232–61; www.ddwei.info/pdf/subsequent/0.pdf.

[112] See *ibid* 238 (cite).

[113] See *ibid*, adding that '[i]n 1600 the bellmen's duty, when a fire broke out was to cry "Fire! Fire! keep in your beds," which was intended no doubt to prevent crowds from collecting and the consequent confusion'. (238–39). Also see: SPB Mais, *Fifty Years of the L.C.C.* (Cambridge University Press, 1939) 136; AI Ogus, 'Vagaries in Liability for the Escape of Fire. "How Great a Matter a Little Fire Kindleth"' (General Epistle of James III. 5) (1969) 27(1) *Cambridge Law Journal* 107–108. The statute is the Act of 1707 (6 Anne, c. 31) 'for the better preventing the mischiefs that may happen by fire'.

[114] Although the process was long and complex; see: S Ewen, *Fighting Fires – Creating the British Fire Service, 1800–1978* (Palgrave Macmillan, 2010); CFT Young, *Fires, Fire Engines, and Fire Brigades* (Lockwood & Co, 1866) and van Lieshout (n 108).

[115] See: Gillespie (n 111) 238 (cite); also: http://eprints.lse.ac.uk/90464/1/WP287.pdf; Ogus (n 113) 107.

[116] See: Ogus (n 113) 104–109 and http://subrogationrecoverylawblog.com/2010/12/06/londons-burning-call-a-subro-lawyer/.

[117] See: G Stephenson, *Sourcebook on Torts*, 2nd edn (Cavendish, 2000) 405–406. 'The word "accidentally" in the Act has always been given a narrow construction and effectively been taken to mean "non-negligent"'. (See: (2005) 22(6) *Construction Law Journal*, T105, T111. It is suggested that this change in law ended up hindering the evolution of common law 'within the expanding nineteenth-century doctrine of negligence'. (See: Ogus (n 113) 104 (cite) and 116–21.)

IV. Re-Regulating the Night?

A. A Roadmap: Where Are We Now? Where Do We Go from Here?

The historical laws and regulations of the night of times past have gone several ways. Some have been *revoked*; the night-thief, for example, is no longer fair-game for homeowners. Some became *obsolete*; nightwalking, for example, is rarely still encountered in law books.[118] Some have *limited use*; for example, in modern democracies, curfews are extraordinary measures principally used in two cases: set on the general population as a temporary, emergency measure, to allow authorities to deal with civil unrest, terror threats or natural disasters *or* set on minors, in an effort to control juvenile night-time activities.[119] Some have *morphed*; burglary is no longer limited to breaking and entering into a dwelling at night,[120] and eavesdropping has taken on a modernised technological content, gaining new relevance for our time.[121] Finally, much of the regulation of night activity has gone mainstream; they have become so conventional that they only trouble the most libertarian inclined. One category seeks to prevent nuisance at night – for example, from high levels of light and sound – in an effort to set a baseline for a good night's sleep. A second category regulates night-time activity, in an effort to prevent misadventure; for example, closing public parks at night or imitating the hours liquor can be sold. Some refer to such limitations as a *curfew*, but this is at best, an analogy; while such regulations limit human activity and choice, they do not require exclusion of all options, disallowing some nightly outdoor activities is not a curfew, strictly speaking.[122] A third category covers the regulation of working hours of individuals and businesses, especially at night. The rationales behind

[118] Massachusetts' criminal code still bans 'Common night walkers, common street walkers, both male and female ….' (MGLA 272 § 53). See Beaumont (n 43) 16.

[119] Juvenile curfew is especially pertinent to the United States where it has been used since the late 19th century and where it is still widely and controversially used today. See: PC Baldwin, '"Nocturnal Habits and Dark Wisdom": The American Response to Children in the Streets at Night' (2002) 35(3) *Journal of Social History* 593; Miller and Wright (n 41) 360–63; LS Urban, *The Deterrent Effect of Curfews – An Evaluation of Juvenile Probationers* (LFB Scholarly Publishing, 2007).

[120] There is empirical data suggesting that daylight is a factor in burglars' decision when to operate. See: T Coupe and L Blake, 'Daylight and Darkness Targeting Strategies and the Risks of Being Seen at Residential Burglaries' (2006) 44(2) *Criminology* 431; also: JL Doleac and NJ Sanders, 'Under the Cover of Darkness: How Ambient Light Influences Criminal Activity' (2015) 97(5) *Review of Economics and Statistics* 1093.

[121] The article on *electronic surveillance and wiretapping* in Miller and Wright (n 41) 487 notes how 'Even prior to the age of electricity, the practice of eavesdropping was a problem. The main difference was that eavesdropping only occurred within earshot'. Modern technology has changed this significantly.

[122] See, eg, 'Park's Dusk Curfew Enforced' (2017) 23 *City Law* 35; César Rodriguez, 'Laredo Bar Owner Arrested for Allegedly Serving Alcohol Past Curfew, Police Say' *Laredo Morning Times*, 22 February 2019. On the regulation of night drinking see, extensively: D Talbot, *Regulating the Night: Race, Culture and Exclusion in the Making of the Night-time Economy* (Ashgate, 2007).

it are public policy concerns, such as the social rights and welfare of employees and worker safety.[123]

Can we expect more regulation in the future? Extrapolating on contemporary law, I envision two possible paths: *regulating light pollution* and securing *a minimal night sleep as a human right.*

B. Regulating Light Pollution

Humans have won over the dark of the night in spectacular fashion. There is so much artificial outdoor lighting at night that we now speak of *light pollution.*[124] Most people live in urban areas where it never gets dark, day or night,[125] and where we can hardly see the night sky.[126]

The question is: why should we care? It turns out that light pollution – poorly aimed and over-bright night light – has considerable negative effects. First, waste of energy and money: worldwide, an estimated 20 per cent of electricity is used for lighting of which at least 30 per cent is wasted.[127] Second, there is environmental degradation which encompasses such phenomena as skyglow,[128] glare,[129] and light intrusion.[130] Third, there is impact on wildlife: '[v]ast numbers of the world's species are nocturnal. The invasion of their world by artificial lighting has consequences that we are only just beginning to investigate and quantify'.[131] Artificial light disrupts the behaviour of plants and animals, which evolved in a world without such lighting.[132] Initial assessments are quite troubling, for both

[123] See, eg, the *Night Work Conventions* of the International Labour Organization (see: https://en.wikipedia.org/wiki/Night_Work_Conventions).

[124] On the term, see: K Narisada and D Schreuder, *Light Pollution Handbook, Vol. 1* (Springer, 2004) XIX–XX; B Mizon, *Light Pollution – Responses and Remedies*, 2nd edn (Springer, 2012) 53.

[125] And the world is getting *even brighter*: See: CCM Kyba et al, 'Artificially Lit Surface of Earth at Night Increasing in Radiance and Extent (2017) 3(11) *Science Advances* e1701528.

[126] See: Brad Plumer, 'The Night Sky is Vanishing: 80 percent of Americans can no Longer see the Milky Way' *Vox*, 10 June 2016; Stephanie Buck, 'During a 1994 Blackout, L.A. Residents called 911 when they saw the Milky Way for the First Time' *Timeline*, 15 February 2017. See also Bogard (n 34) 3–13.

[127] See: https://undark.org/article/light-pollution-health-consequences/; more details in Mizon (n 124) 53–62.

[128] *Skyglow* refers to light that is being carelessly, or sometimes deliberately, projected from the ground or a structure, colouring the night sky and reducing the visibility of astronomical objects. Its cause is upward light from poorly designed lamps. See: Mizon (n 124) 40.

[129] *Glare* is an over-bright, poorly directed light that can dazzle or distract. It can paradoxically hide whatever (or whoever) is in their vicinity, thus defeating the whole purpose of lighting. Mizon (n 124) 62–63 calls it an 'anti-light'.

[130] *Light intrusion* – once called 'light trespass' – refers to 'the unwelcome spilling of light to cause annoyance to people in neighboring premises'. (Mizon (n 124) 65, noting that *it is an actionable nuisance under UK law*).

[131] Mizon (n 123) 65. On the ecology of darkness, see Bogard (n 34) 125–57 also: C Rich and T Longcore (eds), *Ecological Consequences of Artificial Night Lighting* (Island Press, 2006).

[132] See: Denise Chow, 'New Interactive Map Shows how Light Pollution Affects your Hometown' *NBC News*, 11 March 2019.

wildlife and humans.[133] Fourth, the impact on quality of human life and health: glaring lights cause accidents, and neighbours complain of light intrusion disrupting their sleep.[134] Further, light pollution turns out to be a risk factor for serious health conditions – physical,[135] as well as mental.[136]

Given these concerns and the fact that some domestic and international remedial efforts are already underway, it does not look like a stretch to expect regulation of outdoor lighting at night. We already had a first taste; in recent years, the traditional incandescent bulbs 'have been rapidly replaced by more energy-efficient lighting. The shift has driven down electricity demand in American homes, saving consumers money and cutting greenhouse gas emissions'.[137] This happened by law; joining an extensive international trend,[138] the United States Congress set the first national light bulb efficiency standards in 2007, which required that, starting in 2012, new light bulbs use 28 per cent less power than the existing lights, thus 'essentially ending the sale of the older, inefficient bulbs'.[139] It makes considerable sense to expect more regulation to counter light pollution.

C. A Right to Sleep?

Human activity at night has been expanding for centuries, but the trend seems to have accelerated in recent decades. For some, night activity is a matter of personal choice or owed to their nature as 'night owls'.[140] For many, this is the result of modern life; the economy, global and domestic, runs 24/7 and night work is essential when your clients are many time zones away,[141] or when you work night shift in factories, restaurants, hospitals, newsrooms, airports, etc. Add the

[133] See, eg: Lucy Salt, 'Killing us Softly: The Perils of Light Pollution' *Foreground*, 21 February 2019; Liz Perkin, 'Surprising Effect of Light Pollution: It Disrupts Aquatic Insects' *The Revelator*, 7 March 2019; KJ Gaston, ME Visser and Franz Hölker (eds), 'The Biological Impacts of Artificial Light at Night: From Molecules to Communities' (2015) 370(1667) *Philosophical Transactions of The Royal Society B Biological Sciences*; and Mizon (n 123) 67–75.

[134] See: Mizon (n 124) 77.

[135] See, eg: A Haim and BA Portnov, *Light Pollution as a New Risk Factor for Human Breast and Prostate Cancers* (Springer, 2013); https://undark.org/article/light-pollution-health-consequences/.

[136] See: Lisa Conti, 'How Light Deprivation Causes Depression' *Scientific American*, 1 August 2008; Linda Geddes, 'How do Scandinavians Deal with Long, Dark Winters? And What Might this Teach us About the Relationship between our Moods and Sunlight?' *Mosaic Science*, 14 March 2017.

[137] Nadja Popovic, 'America's Light Bulb Revolution' *New York Times*, 8 March 2019.

[138] See, usefully: https://en.wikipedia.org/wiki/Phase-out_of_incandescent_light_bulbs; see also: D Green and L Minhin, *Screw Light Bulbs – Smarter Ways to Save Australians Time and Money* (The University of Western Australia, 2010).

[139] Popovic (n 137), but *cf* Sean Hollister, 'The Incandescent Light Bulb isn't Dead – Reports were Greatly Exaggerated' *The Verge*, 1 January 2014.

[140] On the health risks associated with the disposition, see: Amanda MacMillan, '13 Ways Being a Night Owl Could Hurt Your Health' *health.com*, 8 February 2019. On the typology of human sleep characters, see: Michael J Thorpy and Jan Yager, *The Encyclopedia of Sleep and Sleep Disorders*, 2nd edn (Facts on File Inc, 2001).

[141] See: R Patel, *Working the Night Shift – Women in India's Call Center Industry* (Stanford University Press, 2010).

long commutes, holding multiple low-paying jobs and household chores, and one realises that many people cannot get the sleep they need for socio-economic reasons.

The baseline is that despite modern pressures, humans still operate on a circadian rhythm and need a healthy dose of daily sleep.[142] Science is still unsure exactly *why* we need to sleep at night. It tells us that during sleep important processing, restoration, and strengthening processes occur,[143] and that insufficient sleep time is detrimental to life and spirit[144] (as well as to the economy[145]). Science is not certain about the precise number of sleeping hours needed, or how we can easily fall asleep; the current 'gold standard' is eight hours of nightly sleep and there is plenty of advice on offer on how to create conducive sleep conditions.[146] The bottom line is that the American Centers for Disease Control and Prevention (CDC) declared insufficient sleep a 'public health problem', estimating that more than a third of American adults are not getting enough sleep on a regular basis.[147] Some commentators speak of a 'great sleep recession' in America,[148] suggesting 'time poverty wreaks havoc on sleep for low-income Americans',[149] and that 'a solid night's sleep as luxury good'.[150] Even the venerable *New York Times* calls sleep 'the New Status Symbol', declaring that '[s]leep today is a measure of success, a skill to be cultivated and nourished'.[151]

[142] On the health effects of night shifts on workers, see: W Olds (ed), *Sleep, Circadian Rhythms, and Metabolism* (Apple Academic Press, 2015).

[143] See: www.sleepfoundation.org/articles/why-do-we-need-sleep; Carl Zimmer, 'The Purpose of Sleep? To Forget, Scientists Say' *New York Times*, 2 February 2017; Veronique Greenwood, 'Why Do We Need to Sleep?' *The Atlantic*, 3 January 2018; Olga Khazan, 'Better Sleep Can Build Emotional Resilience' *The Atlantic*, 23 October 2017.

[144] See, eg: Richard A Friedman, 'Yes, Your Sleep Schedule Is Making You Sick' *New York Times*, 10 March 2017; Laura Sanders, 'Poor Sleep can be the Cause of Anxiety, Study Finds' *Washington Post*, 10 November 2018; Hilary Brueck, 'You are Probably not Getting Enough Sleep, and it is Killing You' *Business Insider*, 27 April 2018.

[145] Marco Hafner et al, 'Why Sleep Matters – The Economic Costs of Insufficient Sleep – A Cross-Country Comparative Analysis' (2017) 6(5) *Rand Health Q* (2017) 11 (findings suggest that insufficient sleep can result in large economic costs in terms of lost GDP and lower labour productivity).

[146] See: Friedman (n 144); Georgia Frances King, 'Why Eight Hours a Night isn't Enough, According to a Leading Sleep Scientist' *Quartz*, 10 June 2018; Kate Leaver, 'Ultimate Sleep Hack: I Spent a Month Trying Every Type of Advice' *BBC*, 15 March 2019; April Fulton, 'Teens Sleeping Too Much, Or Not Enough? Parents Can Help' *NPR*, 23 September 2018; Markham Heid, 'Is Blue Light Really What's Keeping You Awake?' *Medium*, 30 August 2018; Jerome Groopman, 'The Secrets of Sleep' *New Yorker*, 23 October 2017; www.nia.nih.gov/health/good-nights-sleep#good. The Centers for Disease Control and Prevention currently recommends seven or more hours of sleep per might for adults aged 18 to 60, see: www.cdc.gov/sleep/about_sleep/how_much_sleep.html.

[147] See: Hafner et al (n 145).

[148] See: Debra Goldschmidt, 'The Great American Sleep Recession' *CNN*, 18 February 2015 (updated 23 June 2017).

[149] See: Nan Mooney, 'Time Poverty Wreaks Havoc on Sleep for Low-Income Americans' *Slate*, 28 November 2017.

[150] See: Jacob Rosenberg, 'A Solid Night's Sleep Is the New Luxury Good. Have Fun Affording It' *Observer*, 31 July 2018.

[151] See: Penelope Green, 'Sleep Is the New Status Symbol' *New York Times*, 8 April 2017.

If you think that this is a socio-economic concern, you could expect, first, a public campaign to urge that there is a crisis and action is needed, and[152] second, a campaign for some form of legal regulation, in order to secure sleep time.

At first blush, this sounds a little outlandish. Israeli Defense Forces' General Staff Order 33.0231 entitled 'Soldiers Sleep and Rest Hours' holds that during a day a soldier *shall be guaranteed six hours of uninterrupted sleep* to ensure that he can function with alertness and efficiency in the remaining hours of the day.[153] Surely, civilians, not subject to 24/7 regulation like soldiers, need no such legal protection and can secure themselves a sufficient block of time for sleep? Some will surely argue that securing sufficient time for sleep is a matter of personal choice and responsibility. Others will argue that modern life and economics deny many a fair chance of securing sufficient sleeping time. This could be translated into the more familiar debate over regulation – necessity, legitimacy, costs and benefits.

Should the public support such a regulation on this issue, what might it look like? It might start with a public education campaign,[154] and it is very unlikely to end with forcing people to find time for sleep – curfew style – as this is much too invasive for liberal democracies. But, in-between, I can discern two potential avenues.

One route traces the history of labour law where it advocated *shortening* work hours and *limiting business operating*. An example of the former is the effort to limit the workday to eight hours in the United States. The '8 hour movement' argued that allowing workers more free time – to study or socialise – would be a benefit for individuals and polity alike.[155] An example of the latter is the resistance that still exists (notably in continental Europe) to extending business hours; despite the benefits of 24 hour supermarkets, such countries are more concerned with its detrimental effects.[156] Such a public policy – siding with tradition and supporting workers' rights – has a price, like a ranking hit in the World Economic

[152] Politics is not my expertise – but there are some indications that a public campaign is likely, see: A Huffington, *The Sleep Revolution: Transforming Your Life, One Night at a Time* (Harmony Books, 2016).

[153] My translations; sections 1, 2 & 4; some exemptions apply. A day is defined as the 24 hours between 6am of one day and the next. See: www.idf.il/media/16807/330213.pdf (Hebrew).

[154] Even bringing up-to-date scientific information to the public might be met with distrust. *Cf* Bastiaan T Rutjens, 'What Makes People Distrust Science? Surprisingly, not Politics' *aeon*, 28 May 2018.

[155] Workers did not have the economic power to win but they had the votes to enlist government on their side. See: P Kens, *Lochner v. New York – Economic Regulation on Trial* (University Press of Kansas, 1998) 15–16; see also: WA Mirola, *Redeeming Time: Protestantism and Chicago's Eight-Hour Movement, 1866–1912* (University of Illinois Press, 2015). Researchers are now reassessing the importance of leisure time in a global era; see, eg, M Shir-Wise, *Time, Freedom and the Self: The Cultural Construction of 'Free' Time* (Palgrave Macmillan, 2019).

[156] See, eg: Sam Judah and Olivia Sorrel-Dejerine, 'UK v France: The Sunday Shopping Difference' *BBC*, 10 October 2013; www.german-way.com/travel-and-tourism/germany-for-tourists/shopping-in-germany/shopping-hours-in-germany.

Forum's annual global competitiveness report.[157] But this is a public policy decision for each nation to take.

That said, the traditional 'nine to five' work model is under serious attack; newer work models offer more flexibility in terms of when and where that work is performed.[158] Such models have positive good effects like easing or even negating the work commute[159] and allowing people to synchronise work with their individual internal body clock.[160] But it is not sure that the new models will leave workers with more free time. In the modern workplace, employees often work long hours, and are expected to remain 'connected' and available at all times. This is why French law intervened, establishing, effective from 2017, '*a right to disconnect*' and requiring companies with over 50 employees to establish hours when staff should neither send nor answer emails.[161] In sum, some form of heightened regulation of working hours is a relevant, viable possibility.

A second potential legal route sounds more dramatic, but may well be within reach; it would legally recognise uninterrupted continuous peaceful sleep as a *human right*. This fits well with the ongoing debate on, and demand for, recognising and securing legal protection for 'new rights'.[162] In recent years, we have seen (at least in liberal democracies) interesting advances on social rights (eg, same sex marriage, support for which built-up amazingly fast)[163] and even on socio-economic rights (such as the right to Internet access).[164] Moreover, there is a strong argument for a right to sleep to be derived from already established human

[157] France was ranked 17th in 2018. The report noted encouragingly that it had recently 'taken on an ambitious reform program that encompassed overhauling labour laws' but then added, less favourably the '[w]hile labour market rigidities are being addressed, there is still a long way to go as the economy currently ranks 53rd on this dimension'. (See: www3.weforum.org/docs/GCR2018/05FullReport/TheGlobalCompetitivenessReport2018.pdf, 30).

[158] See, eg, Jana Kasperkevic, 'Do People even Work 9-to-5 Anymore?' *Marketplace.org*, 28 May 2018; Eric Morgan, 'Whatever Happened to the 9 to 5?' *Inc*, 2 December 2015; Matt Byrom, 'The Death of the Workday: Is 9 to 5 Working Obsolete?' *Business*, 22 February 2017 (last modified).

[159] See, eg, Annette Schaefer, 'Commuting Takes Its Toll' *Scientific American*, 1 October 2005; Fiona M Haslam McKenzie, 'The Socio-economic Impacts of Long-Distance Commuting on People and Communities' in FM Haslam McKenzie (ed), *Labour Force Mobility in the Australian Resources Industry: Socio-Economic and Regional Impacts* (Springer, 2016) 11–28.

[160] There is evidence that '[p]ushing back high school start times not only improves the quality of students' sleep, it also boosts attendance and academic performance' (Alice Klein, 'First Direct Evidence that Later School Day Really does Help Teenagers' *New Scientist*, 12 December 2018). For extensive discussion, see: M Walker, *Why We Sleep: Unlocking the Power of Sleep and Dreams* (Scribner, 2017).

[161] See: David Z Morris, 'New French Law Bars Work Email After Hours' *Fortune*, 1 February 2017; https://en.wikipedia.org/wiki/Right_to_disconnect.

[162] See, eg: Emily Atkin, 'Why Animal Rights Is the Next Frontier for the Left' *New Republic*, 14 March 2019.

[163] See: Chris Cillizza, 'Why Support for Gay Marriage has Risen so Quickly' *Washington Post*, 19 March 2013.

[164] See: https://en.wikipedia.org/wiki/Right_to_Internet_access; M Bagaric, N Fischer and D Hunter, The Hardship that is Internet Deprivation and what it means for Sentencing: Development of the Internet Sanction and Connectivity for Prisoners (2017) 51 *Akron Law Review* 261.

rights, such as the rights to health, life, rest and leisure.[165] Often in the vanguard, the Indian Supreme Court *has already ruled*, in 2016, based on the right to life, that sleep is a fundamental right.[166] Similar logic is behind the gradual recognition in the United States that homeless people have a right to sleep in public places and that authorities cannot ban the practice.[167] I hope that a new, fair, balance will be reached, once the importance of sleep is fully realised.

[165] The latter is enshrined in Article 24 of the Universal Declaration of Human Rights: '[e]veryone has the right to rest and leisure, including reasonable limitation of working hours and periodic holidays with pay'. (See: www.un.org/en/universal-declaration-human-rights/; www.standup4humanrights.org/layout/files/30on30/UDHR70-30on30-article24-eng.pdf.)

[166] See: Jen Mills, 'Sleep is a Fundamental Human Right, India's Supreme Court Rules' *Metro.co.uk*, 28 February 2016. For the holding, see: www.advocatekhoj.com/library/judgments/announcement.php?WID=1841.

[167] See: Mihir Zaveri, 'Laws Punishing Homeless People for Sleeping in Public Are Cruel and Unusual, Court Rules' *New York Times*, 5 September 2018 (decision of the Court of Appeals for the Ninth Circuit).

PART II

Courts and Time

5

Law and Time in Two Dimensions: Legitimate Expectations in the Case Law of the Court of Justice of the European Union

PATRICIA POPELIER

I. Introduction

Laws are captured by time in two dimensions. One dimension is the interval of time between laws coming alive and subsequently being abolished. The other dimension is the interval of time during which the persons affected by laws plan and act. Tensions of time occur where one dimension interferes with the other. At the intersection of both dimensions are 'legitimate expectations'. The leading question addressed is how courts use this legal concept to deal with tensions of time.

This chapter focuses on the case law of one particular court, the Court of Justice of the European Union (CJEU). Three reasons explain this choice. First, the CJEU's case law applies to a broad complex of legal systems, 29 in total: (at the time of writing) 28 Member States and the European Union (EU) itself as a whole. Second, the Court functions in a legal system built on the basic principles of economic freedoms and an economic market, which should make it particularly susceptible to arguments of legitimate expectations which, as explained below,[1] have specific significance in economic behaviour. Third, as the CJEU rules on EU as well as national laws, the question arises whether the concept of legitimate expectations is used in a similar way with regard to both subsets of laws.

In order to address this question, this chapter is structured as follows. After (I) having introduced the leading question, the next sections explain (II) what the two dimensions of time and (III) the concept of legitimate expectations involve.

[1] Section II.A.

These sections already draw on the CJEU's case law to illustrate what otherwise remains abstract reasoning, and refer in particular to older landmark decisions. Subsequently, (IV) the CJEU's case law from 1998 to 2017 is analysed to examine how the CJEU employs the concept of legitimate expectations to deal with tensions of time in the last 20 years. Finally, (V) the conclusion links the findings in section IV with the initial exposition of the two dimensions of time in law.

One preliminary remark must be made before advancing: This chapter is only concerned with legitimate expectations with regard to the effect of time of laws with a general scope. It is not concerned with case law that uses the concept of legitimate expectations to rule on individual administrative decisions or judicial pronouncements. Notices or Guidelines that lay down administrative practice with the effect of a general rule are not touched upon.

II. Law and Time in Two Dimensions

A. The First Dimension: The Legal Timeframe

The first dimension of time for laws is the interval of time that determines the law's course of life. This dimension is static: it is an analytic concept that reduces time to a calculable line with a concrete starting and finishing point. The most important reference points in this dimension are the dates on which the act is adopted, the date on which it is published, the day it enters into force, its temporal function, and the moment on which it expires. Various events can cause expiration: the lawmaker may amend or abolish the law, or a court may annul it. This moment has its own dimension of time: the amending law may replace the old rule by a new rule with retroactive or immediate effect; likewise, a court may annul with retroactive effect or may limit the effects in time of its judgment. This dimension of time organises the temporal framework in which events are placed to become 'legal' events.[2]

The most crucial factor in this dimension of time is the temporal function. This determines whether a new law interferes with the timeframe created by the old law. The temporal function is a technical device used to determine the facts to which the new rule applies.[3] Rules attach legal consequences to legal facts. Moreover, legal facts are facts that become legal because the law identifies them as something that triggers legal consequences. These legal consequences can take the

[2] See further: M Blegvad, 'Time, Society and Law' in *Time, Law and Society* (ARSP N° 64, Stuttgart: Franz Steiner Verlag, 1995) 15; F Ost, 'Les multiples temps du droit' in JJ Austruy et al, *Le droit et le future* (Presse Universitaire de France, 1985) 118, 144.

[3] I summarise earlier findings that appeared in Dutch-language publications: P Popelier, *Toepassing van de wet in de tijd* (*Application of the Law in Time*) (Kluwer, 1999) 31–36, elaborating on G Knigge, *Verandering van wetgeving* (*Alteration of Legislation*) (Gouda Quint, 1984) 687.

shape of a right or a duty: for example, the right to an allowance or a fiscal benefit, or the right to act or not act in a specific way; the duty to pay damages or penalties, or the duty to act or not act in a specific way. A law with immediate effect attaches legal consequences to facts that occur in the present, while a law with future effect attaches legal consequences to facts that occur after the elapse of a certain period of time. Similarly, a law with retroactive effect attaches legal consequences to facts that have occurred in the past. To distinguish the past, present and future, the point of reference is usually the date on which the act enters into force. This is the date on which the act becomes binding for persons, and thus the date on which the legal consequences dictated by the new law can take effect. This implies that laws can invariably operate in both the present and future; although often labelled as 'fiction',[4] or as 'magic' that changes past events,[5] they neither change nor aspire to change what happened in the past.

According to Art 297 Treaty on the Functioning of the European Union (TFEU), 'legislative acts shall be published in the Official Journal of the European Union' and 'enter into force on the date specified in them or, in the absence thereof, on the twentieth day following that of their publication'. This technically means that the legislative act may specify a day *before* its publication as the day on which it enters into force. As retroactivity is a legal-technical figure, meaning that new legal consequences are attached to facts that occurred before the new law entered into force, specifying a date before the publication would mean that the law is not retroactive, although it applies to facts that occurred before the act was made public, or was even adopted. In some legal systems, the constitution prohibits this explicitly, by stating that no law is binding or shall enter into force until it has been published.[6] As the *ratio legis* of the publication requirement is that persons can take cognizance of the law that is applied to them, it may be derived from this that laws should not take effect on a date before its publication.

The CJEU does not always seem to have a clear picture of this timeframe. For example, when stating the principle that laws do not have retroactive effect, the CJEU refers to both the date of publication and the entry into force as reference points to define retroactivity.[7] This possibly serves as a safeguard considering the technical possibility to let the EU law enter into force on a date before its publication.

Furthermore, it often employs a rather confusing mantra to describe the temporal effects of a law. According to this recurring phrase, a new rule 'applies

[4] Amongst many others, P Roubier, *Le droit transitoire. Conflits des lois dans le temps*, 2nd edn (Dalloz/Sirey, 1960) 10.

[5] J Waldron, 'Retroactive Law: How Dodgy was Duynhoven?' (2004) 10 *Otago Law Review* 631.

[6] For example: Art 190 Belgian Constitution; Art 88 Dutch Constitution; Art 112 Luxembourg Constitution; Art 88.1 Polish Constitution. Art 79 formulates it as follows: 'If the Act has not been published by the date provided for its entry into force, it shall enter into force on the day of its publication'.

[7] See, eg, C-37 and 38/02 *Di Lenardo Adriano srl and Dilexport* [2004] ECR I-6911, para 67; C-522/10, *Reichel-Albert*, 19 July 2012, para 25–26; C-120/08, *Bavaria* [2010] ECR I-13393, para 40.

immediately to the future effects of a situation which arose under the old rule'.[8] Technically, however, a new rule does not apply to effects; it applies to facts and *determines* the new effects of these facts. The situation to which the CJEU refers is when a fact transpires over a longer period of time (for example, a contract for a specific duration) in the middle of which a new rule is enacted that changes the legal consequences. The question, then, is: (1) whether this fact can be decomposed in a series of relevant facts; (2) whether one determining fact in a series of events can be identified; or (3) whether this fact, spread over time, is indivisible. In situation (1), the facts that occurred before the entry into force of the new rule still induce the legal consequences stated by the old rule, whereas the new rule applies to facts that occur after that date. In situation (2), everything depends on the question of whether this determining fact occurred before or after the entry into force of the new rule. In situation (3), if the fact spreads over time to a point after the entry into force of the new rule, then the new rule applies and dictates the legal effects that must be attached to this fact. Hence, it is paramount that the new law clearly determines the precise fact that actually brings about the new legal consequences.

The CJEU's case law further becomes somewhat confusing when dealing with time and law in terms of 'acquired rights'. Acquired rights are defined in terms of time: if the event giving rise to them occurred before the legislative amendment.[9] The 'event giving rise to the right' is the legal fact (eg the passing of an internal competition); the 'right' is the legal consequence of that event: if this event occurs, then one obtains 'the right to' (eg eligibility for appointment). The protection of acquired rights thus simply comes down to the prohibition of retroactive laws.

B. The Second Dimension: The Timeframe of Human Action

The second dimension of time refers to the time structure of human action in which laws interfere. Humans plan certain actions for a specific purpose they wish to achieve over time; they make investments and take several actions to produce a certain result. The law is one of many factors that determine either the purpose or the actions taken to achieve this result. Fuller accordingly defined law as 'the enterprise of subjecting human conduct to the governance of rules'.[10] The purpose of the law is to enable persons to make plans and carry them out, by making the conduct of other parties, especially the government, predictable and by encouraging or discouraging specific activities by attaching beneficial or adverse consequences to them. This way, the law provides a framework of action for persons by reducing uncertainties.

[8] See eg C-226/08 *Stadt Papenburg* [2010] ECR I-131, para 46; C-89/14 *A2A*, 3 September 2015, para 37.

[9] C-443/07 P *Centeno Mediavilla* [2008] ECR I-10945, para 63; C-496/08 *Pilar Ange Serrano*, [2010] ECR I-1793, para 84.

[10] L Fuller, *The Morality of Law* (Yale University Press, 1973) 106.

This is especially relevant in an economic context, as Fuller himself pointed out.[11] Traders take risks in order to gain profit. The risk is calculated, amongst other things on the basis of the legal framework that conditions the activities and holds out the prospect of tax reductions or other benefits to encourage a specific activity, or tax increases or penalties discourage other activities. Legal changes in the middle of an economic activity, then, produce costs: the trader loses their investments made in reliance on the previous law and misses benefits because the expected legal consequences do not materialise due to the interference of the new law.[12]

This second dimension is more dynamic than the first. It does not consist of specific points of reference: in the way persons perceive time, the past, present and future flow together as a continuum in which preparations, actions and effects develop.

C. The Intersection of Two Dimensions

Law and time become problematic at the point where the two dimensions of law intersect. If a trader has made investments, the plan of action has started off and is difficult to reverse without costs; yet, a law may interfere and replace the expected legal consequences by other disadvantageous effects.

The problem is most prominent if the law is retroactive. In most legal systems, retroactivity is therefore looked upon with suspicion. Laws are in principle not retroactive and retroactivity needs stronger justification or is, in some domains, forbidden – for example, retroactivity of a stricter penal law.[13] However, retroactivity is not always a problem – for instance, if the new rule attaches more beneficial legal consequences to past facts. Moreover, laws with immediate effect and even laws with future effect can be just as problematic.

Laws with immediate effect that interfere in the timeframe of a human enterprise can also be disruptive, in that they change the legal consequences of an action: even if the action as such has not yet occurred, the person may have made investments and may not be able to adapt their course of actions. The CJEU highlighted this in the famous *Mulder* judgment. Mr Mulder was a farmer who, under a system established by a Council Regulation, agreed to stop delivering milk or dairy products for a period of five years in return for a premium. After that period had passed, he wanted to resume his dairy production and made investments accordingly. Meanwhile, however, a new Regulation entered into force, allocating

[11] *Ibid* 24.

[12] A Portuese, 'The Case for a Principled Approach to Law and Economics: Efficiency Analysis and General Principles of Law' in K Mathis (ed), *Law and Economics in Europe. Foundations and Applications* (Springer, 2014) 311.

[13] Art 7 European Convention of Human Rights; Art 15 International Covenant on Civil and Political Rights.

reference quantities on the basis of production in a reference year that fell precisely in the five-year period during which Mr Mulder had stopped dairy production. This new Regulation took immediate effect; however, when Mr Mulder, encouraged by a previous Regulation that aimed at limiting surplus of milk and dairy products, decided to stop dairy production for a limited period, he already envisaged resuming production after that period – to him, it was all part of one continuum so that, had he known the consequences dictated by a later Regulation, he would not have entered into the agreement to stop dairy production. The Court decided that the Regulation was invalid, given that it expanded the effects of the system to which Mr Mulder and farmers in similar circumstances had rendered themselves to an extent that could not have been foreseen.[14]

However, even laws with future effects are not by definition harmless. They may surprise individuals that counted on the new law to take effect immediately. Laws that leave applicable the old rule with regard to a category of persons that specifically relied on that rule may likewise be open to criticism. Such laws may protect the investments made by one category of persons while making it difficult for another category to enter the market under equal conditions. For example, the CJEU had to rule on national legislation that implemented a new Directive under which copyrights revived. The national law permitted persons who were reproducing and marketing sound-recording media in respect of which the rights of use had expired under the previous legislation but had now revived, to distribute those media for a limited period from its entry into force. The requesting producer, Butterfly, argued that this transitional period was too short. The CJEU refuted that argument, inter alia, by reminding of the Directive's principle objective of protecting copyright and related rights.[15] In other words, postponing the effect of a new rule is beneficial for these producers, but harms the interest of the holders of copyright. It further harms the interests of another category, not mentioned in the judgment: competitors that are new in the market, operate under more restrictive conditions, since they immediately had to calculate the costs of those copyrights. In another judgment, the CJEU acknowledged that legitimate expectations justified a transitional period, but that maintaining the old situation permanently for a specific group of civil servants, created a discrimination that was no longer justified.[16]

In some cases, the CJEU confuses both dimensions of time. In the often quoted *Crispoltoni* case, the CJEU ruled on Regulations that reduced intervention prices and premiums if the maximum guaranteed quantity for tobacco of the 'Bright' variety harvested in 1988 had been exceeded. They were published on 29 April 1988 and 26 July 1988 respectively and entered into force the day after respectively of their publication. Clearly, the relevant legal fact was the exceeding of the

[14] C-120/86 *Mulder* 28 April 1988, *Rep* [1988] ECR I-2321, para 25–27.
[15] C-60/98 *Butterfly Music* [1999] ECR I-3939.
[16] C-530/13 *Leopold Schmitzer*, 11 November 2014, para 44.

maximum guaranteed quantity for tobacco harvested in 1988. The Regulations therefore had immediate effect: they applied to the tobacco harvested and exceeding the maximum guaranteed quantity *after* their entry into force. The Court, nonetheless, stated that the contested Regulations had retroactive effect *because* they were published after a date that was relevant for the affected farmers.[17] It therefore confused the first dimension of law, according to which the temporal effect of a law is determined, with the second dimension of law, on the basis of which the validity of the temporal effect of that law is judged. In other words: the CJEU should have established that the Regulations had immediate effect but that they were nevertheless invalid because they frustrated legitimate expectations. Instead, the CJEU established that the Regulations frustrated legitimate expectations and that they were *therefore* retroactive. As *Crispoltoni* became a reference case, the CJEU has repeated this mistake later on, even where it argued that the Regulation was nevertheless valid considering the reasons for interference.[18]

III. The Concept of 'Legitimate Expectations' at the Intersection of Two Dimensions of Time

'Legitimate expectations' is a notion that courts use to assess the validity of laws that interfere in the second dimension of time. It is derived from the principle of legal certainty[19] and implies a balance of interests that takes into consideration the purpose of the law and the evolving circumstances under which the lawmaker operates, on the one hand, and the costs of persons that relied on the previous law, on the other. As such, it searches for optimality in the promotion of efficiency in legal change: in a first step, the affected person's costs are acknowledged only if reliance is legitimate; if it is, then, in a second step, these costs are weighed against the social benefits that are expected to follow from legal change.[20]

The rules of thumb that European courts and national supreme or constitutional courts use in this balancing exercise have already been identified in existing scholarship.[21] They derive from a dynamic concept of legal certainty, which starts from the observation that persons as well as legislative authorities operate in a dynamic environment that induces them to be flexible so as to adapt to new circumstances, to the extent that persons may even rely on the legislator to change

[17] C-368/89 *Crispoltoni*, [1991] ECR I-03695.

[18] C-324/96 *Odette Nikou Petridi Anonymos Kapnemboriki AE* [1998] ECR I-1333.

[19] More on the concept of legal certainty and legitimate expectations can be found in P Popelier, 'Legal Certainty and the European Courts: Accessibility and Legitimate Expectations as Standards of Reasonableness' in M Fenwick et al (eds), *The Shifting Meaning of Legal Certainty in Comparative and Transnational Law* (Hart Publishing, 2017) 29–54.

[20] Portuese (n 12) 314.

[21] Most recently in Popelier (n 19).

a rule so as to keep it efficient, or to allow them to adjust to new developments.[22] These rules of thumb are:[23]

Costs resulting from the disappointment of expectations gain weight as:

1. Rules are explicitly motivational, making governmental action a decisive factor for human behaviour;
2. It is difficult for persons to adapt their behaviour to new rules without costs or a loss of investments;
3. They are based on clear and coherent rules or established practice or jurisdiction;
4. Legal changes were unpredictable, for example because they do not respond to new challenges; or the former rules were not the object of public debate; or administrative practice or judicial decisions created different expectations;
5. Persons acted in good faith.

Conversely, these costs lose weight as:

1. The social benefits are so strong that they outweigh individual costs;
2. It is easy for persons to adapt their behaviour to new rules without costs or a loss of investments;
3. The previous situation did not clearly predict the realisation of the expected legal consequences;
4. Legal changes were predictable, for example because they were announced, or they responded to new challenges or changing circumstances of which persons should be aware, or they did not really bring anything new compared to the situation under the old legal framework or at least preserve the original intent, or they are a coherent part of a larger reform that gradually takes place;
5. Persons acted in bad faith.

The question remains which arguments prevail and how the Court makes them concrete. Answering this question enables us to assess the practical value of the notion of legitimate expectations. The next section reports on this exercise.

IV. The Concept of Legitimate Expectation in the Case Law of the CJEU

A. Preliminary Observations

For the purposes of this chapter, all the judgments where the CJEU deals with the effects of general (national or EU) laws in time, under the notion of 'legitimate

[22] In more detail: Popelier (n 19) 30–36.
[23] Popelier (n 19) 37–38

expectations' in the last 20 years (1998–2017), were analysed. Three observations are immediately apparent.

First, the Court is rarely confronted with this type of problem: after filtering out judgments that concerned individual administrative decisions or judicial decisions and those that concerned the accessibility of laws rather than their effects in time, the Curia search engine produced 64 judgments: 33 concerning EU laws and 31 concerning national legislation.

Second, the claim that legitimate expectations are frustrated only rarely leads to the invalidation of an act: only once with respect to EU laws (three per cent).[24] When ruling on national laws, the CJEU does not always assess the national law but refers this matter back to national courts. Even then, the CJEU sometimes makes clear what the outcome should be, in the phrasing of its preliminary decision, or when the circumstances that the national court must take into consideration all point to one result. In four cases the CJEU left it completely up to the national court to decide whether the national law breached legitimate expectations.[25] Apart from this, the CJEU found the national legislation to be susceptible to criticism in light of the principle of legitimate expectations in five cases (16 per cent). Either the legislators are perfectly aware of the problems of law in time and make careful decisions, taking into account the timeframe of human action, or the CJEU is very strict in the concretisation of the argumentation scheme of legitimate expectations, depriving it of practical value.

This leads us to a third immediate observation: the odds of success for the claim that a law with general scope violates the principle of legitimate expectations are higher where national laws are at issue compared to cases concerning EU laws. Does this mean that the CJEU has a bias in favour of the EU legislator, or are national legislators simply less concerned with legitimate expectations when enacting legislation? The latter possibility cannot be ruled out, especially when we compare the number of retroactive laws, which are more likely to harm legitimate expectations: retroactive laws were at issue in only four cases in the category of EU laws – and in one case the Regulations at stake were, technically, not retroactive at all[26] – compared to eight cases in the category of national laws.

A further analysis establishes which arguments are determining for the outcome. The judgments with regards to EU laws are discussed separately from the ones where national legislation is at stake, considering the fact that the claim based on legitimate expectations is more successful in the latter category.

[24] C-17/01 *Finanzamt Sulingen* [2004] ECR I-4243.

[25] C-381/97 *Belgocodex SA vs Belgian State* [1998] ECR I-8153; C-376/02 *Stichting 'Goed Wonen'* [2005] ECR I-3445; C-201/08 *Plantanol* [2009] I-8343; C-98/14 *Berlington Hungary Tanácsadó és Szolgáltató*, 11 June 2015.

[26] In *Crispoltoni* (n 17), discussed below.

B. Predominant Arguments used to Decide on EU Laws and Legitimate Expectations

As mentioned above, in all but one case, the CJEU rejected the claim that legitimate expectations were frustrated. Above, five categories of arguments were mentioned in rejecting such claims:

1. the social benefits of the new law outweigh individual costs;
2. persons can adapt to the new rules without costs or loss of investments;
3. the previous situation did not clearly predict the realization of the expected legal consequences;
4. legal changes were predictable; and
5. persons acted in bad faith.

They are discussed in order of relevance, ie the frequency of arguments used by the CJEU to refute the claim based on legitimate expectations: (4) – (3) – (2) – (1) – (5). Subsequently, the one judgment in the reference period where the CJEU did find that legitimate expectations were violated, is discussed.

In most judgments, several arguments are taken into consideration to refute or accept legitimate expectations. Occasionally, the CJEU uses only one argument to put an end to the discussion.[27] Exceptionally, the CJEU ends the discussion with a 'tis-'tisn't-argument, for example, where it simply states that the plaintiff complains that the Regulation, which converts, without compensation, temporary suspension of a percentage of the additional reference quantity granted in return for payment into a definitive reduction, but that it saw no factor to enable producers in these circumstances to entertain legitimate expectations that such an additional reference quantity would be treated differently.[28]

Noticeably, the CJEU rarely delves into a nuanced balancing exercise. It merely lists the arguments that support the outcome. Only exceptionally, the Court acknowledges factors to sustain that legitimate expectations were created, but remonstrates with counter-arguments that outweigh these factors. This was the case where the CJEU admitted that the contested Regulation determining the prices and premiums for tobacco varieties for the 1991 harvest was published subsequent to the date on which producers had to take their decisions for the 1991 harvest. It then proceeded to distinguish the situation from the one in *Crispoltoni*, discussed above, and argued why producers could not rely on legitimate expectations. The reason given in this case was that producers knew that prices and premiums were to be determined annually, had witnessed a reduction before, and should thus have expected a second reduction.[29]

[27] C-315/96 *Lopex Export* [1998] ECR I-317; C-149/96 *Portuguese Republic vs Council* [1999] ECR I-8395; C-162-163/01P *Bouma and Beusmans* [2004] ECR I-4509.

[28] C-186/96 *Demand* [1998] ECR I-8529, para 43–44.

[29] C-372/96 *Pontillo and Donatab* [1998] ECR I-5091, para 27–28.

A recurrent phrase used in these judgments is that a person cannot have a legitimate expectation that an existing situation which is capable of being altered by the EU institutions in the exercise of their discretionary power, will be maintained'.[30] Consequently, the CJEU strongly emphasises that the starting position is that EU laws can change, at least with immediate effect; hence, strong arguments are required to convince the Court that legitimate expectations limit the lawmaker's discretionary powers.

i. Legal Changes were Predictable

To reject the claim, the CJEU used arguments for each of these categories, but mostly relied upon arguments that can be placed under category (4): according to the CJEU, the legal changes were predictable and the affected persons could therefore not rely on the previous law but should have adapted to the new situation.

Within this category, the argument that changes in the law were announced, returns regularly. Several circumstances fall within this category. One is the fact that previous laws already predicted the changes by, for example, containing a transitional arrangement,[31] prescribed annual revisions[32] or authorised EU institutions to take measures to end the programme early or to adjust or derogate from it.[33] The 'previous law' may even refer to an Act of Accession, empowering the Commission to adopt transitional measures, on the basis of which economic operators should foresee possible repercussions for their situation.[34] In one judgment, the Court admitted that the possible change in criteria for the classification of staff was not set out in detail in the Regulation, but it was of the opinion that it was foreseeable 'in the light' of the provisions of that Regulation.[35] This argument was complemented with the point that the change of regime was also announced by a Commission's Notice: while this Notice did not set out in detail the forthcoming changes, it omitted stating that the current situation would be continued, so that the affected persons could deduce from that notice that the old Regulation

[30] *Ibid* para 22; C-284/94 *Kingdom of Spain vs Council* [1998] ECR 1-7309, para 43; C-104/97 *Atlanta* [1999] ECR I-6983, para 52; C-402/98 *ATB* [2000] ECR I-5501, para 37; C-110/97 *Kingdom of the Netherlands vs Council* [2001] ECR I-8763, para 115; C-340/98 *Italian Republic v Council* [2002] I-2663, para 42; *Di Lenardo Adriano srl and Dilexport* (n 7) para 70; C-342/03 *Spain v Council* [2005] ECR I-1975, para 49; C-189/02 P, C-202/02 P, C-205/02 P to C-208/02 P and C-213/02 P, *Dansk Rørindustri* [2005] ECR I-5425, para 171; C-310/04 *Kingdom of Spain v Council* [2006] I-7285, para 81; *Pilar Ange Serrano* (n 9) para 93.

[31] *ATB* (n 30) para 42; C-352/09 P *ThyssenKrupp Nirosta* [2011] ECR I-2359, para 77.

[32] *Pontillo and Donatab* (n 29) para 28.

[33] *Kingdom of Spain vs Council* (n 30) para 44; *Dansk Rørindustri* (n 30) para 29–31; C-221/09 *AJD Tuna Ltd* [2011] I-1655, para 75; C-288 and 289/09 *British Sky Broadcasting Group* [2011] ECR I-2851, para 108–111; C-335/09 *Republic of Poland vs Commission*, 26 June 2012, para 182; C-545/11 *Agrargenossenschaft Neuzelle*, 14 March 2013, para 30–33; C-611/12 *Giordano*, 14 October 2014, para 52.

[34] C-179/00 *Gerald Weidacher* [2002] ECR I-501, para 33.

[35] C-373/07 *Membrom*, 2 April 2009, para 92.

would no longer be applied as before.[36] This type of announcement was also taken into consideration in two other judgments, where the Court pointed out that a Commission Communication that contained a proposal for amendment was published in the Official Journal[37] or that information on the Commission's intention to adopt measures had been otherwise spread.[38]

An argument that recurs even more frequently to sustain that the change in regime was foreseeable, refers to external circumstances. The phrase that is most often used is that economic operators are not justified in having a legitimate expectation that an existing situation which is capable of being altered by the EU institutions in the exercise of their discretion will be maintained, 'particularly in a field whose subject matter involves constant adjustment to reflect changes in the economic situation'. The fields that have been mentioned that fall under this category are: state aid in the banking sector;[39] textile imports;[40] or simply the common organisation of the markets.[41] Further, consultations and negotiations between the EU and third countries are a warning sign for traders that changes, for instance in customs rates, are pending.[42] Judicial activity constitutes another warning sign. According to the CJEU, operators in the sugar sector should have foreseen that, after 30 years, no derived intervention price would be fixed for Italy, considering the various actions brought by sugar producers before the national courts and the Court of First Instance, challenging the classification of Italy as a deficit area requiring such interventions.[43] Lastly, economic operators should follow political debate. If a new measure forms part of a more fundamental reform, economic operators should have been prepared,[44] especially if this reform had been discussed at the political level.[45]

ii. The Previous Situation did not Clearly Predict the Realisation of the Expected Legal Consequences

The second most used argument given by the CJEU to reject the claim that legitimate expectations were frustrated, is that it was not clear from the start that the previous situation actually predicted that the legal consequences, relied upon by the petitioners, would materialise.

[36] *Ibid* para 93.

[37] *Kingdom of Spain v Council* (n 30) para 84.

[38] *Kingdom of the Netherlands vs Council* (n 30) para 117.

[39] C-526/14 *Tadej Kotnik*, 19 July 2016, para 66.

[40] *Portuguese Republic vs Council* (n 27) para 75.

[41] *Pontillo and Donatab* (n 29), para 22; C-375/96, *Galileo Zaninotto* 29 October 1998, para 50; *Atlanta* (n 30) para 52; *ATB* (n 30) para 37; *Italian Republic v Council* (n 30) para 42; C-14/01 *Molkerei Wagenfeld Karl Niemann* [2003] I-2279, para 56; *Di Lenardo Adriano srl and Dilexport* (n 7) para 70; *Kingdom of Spain v Council* (n 30) para 81; *Republic of Poland vs Commission* (n 33) para 180.

[42] *Spain v Council* (n 30) para 50.

[43] *Italian Republic v Council* (n 30) para 47–48.

[44] *Odette Nikou Petridi Anonymos Kapnemboriki* (n 18) para 46.

[45] *Kingdom of Spain v Council* (n 30) para 84.

Within this category, the argument commonly used is that no competent EU authority has given precise, unconditional and consistent assurances that the previous situation would be maintained. This is a valid argument where it concerns the behaviour of an administrative body towards individuals, but at first sight somewhat cramped when the expectations rely on legislation with a general scope. Why should a person require explicit statements by the Commission that the legal consequences held out by a legislative act are indeed assured? The argument, however, makes more sense if the implementation of a rule depends on Commission action,[46] or when combined with a previously discussed argument: if the previous law authorised EU institutions to end a programme prematurely or make adjustments or transitional arrangements, then persons can only rely on the continuation of the programme if these institutions have given assurances to that end.[47] For example, in *AJD Tuna* the operators relied on a previous Regulation that laid down Bluefin tuna fishing quotas and concluded contracts with fishermen on that basis. The Regulation, however, authorised the Commission to end the fishing season at an earlier date, and the Commission did so before the full quota had been reached. Given this context, the CJEU held that the operators could not rely on the delivery of the full quantity, since the Commission had not given any assurance that they would not make use of the possibility to take measures.[48] Similarly, if the legal, economic or political context gave reasons to expect a change in the regime, economic operators can only rely on its continuation if EU institutions have given clear assurances to that effect.[49] Moreover, the argument is used to explain, when new regulations are applied immediately, why persons could not expect that the old regulations would still be applied to their specific situation.[50] Less convincingly, it is used additional to the argument that the common organisation of the markets requires constant adjustment, so that economic operators can only rely on existing legislation if EU institutions gave explicit reasons to believe that it would not be amended.[51]

Contention may rise regarding what can amount to precise assurances by competent authorities. In *Lopex Export*, the trader had relied on binding tariff information to conclude contracts to export certain goods, and had obtained an export licence to carry out the transaction. By the time the goods were actually exported, however, a new Regulation adopted new tariff headings. *Lopex* maintained that the administrative measures that were taken had vested rights that created legitimate expectations concerning the tariffs applicable to the exportation of the goods. The CJEU, however, pointed out that the aim of binding tariff

[46] *Tadej Kotnik* (n 39) para 62.
[47] *Gerald Weidacher* (n 34) para 31–32; *AJD Tuna Ltd* (n 33) para 74; *Agrargenossenschaft Neuzelle* (n 33) para 25.
[48] *AJD Tuna Ltd* (n 33) para 74.
[49] *Membrom* (n 35) para 92–94.
[50] *Centeno Mediavilla*(n 9) para 89–90.
[51] *Galileo Zaninotto* (n 41) para 50.

information is to protect traders against any changes in the position adopted by the customs authorities, not against any changes in regulations.[52] In another judgment, the CJEU made clear that, in any case, no legitimate expectations can arise from assurances by national authorities, especially not if they were contrary to European law.[53]

Another argument used to refute the claim based on legitimate expectations on the grounds that the previous legal situation could not create such expectations, is the finding of legal uncertainty in the first place. In particular, if an arrangement has been the subject of constant reforms and adjustments in the past, the affected persons cannot hold the expectation that it will last this time.[54] Conversely, however, as established above, the fact that the legal situation remained the same for several decades, does not in itself create legitimate expectations if external events – for example judicial activities – point to new developments.[55]

iii. The Possibility to Adapt without Costs or Loss of Investments

Third, the CJEU takes into consideration the costs of a change in the legal regime. Within this category falls the *lex mitior* principle, applicable in penal law and applied by the CJEU to the imposition of fines.[56] Mostly, however, the question revolves around the loss of investments, the loss of future gains, and the costs of adapting to the new rules. If these costs are low or absent, this is an argument to reject the claim based on legitimate expectations.

The costs of adapting to new rules are considered non-existent should the CJEU find that the new rule did not really bring about a change in the legal situation. In the reference period, the CJEU noted in three judgments that the new rule was a continuation of the situation under the old legal framework.[57] For example, it pointed out that the rule under the challenged Regulation, excluding free re-entry to third-country nationals with a temporary residence permit who left the territory of an EU Member State, was already included in the Schengen Convention.[58] Interpretative provisions may, for this reason, even have retroactive effect.[59]

If the new rule did bring an innovation, the question is whether sufficient time was given to adjust to the new rule. Here, the reference point is not the date of publication or entry into effect, but the date on which the person affected

[52] *Lopex Export* (n 27) para 27–28.

[53] C-164/01 P *van den Berg* [2004] ECR I-10225, para 69; C-606/10 *ANAFE*, 14 June 2012, para 70, 79.

[54] *Di Lenardo Adriano srl and Dilexport* (n 7) para 71; *Kingdom of Spain v Council* (n 30) para 84.

[55] *Italian Republic v Council* (n 30) para 47–48. See section IV.B.i.

[56] *Thyssen Krupp Nirosta* (n 31) para 85.

[57] C-293/04 *Beemsterboer Coldstore Services* [2006] ECR I-2263, para 22–23; C-4/10 and C-27/10 *Bureau national interprofessionnel du Cognac* [2011] ECR I-6131, para 32; *ANAFE* (n 53) para 65.

[58] *ANAFE* (n 53) para 65.

[59] *Beemsterboer Coldstore Services* (n 57) para 22–23.

knew the change that was to come about. For instance, the CJEU agreed with the Advocate General that the importer of rice knew of the Commission's intention to adopt safeguard measures and could therefore have obtained import licences before the entry into force of those measures.[60] Apart from this judgment, the argument was not explicitly raised with regard to EU rules. It is, however, implicit in the argument discussed above, claiming that the change of law was foreseeable: if the person affected could have foreseen that a new rule was going to be adopted, they could already act accordingly.

The last step consists in assessing the costs of adapting to the new rule. If these costs are low, this is an argument that pleads against a breach of legitimate expectations. For example, in one judgment, the CJEU pointed out that the new criteria for classification in grade on recruitment did not seem to result in a downward alteration of those grades.[61] A similar counter-argument is the fact that the new rule improves the situation of the affected persons, for instance if it gives better protection to traders in the event of errors on the part of customs authorities.[62]

Importantly, the costs must be a consequence of the new Regulation. This was the central argument in *Bouma and Beusmans*. The affected parties referred to the *Mulder* case to argue that the new Regulation prevented them from having temporarily terminated milk production in a reference period. The CJEU, however, stated that their loss of income was the consequence of freely deciding to give up milk production at the time the non-marketing undertaking expired. Unlike *Mulder*, the traders had not manifested in an appropriate manner their intention to resume milk production, for example by making investments or repairs, or by maintaining the equipment necessary for such production.[63]

In some judgments, losses are phrased in terms of 'acquired rights'. As explained above, in the CJEU's definition, acquired rights come down to the occurrence of legal facts before the entry into force of the new rule, with the prospect of legal consequences defined under the old rule. The claimed infringement of an acquired right, then, comes down to the complaint that the expected benefits are lost. This evolves around which legal consequences could be expected. In one judgment the petitioners hoped to be classified in a specific grade after having passed internal competitions, whereas the CJEU denied that this event automatically led to that result.[64] However, where affected parties seem to equate acquired rights with the right to obtain the same legal consequences as before, the CJEU has promptly pointed out that parties cannot claim a vested right to the maintenance of a certain advantage under the old rule: this is merely a situation capable of being altered

[60] *Kingdom of the Netherlands vs Council* (n 30) para 117.
[61] *Centeno Mediavilla* (n 9) para 36.
[62] *Beemsterboer Coldstore Services* (n 57) para 25.
[63] *Bouma and Beusmans* (n 27) para 87–89.
[64] *Pilar Ange Serrano* (n 9) para 85–86.

by the EU legislator in the exercise of its discretionary powers.[65] For this reason, the CJEU refuses to take into consideration the damage which the affected party claims to have suffered as a result of the interference of a new rule.[66]

iv. Social Benefits

Even if a new rule infringes upon legitimate expectations, the social benefits advanced by the new rule may still outweigh this fact. In most judgments, however, the CJEU focuses mainly on explaining why no legitimate expectations were frustrated. The fact that the new rule was necessary for reasons of public interest, is only mentioned occasionally and as a secondary consideration. The public interest is invoked to explain why the new rule was adopted – eg to guarantee free competition, financial stability, or the fight against overproduction – however, the CJEU nonetheless argues that it did not infringe upon legitimate expectations.[67] For example, in one judgment the CJEU noted that an overriding public interest – in this case: ensuring the stability of the financial system – may preclude transitional measures from being adopted, even if the EU had created legitimate expectations, subsequently hurrying to assure that in that case, the EU had not.[68]

v. Bad Faith

In none of the judgments pronounced in the last 20 years did the CJEU suspect the plaintiff of bad faith. It came closest to this where it judged that the plaintiffs 'clearly did not behave as prudent and well-informed traders'.[69] 'Prudent and well-informed traders' would, according to the Court, not have planned a 1989 production largely exceeding the 1988 quota.

The concept of 'prudent and well-informed traders', by contrast, returns frequently and in many variations: foreseeable and conscientious economic operators, prudent and circumspect operator, normally diligent economic traders, etc. This is used as a benchmark: the prudent trader examines the regulations that apply to them, reads Commissions Notices, thinks along with the EU legislator to predict reactions to changes in the economic context, and follows up public debates and judicial activity. Traders that do not live up to this ideal economic operator are not accused of bad faith; however, they cannot assume that their expectations are considered legitimate.

[65] *Pontillo and Donatab* (n 29) para 22–23; *Atlanta* (n 30) para 53.

[66] *Atlanta* (n 30) para 54.

[67] *Odette Nikou Petridi Anonymos Kapnemboriki* (n 18) para 53; *Dansk Rørindustri* (n 30) para 169; *Thyssen Krupp Nirosta* (n 31) para 77–78; *Tadej Kotnik* (n 39) para 68–69.

[68] *Tadej Kotnik* (n 39) para 68–69.

[69] *Odette Nikou Petridi Anonymos Kapnemboriki* (n 18) para 51.

vi. Acknowledgement of Legitimate Expectations

As mentioned, in the last 20 years the CJEU accepted in one case that the EU institutions had unlawfully frustrated legitimate expectations.

What distinguishes this case from most others, is that the contested rule had retroactive effect. Only in four other judgments has the CJEU qualified a rule as retroactive. In one case, however, it concerned an interpretative rule, considered to bring no innovation.[70] Similarly, in a second case, the retroactive rule was in fact a continuation of the existing situation.[71] In another case, a Notice that was applied retroactively introduced a new practice to determine fines, but stayed within the limits dictated by the Regulation that was in force when the relevant facts occurred.[72] In the final judgment, the contested rule was, from a technical point of view, not actually retroactive.[73]

However, what distinguishes this case from all other cases discussed in this section, is the ambiguity of the retroactive measure as to its source: while technically a Commission's measure, its effect was to give retroactive effect to a national measure. In the present case, Mr Sudholz had chosen to treat goods used both for business and private purposes as business goods – in this case: a passenger car – with the consequence that the VAT due as input tax in the acquisition of those goods was wholly and immediately deductible. A Directive allowed for derogating national measures, provided they received the Commission's authorisation. On 24 March, the German authorities imposed a limit on the deduction of input VAT paid on passenger cars purchased after 1 April 1999. The Commission's authorisation, however, was adopted and published almost a year later – which was untimely, according to the German authorities – with retroactive effect from 1 April 1999. The CJEU reminded that the principle of legal certainty in principle precludes retroactive measures, and that the principle of legal certainty must in particular be observed all the more strictly in the case of rules liable to entail financial consequences.[74] In this case, the new rule was adopted by the national legislature with immediate effect; however, it was the authorisation of the Commission that gave it retroactive effect. Neither the fact that the national measure was already in force at the time of the purchase, nor the fact that the Commission's authorisation was untimely could convince the CJEU. It reminded that 'the Court has already held that a national legislative amendment retroactively depriving a taxable person of a right to deduct which he has derived from the Sixth Directive, is

[70] *Beemsterboer Coldstore Services* (n 57) para 23.
[71] *Bureau national interprofessionnel du Cognac* (n 57) para 32.
[72] *Di Lenardo Adriano srl and Dilexport* (n 7) para 191.
[73] *Odette Nikou Petridi Anonymos Kapnemboriki* (n 18) as discussed above, section II.C.
[74] *Finanzamt Sulingen* (n 24) para 33–34.

incompatible with the principle of the protection of legitimate expectations',[75] thereby equating the present situation with retroactive national measures.

Both the retroactive nature *and* the fact that the measure was essentially a national one, not one adopted by the EU legislator for the entire Union, may thus explain why the CJEU in this one judgment was of the view that the legitimate expectations of the affected person were violated.

vii. Interim Conclusion

The claim that the temporal effect of an EU law frustrates legitimate expectations, is not often brought before the CJEU. We can assume that the EU legislator is well aware of the problems of time and law and acts carefully, especially with some milestone decisions such as *Muller* and *Crispoltoni* in mind.

Still, in the judgments where the CJEU did have to decide on such issues, the CJEU appears very demanding of its legal subjects, in particular if they are traders. They have to follow up on the details of regulations, as well as political, administrative, economic and judicial developments. They have to think along with the EU lawmaker and adapt to possible changes in the law, even if they do not have a precise picture of what the amendments will involve. The CJEU appears to sympathise more with EU institutions, which are given great latitude to change legal rules.

What is most striking, is that the EU institutions can cover themselves against claims based on legitimate expectations by creating legal uncertainty: if they include in a legislative act that the legal regime can change, or empowers the Commission to amend or even stop the legislative programme, then the persons affected cannot rely on the continuation of the rule, even if they made investments in view of the benefits held out by that rule. If this is taken further, then EU law loses an important function, crucial to the principle of legal certainty, which is to act as a framework that guides the actions of legal subjects.

An example of this is *Pontillo and Donatab*, where the CJEU pointed out that the producers of tobacco knew that the prices and premiums were to be determined annually and could have foreseen a further reduction.[76] The question remains why the Commission, in its annual decision, was not urged to take such annual decisions *before*, and not *after* the moment when producers had to take their decisions concerning the next harvest. In its eagerness to underpin the judgment that there is no violation of the principle of legitimate expectations, the CJEU seems to neglect a more nuanced weighing of arguments and counter-arguments.

[75] *Finanzamt Sulingen* (n 24) para 40.
[76] *Pontillo and Donatab* (n 29) para 27–28.

C. Predominant Arguments to Decide on National Legislation and Legitimate Expectations

Compared to the judgments on EU law, the case law on the compatibility of national laws against the principle of legitimate expectations stands out in three ways. First, as already mentioned, the CJEU finds more violations of the principle. Second, it more often identifies both arguments and counter-arguments. This is especially the case where the CJEU gives criteria to which the referring national court should give specific attention when it ultimately decides on the validity of national law. Third, the hierarchy of arguments is different. With regard to EU laws, the argument that legal changes were predictable and that the old rule did not clearly predict that the expected legal consequences would follow were the most used arguments to deal with the claim based on legitimate expectations. By contrast, with regard to national laws, the question of whether persons can easily adapt their behaviour to new rules without costs or loss of investment, is the most dominant argument.

However, the CJEU is consistent in its starting point that the law can change with immediate effect. Here as well, the starting point is that persons cannot have a legitimate expectation that an existing situation which is capable of being altered by the national authorities in the exercise of their discretionary powers, will be maintained.[77] Consequently, the CJEU only rarely finds that laws with immediate effect violate the principle of legitimate expectations.[78] The counterpart is that laws with retroactive effect harbour more distrust – even, according to the CJEU, irrespective of whether this may produce favourable or unfavourable effects for the person concerned.[79] This is in particular the case with regard to penalties.[80] Four out of eight retroactive national laws were considered an infringement of legitimate expectations;[81] in two other cases, the verdict was left to the national court.[82] In one of these cases the CJEU found that the retroactive law in principle did not violate legitimate expectations, but left it to the national court to decide whether press releases announcing the retroactive amendment had been sufficiently clear for economic operators to understand the consequences.[83] In only two judgments was the Court unambiguously of the opinion that the retroactive

[77] C-487/01 and C-7/02 *Gemeente Heusen and Holin Groep BV* [2004] ECR I-5337, para 66; C-241/07 *JK Otsa Talu OÜ* [2009] ECR I-4323, para 51; *Plantanol* (n 25) para 53; C-449/08 *Elbertsen* [2009] ECR I-10241, para 5.

[78] C-286/12 *Commission v Hungary*, 6 November 2012.

[79] *A2A* (n 8) para 37.

[80] *Gerekens and Procola* (n 80) para 35. Additional levies, however, are not regarded as penalties: para 36.

[81] C-62/00 *Marks & Spencer* [2002] ECR I-6325; C-107/10 *Enel Maritsa Iztok 3 AD* [2011] ECR I-3873; C-362/12 *Test Claimants in the Franked Investment Income Group Litigation*, 12 December 2013; C-640/13 *European Commission v the UK*, 18 December 2014.

[82] *Belgocodex SA vs Belgian State* (n 25); *Stichting 'Goed Wonen'* (n 25).

[83] *Stichting 'Goed Wonen'* (n 25) para 44–45.

law did not frustrate legitimate expectations.[84] Conversely, laws with future effects can be problematic for other reasons, if the transitional term is too long to be justified in terms of legitimate expectations.[85]

i. The Possibility to Adapt without Costs or Loss of Investment

This argument that is most used with regard to national laws, works both ways: as an argument in favour and against the claim that legitimate expectations are frustrated.

It is not difficult to comply with the rule, and the costs thereof are non-existent, if the new rule entails a continuation of the previous legal situation. In that case, the claim is refuted even if the law is enacted with retroactive effect.[86]

Conversely, the CJEU has laid down that while national authorities are in principle entitled to adopt rules with immediate effect, without transitional arrangement, the principle of legitimate expectations prohibits that new laws are suddenly and unexpectedly adopted, withdrawing rights that persons enjoyed until then without allowing them the time necessary to adjust. This applies especially if the new law obliges them to carry out consequential economic adjustments.[87] The CJEU thus examines whether these conditions are satisfied.[88] Transitional periods are in particular called for, when the new legislation revokes licences that allow their holders to exercise an economic activity.[89] An additional argument in favour of legitimate expectations is if the person affected made costly investments with the exact purpose to comply with the scheme adopted previously and these investments are lost.[90] On the other hand, if the legislator took precautions to enable the affected persons to adapt to the new situation, this is an argument against the claim of legitimate expectations. When calculating whether a transitional period was sufficient to enable the persons affected to adapt to the new rule,[91] the CJEU takes into account the transitional arrangement as well as the period before the law entered into force, for example when legislative amendments were announced or otherwise foreseeable.[92]

[84] C-216 and 222/99 *Riccardo Prisco* [2002] ECR I-6761; *Gerekens and Procola* (n 80).

[85] C-168/09 *Flos SpA* [2011] ECR I-181; C-501/12 *Specht and others v Land Berlin*, 19 June 2014.

[86] *Riccardo Prisco* (n 84) para 69.

[87] C-332/14 *Wolfgang und Dr. Wilfried Rey Grundstücksgemeinschaft*, 9 June 2016, para 56, 58–59.

[88] In *ibid* para 60–62, the conditions were not fulfilled, but they were in *Commission v Hungary* (n 78) para 68–70; *Test Claimants in the Franked Investment Income Group Litigation* (n 81) para 47; *Berlington Hungary Tanácsadó és Szolgáltató* (n 25) para 85–87; C-322/16 *Global Starnet Ltd*, 20 December 2017, para 49. In *Plantanol* (n 25) para 47, 52, 57 the CJEU notes that the national authorities withdrew a tax exemption scheme suddenly and unexpectedly, without time to adjust, but leaves it to the national court to consider whether a prudent and circumspect economic operator could nevertheless have foreseen the withdrawal.

[89] *Berlington Hungary Tanácsadó és Szolgáltató* (n 25) para 85.

[90] *Berlington Hungary Tanácsadó és Szolgáltató* (n 25) para 87.

[91] *Flos SpA* (n 85) para 62; *Global Starnet Ltd* (n 88) para 48–49.

[92] *Butterfly Music* (n 15) para 27; *Gemeente Heusen and Holin Groep BV* (n 77) para 81.

If the law is retroactive, it is in particular difficult for persons affected to comply with the new rule, unless the new rule comes down to the continuation of the previous one. The CJEU aims in particular at retroactive laws that entail financial costs, for example in tax law.[93] Conversely, avoiding costs such as a loss of income can justify transitional measures, even if they have a discriminatory effect, given that the transitional period remains proportional.[94] National transposition laws that provide for an indefinite transitional scheme are unacceptable, should this prevent the application of the EU law for this group of people or preserves, more generally, a violation of EU law.[95]

Another question concerns which costs are taken into consideration. In *Gemeente Heusden and Holin Groep*, the plaintiff argued that it took account of the possibility of making a tax deduction when it set the amount of the rent charged to its lessee, a sports club. It held that it could not increase the rent as the sports club would find it difficult to finance that. The CJEU considered that this argument was not relevant because it referred to the lessee instead of the taxable person affected by the law itself.[96] In any case, the basic requirement is the existence of legitimate expectations: costs such as investments to raise production capacity do not impress the CJEU if the economic operators could not legitimately expect that the law would remain unaltered.[97]

ii. Legal Changes were (not) Predictable

The argument that changes in the law were announced by EU institutions or in previous laws, is most frequently used to reject the claim based on legitimate expectations. The CJEU, however, has sometimes imposed quality requirements regarding the preciseness of these announcements.

The CJEU expects individuals to not only take notice of legislation that directly applies to them, but to also examine Directives addressed to Member States. If this Directive sets out a list of measures that have to be implemented by the Member States, individuals affected by the transposition laws cannot entertain legitimate expectations that they will not be subjected to such measures.[98] The same expectation applies to Regulations.[99] The fact that the Member States for many years had not taken such measures, does not make legitimate the expectation that this will remain the case in the future.[100] On the contrary, no legitimate expectations

[93] *Marks & Spencer* (n 81) para 45–46; *Flos SpA* (n 85) para 62–63; *Enel Maritsa Iztok 3 AD* (n 81) para 40–41; *Test Claimants in the Franked Investment Income Group Litigation* (n 81) para 45; *European Commission v the UK* (n 81) para 38.

[94] C-297–298/10 *Hennigs v Eisenbahn-Bundesamt* [2011] ECR I-7965, para 98.

[95] *Flos SpA* (n 85) para 64; *Leopold Schmitzer* (n 16) para 44.

[96] *Gemeente Heusen and Holin Groep BV* (n 77) para 73.

[97] *Elbertsen* (n 77) para 5.

[98] C-496/04 *Slob v Productschap Zuivel* [2006] ECR I-8257, para 47; C-277/13 *European Commission v Portuguese Republic*, 11 September 2014, para 57.

[99] *Gerekens and Procola* (n 80) para 33.

[100] C-495/00 *Azienda Agricola Giorgio, Giovanni e Luciano Visentin* [2004] ECR I-2993, para 55.

can be entertained that a situation which is clearly in violation of EU law will be continued.[101] Even when the Directive allows but does not oblige Member States to take certain measures, individuals should foresee legislative amendments.[102] However, if Member States claim that individuals were in a position to anticipate to changes given developments in national legislation, the CJEU requires that these provisions are sufficiently clear to allow individuals to anticipate the amendments with 'a sufficient degree of certainty'.[103]

Furthermore, announcements by the national authorities that certain measures would be taken, even with retroactive effect, serve to rebut the claim of legitimate expectations.[104] Announcements can take the form of, for example, coalition agreements or draft laws.[105] It was observed above that with regard to EU announcements, changes to the legal regime do not have to be spelled out in detail.[106] With regard to national measures, however, the CJEU explicitly requires that the announcements are 'sufficiently clear' to enable an economic operator to understand the consequences of the proposed legislative amendment.[107]

In the reference period, the argument that amendments in the law were foreseeable due to external circumstances, has not been used so often. In *Gerekens and Procola*, the CJEU used it to strengthen the argument that given existing EU Regulations, the affected parties should have foreseen that new rules would be adopted with retroactive effect. The purpose of the Regulation was to re-establish a balance between supply and demand in the milk market. Considering the economic situation, with continuing production surpluses on the milk market, the affected parties should have been aware of the urgency to implement that Regulation.[108] Likewise, the fact that discussions with stakeholders had taken place reinforced the argument that announcements made legislative changes foreseeable.[109]

iii. The Previous Situation did not Clearly Predict the Realisation of the Expected Legal Consequences

The counterpart of the situation in which legislative provisions and announcements predicted the change of a legal regime, is the situation in which the continuation of the programme in force is assured, at least until a specific date.

[101] *Ibid* para 55; C-480/00 to 482/00, C-484/00, C-489/00 to C-491/00 and C-497/00 to C-499/00 *Azienda Agricola Ettore Ribaldi* [2004] I-2943; C-231/00, C-303/00 and C-451/00 *Cooperativa Lattepiù* [2004] I-2869; *Plantanol* (n 25) para 54. See also C-150/97 *Commission vs Portuguese Republic* [1999] ECR I-259, para 21; C-516/16 *Erzeugerorganisation Tiefkühlgemüse eGen v Agrarmarkt Austria*, 20 December 2017, para 69–73.

[102] *Gemeente Heusen and Holin Groep BV* (n 77) para 66.

[103] *Commission v Hungary* (n 78) para 72.

[104] *Gerekens and Procola* (n 80) para 32; *Plantanol* (n 25) para 63.

[105] *Plantanol* (n 25) para 63.

[106] *Membrom* (n 35) para 93. See above, section IV.B.i.

[107] *Stichting 'Goed Wonen'* (n 25) para 44.

[108] *Gerekens and Procola* (n 80) para 33.

[109] *Plantanol* (n 25) para 63.

This is used in the CJEU case law to support the claim of legitimate expectations. If express provisions and announcements were made that a legislative programme would remain in force until a specific date, the earlier withdrawal of that programme supports the claim that expectations were legitimate.[110] Similarly, if a specific tax regime has not been the subject of any legislative change for nearly 20 years, the CJEU asks the referring national court to take into consideration the claim that a change should enter into force only after an appropriate adaptation period[111] – unless the previous situation violated EU law.[112]

iv. Social Benefits

The CJEU refers more often to the public interest with regard to national measures compared to EU laws. The CJEU acknowledges the Member States' purpose of protecting acquired rights,[113] the prevention of tax evasion,[114] the protection of copyright,[115] or finding a balance between supply and demand of milk.[116] Obviously, if the national law implements an EU Directive or Regulation, the national legislator's purpose coincides with the public interest pursued by the EU legislator.[117] Conversely, the purpose of national measures – eg to protect acquired rights by putting in place transitional schemes – is of secondary importance if that arrangement delays the implementation of a Directive beyond the deadline for transposition[118] or maintains a situation that breaches EU law.[119]

v. Bad or Good Faith

In this set of judgments, the CJEU similarly refrained from accusing any of the plaintiffs of bad faith. In one judgment, it did notice that the plaintiff, a lessor affected by a tax amendment, was not covered by the transitional provisions of the law because the amount of the rent was not high enough compared with the cost of the investment, which the national authorities presumed to be a legal arrangement intended to avoid taxation.[120] The Court, however, considers the fact that

[110] *Plantanol* (n 25) para 51.
[111] *Berlington Hungary Tanácsadó és Szolgáltató* (n 25) para 83.
[112] *Azienda Agricola Giorgio, Giovanni e Luciano Visentin* (n 100) para 55. Also: *Azienda Agricola Ettore Ribaldi* (n 101) and *Cooperativa Lattepiù* (n 101) 25 March 2004.
[113] *Commission vs Portuguese Republic* (n 101) para 21; *Flos SpA* (n 85) para 55; *Hennigs v Eisenbahn-Bundesamt* (n 94) para 91, 98; *Specht and others v Land Berlin* (n 85) para 64, 68; *European Commission v Portuguese Republic* (n 98) para 56; *Leopold Schmitzer* (n 16) para 42.
[114] *Gemeente Heusen and Holin Groep BV* (n 77) para 76; *Stichting 'Goed Wonen'* (n 25) para 45.
[115] *Butterfly Music* (n 15) para 28.
[116] *Gerekens and Procola* (n 80) para 25–26.
[117] *Butterfly Music* (n 15) para 28; *Gerekens and Procola* (n 80) para 25–26. See more implicitly also *A2A* (n 8) para 42.
[118] *Commission vs Portuguese Republic* (n 101) para 21; *European Commission v Portuguese Republic* (n 98) para 56–57.
[119] *Leopold Schmitzer* (n 16) para 44.
[120] *Gemeente Heusen and Holin Groep BV* (n 77) para 75.

the plaintiff acted in good faith irrelevant, where expectations are drawn from a national authorities' action that was contrary to EU law.[121]

The concept of 'prudent' or 'diligent' economic operators is considerably less employed compared to the judgments regarding EU laws. Where it is used – in only two judgments – the CJEU seems less demanding of the trader when national law is at stake.

In the first judgment, it is used with regard to EU law: the CJEU states that where 'a prudent and discriminating trader' could have foreseen the adoption of EU law likely to affect his interest, he cannot plead the principle of legitimate expectations if the measure is adopted and national law implements it.[122]

In the second judgment, the concept is used in the same way with regard to national law.[123] Here, however, the CJEU leaves it to the referring national court to establish whether the national law was indeed foreseeable for a 'prudent and circumspect' trader.[124] To that end, the CJEU reminded the court of several relevant facts in the decision (within the discretionary power of the national authorities in implementing the EU Directive) to first grant and later withdraw a tax exemption. On the one hand, there were good reasons to end the exemption, and the withdrawal had been announced in the coalition agreement and by a draft law. On the other hand, the EU regulatory framework had not changed, and after the announcements that the tax exemption would be withdrawn, the legislator maintained the exemption system until the expiry date initially laid down.[125]

vi. Interim Conclusion

The claim that the temporal effect of a national law frustrates legitimate expectations is as occasional as when EU laws are at stake; however, although it is not often successful, the CJEU seems less reluctant to sustain the claim.

The CJEU is still demanding of individuals, but the use of the term 'prudent trader'[126] arouses the suspicion that the CJEU is much more sympathetic with the trader when national laws are at stake. This is endorsed by judgments where the CJEU seems to set higher standards for the national authorities to decide whether the new law was foreseeable. As mentioned above, the CJEU is not satisfied that a new law was announced in the law or by press releases; the referring national court is required to determine whether these announcements were 'sufficiently clear' for an economic operator to understand the consequences of the law to be adopted.[127]

[121] *Erzeugerorganisation Tiefkühlgemüse eGen v Agrarmarkt Austria* (n 101) para 74.
[122] *Gerekens and Procola* (n 80) para 29.
[123] *Plantanol* (n 25) para 53.
[124] *Plantanol* (n 25) para 57.
[125] *Plantanol* (n 25) para 62–67.
[126] Above, section IV.C.vii.
[127] *Stichting 'Goed Wonen'* (n 25) para 44; *Commission v Hungary* (n 78) para 72. See section IV.C.ii.

The CJEU is in particular sympathetic with the plaintiff, where the national law derogates from EU law, or defers its implementation. Hence, transitional schemes to protect the rights of individuals are not appreciated, unless the EU Directive gave explicit scope for such arrangements.[128] Conversely, the CJEU keeps a close watch on the implementation of EU law, finding national measures foreseeable, even if they have retroactive effect, if they are aimed at transposing an EU Directive.[129]

V. Conclusion

The research question in this article concerns how the CJEU uses the principle of legitimate expectations to deal with tensions between the two dimensions of time and law, ie when the temporal effect of a law interferes with the time schedule guiding the actions of individuals.

A first conclusion is that the CJEU clearly distinguishes both dimensions, although it sometimes has trouble clearly defining retroactive vs immediate effect.[130] The principle of legitimate expectations is there to protect the second dimension; the time structure of human action. The consequence thereof is that the foreseeability of laws is the most crucial criterion, irrespective of whether the law has immediate or retroactive effect. This manifests itself in the CJEU's case law, where laws with immediate effect may frustrate legitimate expectations, whereas, in some circumstances, retroactive laws do not. Still, the first dimension greatly impacts the Court's reasoning, where retroactivity is looked upon with more suspicion, even as immediate effect and the discretionary power of authorities to act are regarded as the starting point. While individuals should, as a rule, foresee that changes can be brought about at any time, the claim that in this case the legislative change was too abrupt, necessitates intense substantiation.

A second conclusion is that claims of legitimate expectations are rarely successful. This is especially the case when EU laws are at stake. The full implementation of EU law is the CJEU's first concern. As a result, the claim is more successful when national laws are at stake. Individuals are expected to foresee legislative changes on the basis of economic developments, the regulatory framework, judicial activity, political debate and announcements; nonetheless, the CJEU is more demanding of traders with regards to EU law compared to national law.[131]

A third conclusion is that while the general interest justifies priority for the law over individual interests, the Court should keep in mind that creating a climate

[128] See section IV.C.iv.
[129] *European Commission v Portuguese Republic* (n 98) para 57.
[130] See section II.B.i.
[131] See section IV.C.vi.

of trust and legal certainty also serves the general interest. The CJEU seems more nuanced when national laws are at stake, where it sets higher standards as to the quality of announcements.[132] By contrast, the way in which legitimate expectations are dismissed when EU laws are at stake sometimes risks undermining the very purpose of the law to serve as a general framework for human action.[133]

[132] See section IV.C.ii. and vi.
[133] See section IV.B.vii.

6

The Timing of Judicial Review of Constitutional Amendments – Towards a 'Time Sensitivity Test' Following the Moldovan Constitutional Court's Decision on the Modality of Electing the President

ZOLTÁN POZSÁR-SZENTMIKLÓSY AND YANIV ROZNAI

I. Introduction

In 2004, the Ukrainian Parliament considered an important constitutional amendment that aimed to make substantial changes to the organisation of the executive branch. As required by Article 159 of the Ukrainian Constitution, the draft amendment was submitted to the Constitutional Court for a preliminary opinion regarding the conformity of the proposed amendment with the requirements of Article 157 of the Constitution.[1] The Constitutional Court confirmed the admissibility of the amendment. In the course of the parliamentary debates, the original text of the amendment was substantially modified, and the final version was adopted without being resubmitted to the Constitutional Court for its opinion. In 2010, six years later, 252 People's Deputies filed a constitutional complaint challenging the procedural validity of the 2004 amendment before the Constitutional Court. In its judgment, the Constitutional Court held that the amendment was adopted in a procedural violation of Article 159 of the Constitution on the grounds that it was revised and approved by the Verkhovna Rada (the unicameral parliament of Ukraine) without the obligatory opinion

[1] Article 157 of the 1996 Constitution stipulates: 'The Constitution of Ukraine shall not be amended if the amendments foresee the abolition or restriction of human and citizens' rights and freedoms, or if they are oriented toward the liquidation of the independence or violation of the territorial indivisibility of Ukraine.'

of the Constitutional Court. Accordingly, the Constitutional Court declared it unconstitutional and void.[2]

The Venice Commission questioned the legality and legitimacy of that decision, mainly as it is unclear whether the Constitutional Court had the authority to review the constitutional amendment once it had entered into force. The Venice Commission also made a special reference on the time that had passed since the adoption of the amendment and its judicial annulment by the Constitutional Court:

> It also considers highly unusual that far-reaching constitutional amendments, including the change of the political system of the country – from a parliamentary system to a parliamentary-presidential one – are declared unconstitutional by a decision of the Constitutional Court after a period of six years. The Commission notes however, that neither the Constitution of Ukraine nor the Law on the Constitutional Court provide for a time-limit for contesting the constitutionality of a law before the Constitutional Court. As constitutional courts are bound by the Constitution and do not stand above it, such decisions raise important questions of democratic legitimacy and the rule of law.[3]

In the Venice Commission's opinion, 'The principle of legal certainty, being one of the key elements of the rule of law … requires that when declaring a constitutional amendment unconstitutional the time elapsed since its adoption is taken into account.'[4] This final point – the taking into account the time since an adoption of a constitutional amendment and its invalidation by a court – is the focus of our chapter.

Should the timing of judicial review matter? Should it be integrated as part of the analysis? Should the court take it into consideration – and if so – how?

In judicial review, mainly of administrative acts, there is the well-known doctrine of laches or time-limit for judicial review.[5] For example, in British law, there is a three-month limit in regards to filing a claim form with the Administrative Court, after the grounds to make the claim first arose.[6] The rationale behind

[2] Decision of the Constitutional Court of Ukraine, No 20-rp/2010 (30 September 2010) (concerning the constitutionality of the Law of Ukraine 'On Introducing Amendments to the Constitution of Ukraine' No 2222-IV (8 December 2004), English translation available at 'Decision of the Constitutional Court of Ukraine No. 20-rp of 30 September 2010 (Case of Observing the Procedure for Amending the Constitution of Ukraine)' (2011) 46 *Statutes & Decisions – The Laws of the USSR and Its Successor States* 41–52; see Yaniv Roznai and Silvia Suteu, 'The Eternal Territory? The Crimean Crisis and Ukraine's Territorial Integrity as an Unamendable Constitutional Principle' (2015) 16(3) *German Law Journal* 542, 559–60.

[3] Opinion on the Constitutional Situation in Ukraine, Opinion of the European Commission for Democracy Through Law (Venice Commission) No 599/2010, paras 35–36 (17–18 December 2010), www.venice.coe.int/webforms/documents/default.aspx?pdffile=CDL-AD%282010%29044-e.

[4] *Ibid* para 38.

[5] The doctrine of laches 'permits a court of equity to cut off a right of action when the plaintiff has delayed unreasonably and inexcusably in instituting litigation and thereby has prejudiced the defendant significantly'. See David D Higgins, 'The Application of the Doctrine of Laches in Public Interest Litigation' (1976) *Boston University Law Review* 56, 181.

[6] See Gregory Jones, 'Some Practical Impacts of European Union Law upon the Procedure for Judicial Review' (2010) 15:2 *Judicial Review* 139.

such a time-limit usually rests on efficiency, certainty and stability grounds. As Lord Diplock remarked:

> The public interest in good administration requires that public authorities and third parties should not be kept in suspense as to the legal validity of a decision the authority has reached in purported exercise of decision-making powers for any longer period than is absolutely necessary in fairness to the person affected by the decision.[7]

Such considerations of stability seem to be strengthened when it comes to judicial review – not of administrative acts – but of constitutional norms themselves.

Imagine the decision by the Constitutional Chamber of the Supreme Court of Honduras, of 2015, declaring the original constitutional provisions on presidential term-limits, which existed in the original Constitution of 1982, as unconstitutional.[8] In other words, 33 years after these constitutional provisions were applied and relied upon, the Court declared them to be unconstitutional and inapplicable. Such a decision surely has a destabilising effect. As Landau, Dixon and Roznai note concerning the Honduran decision:

> Review of constitutional amendments may be particularly destabilising because it attacks the basic norms of the system at the constitutional level, not simply laws or regulations issued under those norms. It thus threatens to change the ground rules upon which political actors may be coordinating around and on which they may rely. The stability critique, of course, poses particularly substantial difficulties if deployed well after the amendment was adopted. When there is a long lag between adoption and invalidation, a number of lower-level norms may have been adopted, based on the invalidated constitutional provision, and actors may have changed their interactions based on it.[9]

The aim of this chapter is to bring to the fore the temporal aspect in judicial review of constitutional amendment. As a general backdrop for our analysis, we use a decision by the Moldovan Constitutional Court[10] on the modality of electing the President, in which the Court decided the unconstitutionality of a constitutional amendment 16 years after its enactment, without assessing in its reasoning the temporal question.

The structure of the chapter is as follows: In section II we provide the background for that decision, the reasoning of which is summarised in section III.

[7] *O'Reilly v Mackman* [1983] UKHL 1; [1983] 2 AC 237, 280–81. On this question, see also William I Jr Marschall, 'Timing of Judicial Review' (1955) 33 *Texas Law Review* 701; Alistair Lindsay, 'Delay in Judicial Review Cases: A Conundrum Solved?' [1995] *Public Law* 417; Enid Campbell and Matthew Groves, 'Time Limitations on Applications for Judicial Review' (2004) 32(1) *Federal Law Review* 29.

[8] Decision of 22 April 2015 (Constitutional Chamber, Supreme Court).

[9] David E Landau, Rosalind Dixon and Yaniv Roznai, 'From an Unconstitutional Constitutional Amendment to an Unconstitutional Constitution?' (2019) 8(1) *Global Constitutionalism* 40, 45.

[10] Decision 7/2016 (4 March 2016). Thus, Serkan Yolcu calls this decision, the 'time travel of Moldova'; in Dr Öğr. Üyesi Serkan Yolcu, 'Siyasetin Yargısallaşmasının Hükümet Sistemine Etkisi Üzerine Bir İnceleme: Moldova'nın Anayasal Zaman Yolculuğu' (2018) 2018/2 *Galatasaray Üniversitesi Hukuk Fakültesi Dergisi* 154.

In section IV we analyse the Court's reasoning, and in section V we assess the question of time. We propose a 'time sensitivity test' to be considered when adjudicating constitutional amendments. We do not claim that a lengthy time should bar judicial review of amendments, but that courts cannot ignore the time element and incorporate it in their analysis. Section VI concludes.

II. Background

A. The Foundations of the Constitutional System

The Republic of Moldova became an independent country after the collapse of the Soviet Union. The period of modern statehood of this multi-ethnic country[11] officially began with the adoption of the Declaration of Independence by the Parliament of Moldova on the 27th of August 1991. The country's new Constitution was adopted on the 28th of July 1994. Since then, the country faced various challenges from the point of view of internal[12] and external politics.[13]

The Constitution is based on classic constitutional principles, contains safeguards for fundamental rights, as well as structures and limits for the exercise of state power. As for constitutional amendment rules, the Constitution prescribes: (a) substantive; (b) procedural; (c) temporal; and (d) emergency constraints.[14] (a) Similar to 'eternity clauses' of other constitutions, the Constitution contains – as a substantive constraint – a general prohibition of amendments that infringe fundamental rights and their safeguards.[15] (b) In the case of amendments that affect the sovereignty, independence and unity of the state, as well as its 'permanent neutrality', it is prescribed a special amendment procedure (different from the regular one), namely a national referendum and the majority vote of the registered voters.[16] (c) As a temporal limit, the proposals for constitutional amendments may

[11] The majority of the population speaks Moldovan (a dialect of the Romanian language), while a significant minority speaks Russian, Ukrainian and the Gagauz language (a Turkic language).

[12] Even the 1994 Constitution provided autonomy for the Transnistrian and Gagauzian regions of the country, the de facto exercise of state powers is limited in Transnistria as this region proclaimed its independence – an act not officially recognised but supported by the Russian Federation.

[13] Due to different, but simultaneously present economic interests, as well as cultural differences, there is a tension between the possible European and Russian orientation of the country. As a move towards the European direction, the EU-Moldova Association Agreement was signed on the 27th of June 2014.

[14] On different constraints on constitutional amendments, see: Oran Doyle, 'Constraints on Constitutional Amendment Powers' in Richard Albert, Xenophon Contiades, Alkmene Fotiadou (eds), *The Foundations and Traditions of Constitutional Amendment* (Hart Publishing, 2017) 73; Yaniv Roznai, *Unconstitutional Constitutional Amendments – The Limits of Amendment Powers* (Oxford University Press, 2017); Richard Albert, *Constitutional Amendments: Making, Breaking, and Changing Constitutions* (Oxford University Press, 2019).

[15] Constitution of the Republic of Moldova, Article 142, para 2.

[16] *Ibid*, Article 142, para 1.

be passed six months after their initiation at the earliest and one year after at the latest.[17] (d) The Constitution also prohibits its revision under a state of national emergency, martial law or war.[18]

Taking into consideration the above presented constraints, the regular procedure for the amendment of the Constitution can be initiated by 200,000 voters, one third of the members of Parliament or the Government.[19] The advisory opinion of the Constitutional Court on the draft amendment is a necessary condition for the submission of the proposal to the Parliament.[20] Constitutional amendments can pass in Parliament with the support of a two-third majority of MPs.[21]

The Constitutional Court of the Republic of Moldova is a Kelsenian-type court, its legal status and independence is supported by sufficient safeguards prescribed in the Constitution.[22] The Constitution accords a comprehensive set of powers for the Constitutional Court based on which it exercises norm control and acts as the final instance in legal disputes related to the exercise of competences of other state organs, as well as disputes related to elections and referenda.[23] One can add that the Constitution – as noted above – expressly accords to the Constitutional Court the power for the *ex-ante* review of constitutional amendments.[24]

The Constitution acknowledges the institution of (national) referendum as a tool to solve 'problems of utmost importance confronting the Moldavian society'.[25] The initiation of national referenda is a power of the President (the head of state)[26] on a discretionary basis. However, as mentioned above, in the case of an amendment to the Constitution which affects certain constitutional principles, a national referendum must be organised.

B. The Foundations of the Governmental System

The Constitution of the Republic of Moldova regulates in detail the status and competences of the Parliament, the President and the Government.[27] Based on these provisions, the Republic of Moldova can be considered to have a parliamentary form of government, in which the President has significant powers relating to the executive branch, therefore, the characteristics of semi-presidentialism are also present.

[17] *Ibid*, Article 143, para 1–2.
[18] *Ibid*, Article 142, para 3.
[19] *Ibid*, Article 141, para 1.
[20] *Ibid*, Article 141, para 2.
[21] *Ibid*, Article 143, para 1.
[22] See *ibid*, Articles 136–39 on the appointment conditions, appointment procedure, status and incompatibilities of the judges of the Constitutional Court.
[23] See *ibid*, Article 135.
[24] *Ibid*, Article 135, para 1. (c).
[25] *Ibid*, Article 75.
[26] *Ibid*, Article 88. (f).
[27] *Ibid*, Articles 60–106.

The Parliament, as the legislative branch, is composed of 101 members,[28] elected based on universal, equal, direct, secret and freely expressed suffrage,[29] for a four-year term.[30] The Government, as part of the executive branch, is composed of the prime minister, deputy ministers and ministers.[31] As a basic characteristic of parliamentary systems,[32] the Parliament has the power to grant its vote of confidence for the designated prime minister, as well as the programme and composition of the Government with the majority vote of the MPs – a precondition for the appointment of the Government.[33] From the other perspective, the Parliament has the power to end the mandate of the Government based on a vote of no confidence[34] or motion of censure relating to the proposals of the Government upon which it assumes responsibility before the Parliament.[35]

Based on the provisions of the Constitution, the President is not expressly part of the executive branch, however, alongside the traditional powers which belong to the head of state, he/she has significant powers which can influence the room of manoeuvre of other state organs: he/she may address messages to the Parliament,[36] initiate national referenda,[37] and repeal the acts of the Government which run contrary to the legislation until the final judgment of the Constitutional Court.[38] Moreover, he/she has a strong position in external policy[39] and national defence,[40] also supported by his/her function of being 'the guarantor of the national sovereignty and independence, as well as of the territorial unity and integrity of the State'.[41] Based on these particularities, the role of the President of the Republic of Moldova is close to the role of the head of state in semi-presidential governmental systems.[42]

Though the provisions of the Constitution on the powers of the President were not the subject of constitutional amendments, the modality of the election of the President was changed. Since the enactment of the Constitution (1994), the fundamental law prescribed the direct election of the head of state. The amendment

[28] *Ibid*, Article 60, para 2.
[29] *Ibid*, Article 61, para 1.
[30] *Ibid*, Article 63, para 1.
[31] *Ibid*, Article 97.
[32] Anthony W Bradley and Cesare Pinelli, 'Parliamentarism' in Michel Rosenfeld and András Sajó (eds), *The Oxford Handbook of Comparative Constitutional Law* (Oxford University Press, 2012) 650–70, 651.
[33] Constitution of the Republic of Moldova, Article 98, para 1–4.
[34] *Ibid*, Article 106, para 1.
[35] *Ibid*, Article 106a, para 1–2.
[36] *Ibid*, Article 84, para 2.
[37] *Ibid*, Article 88, f.
[38] *Ibid*, Article 88, i.
[39] *Ibid*, Article 86.
[40] *Ibid*, Article 87.
[41] *Ibid*, Article 77, para 2.
[42] See Maurice Duverger, 'A New Political System Model: Semi-presidential Government' (1980) 8:2 *European Journal of Political Research* 165–87.

to the Constitution in 2000[43] changed this modality, prescribing the indirect election of the President (election by the Parliament). These provisions were in force until 2016, when the Constitutional Court declared these unconstitutional and repealed them from the text of the Constitution, recognising at the same time the former version of the provisions on the election of the President as part of the Constitution.[44]

III. The Case before the Constitutional Court

A. The Facts

The case originated with a motion for posterior norm-control, submitted by 18 MPs in 2015. The claim challenged the provisions of the Law that amended the Constitution in 2000[45] with regard to the modality of electing the President. The petitioners based their motion both on formal and substantive arguments: on one hand, questioning the procedure related to the enactment of the amendment; on the other hand, arguing that the substance of the amendment is in contradiction with the rule of law principle.

As to the background of this claim, the following circumstances can be mentioned – which were also presented in detail in the Constitutional Court's decision.

In 1999 a consultative national referendum was held on the question of establishing a presidential system of government, in which the President would bear the responsibility for the exercise of the executive power. The majority of the voters took part in the referendum and the majority of the votes supported the proposed change to the governmental system.

The next year saw the Parliament enact an amendment to the Constitution (the challenged piece of legislation) which differed to the opinion of the people expressed in the referendum. According to the amendment, the President has to be elected by the Parliament, with a three-fifth majority of its members (and, accordingly, the system of government was not changed).

In the period between the enactment of the amendment to the Constitution (2000) and the submission of the motion of the MPs for posterior norm control (2015), three times a situation occurred in which Parliament failed to elect a President as the required majority was not met, therefore, the Parliament was dissolved and early parliamentary elections took place (2000, 2009, 2010). On the whole, this period can be characterised by instability of the political and the governmental system.

[43] Law 1115-XIV of 5 July 2000.
[44] Decision 7/2016 (4 March 2016).
[45] Law 1115-XIV of 5 July 2000.

B. Formal Review

As mentioned above, the advisory opinion of the Constitutional Court on the draft amendment to the Constitution is a necessary condition for the submission of the proposal to the Parliament.[46] Contrary to this provision, in the recent case, the Parliament adopted substantive amendments to the Constitution that had not been entirely submitted to the Constitutional Court for an opinion.

Based on the proposal of a group of MPs, the Constitutional Court delivered its positive opinion on the initiative to amend the Constitution on the 16th of November 1999.[47] The initiative was related to various provisions of the Constitution including the provisions on the election of the President. As the Parliament has two readings on draft laws, the recent draft was voted by the Parliament in the first reading, alongside two other proposals, previously also examined by the Constitutional Court. These three proposals were merged into a single draft constitutional law which was the subject of substantive amendment in the second reading.[48] As a result, the original initiative examined by the Constitutional Court and the enacted draft differed in key elements relating to the modality of electing the President, eg the age requirement of candidates was changed from 35 years to 40 years, the required majority vote in Parliament was replaced by the vote of three-fifths of MPs, as well as those relating to some of the powers of the President, eg the dissolution of Parliament.[49]

The detailed rules on the readings of constitutional amendments are prescribed in the Rules of Parliaments (the Standing Orders). These provisions specify that in the second reading the amendments to the initiative could not refer to any supplementary elements that were not covered by the draft law upon which the Constitutional Court delivered its opinion, except the editing of the text, modifications (or omissions) that do not alter the essence of the original provisions.[50] In the Court's view, the above mentioned amendments in the second reading were substantive in nature, therefore the procedural requirements were not met.[51] The Court also expressed its position on the function of its opinions on the draft laws amending the Constitution, as these 'aim at safeguarding the fundamental values of the Constitution from abusive practices of political, social and institutional actors'. Moreover, the Court declared that 'a substantially amended draft law amending the Constitution shall be considered as a new draft law'[52] – which may be initiated together with the opinion of the Constitutional Court. As a consequence, the Court stated that 'ignoring or overstepping these opinions may constitute

[46] Constitution of the Republic of Moldova, Article 141, para 2.
[47] Decision 7/2016 (4 March 2016) CC 110–113.
[48] *Ibid*, 116–20.
[49] *Ibid*, 127.
[50] *Ibid*, 138.
[51] *Ibid*, 158–59.
[52] *Ibid*, 93.

grounds for the nullity of such modifications'[53] – accordingly, the violation of the procedural requirements may be basis for the declaration of unconstitutionality separately from the substantive review.

C. Substantive Review

The Constitutional Court expressed some of the arguments which support the substantive review of the constitutional amendment within the frame of examination of admissibility. On the formal ground, the Court declared that it is competent to examine the constitutionality of a law amending the Constitution as the basic law accords it the general competence of reviewing the constitutionality of laws, as well as the specific competence to deliver its opinion upon the initiatives to amend the Constitution.[54] One can note that the Court did not examine in detail the relation between the general competence of reviewing constitutionality of laws and its applicability in the case of constitutional amendments which are already enacted (differing from the ex-ante review of these).

The Court added an additional argument by stating that constitutional review of certain legal acts inevitably includes the *interpretation of the related norms of the Constitution*.[55] Even the Court did not express this relation explicitly, in the recent case this argument led to the examination of the effects of the legal act in question (the act on the amendment of the Constitution) on the integrity of the provisions of the Constitution. The *integrity argument* also plays an important role relating to other elements of the substantive review.

Also within the frame of the examination of admissibility, the Court made important statements relating to its approach to the question of time: (a) First, the Court stated that 'the state of constitutionality evolves along with the transformation of socio-economic, political and moral circumstances within the society'[56] – an emphasis on contextual elements of constitutional disputes that change over time; (b) Second, the Court created the link between legitimacy and future amendments of a legal provision which is the subject of constitutional disputes, when stating that 'the challenged provisions have not been amended since their adoption in 2000, and thus have never been validated by a successive modification'.[57]

The petitioners based their complaint – alongside the arguments relating to the procedure, summarised above – on the possible infringement of the constitutional principle of the rule of law.[58] The petitioners also referred to the requirement of

[53] *Ibid*, 204.
[54] *Ibid*, 36.
[55] *Ibid*, 37–38.
[56] *Ibid*, 32.
[57] *Ibid*, 34.
[58] *Ibid*, 2.

constitutional balance in the relationship of state organs, as the challenged provisions generated deadlocks and institutional imbalances.[59]

The Court took these requirements into consideration when examining the merits of the case, however, it took a much broader perspective, a mixture of more substantive considerations.

The Court emphasised that the Constitution is founded on *universal and undeniable values* – among these, the principle of the rule of law and the system of checks and balances.[60] Elsewhere, the Court referred explicitly to substantial limitations of constitutional amendments, by taking into consideration *fundamental values* that are 'inseparably interlinked and form the foundation of the state'. In the Court's view the 'independence of the State, democracy, the republic as a form of government and the intrinsic character of human rights and fundamental freedoms' formulate these fundamental values.

The Court also made an imperative statement by declaring that 'no amendment of the Constitution shall disturb the harmony of the Constitution or the *harmony of values* enshrined in the Constitution'.[61] Accordingly, the Court declared that the provisions of the Constitution make up a harmonious system.[62] The Court referred to harmony as 'the unity of matter and the balance of values enshrined in the Constitution'.[63] Later unity is described as the 'indissoluble logical and legal connection' of the constitutional provisions.[64]

According to the Court, this unity implies a ban on the adoption of legislative steps (including constitutional amendments) that would lead to conflicts between state organs and institutional deadlocks. The Court emphasised that 'almost all political and institutional stalemates in the Republic of Moldova are related to the modified provisions referring to the procedure to elect the President and the mechanism to dissolve the Parliament'.[65] These provisions have led to 'a rupture' of the constitutional unity as these failed to 'secure the normal functioning of the presidential institute',[66] as the Parliament could not elect compromise candidates, and, accordingly, political and institutional deadlocks occurred.

According to the Court, the harmony of the provisions of the Constitution was also challenged by the fact that – due to the stalemates described above – the Speaker of the Parliament cumulated offices when acting as an interim president.[67] Such situation is in contradiction with the principle of separation of powers.[68]

[59] *Ibid*, 46–47.
[60] *Ibid*, 59.
[61] *Ibid*, 74–75.
[62] *Ibid*, 69–70.
[63] *Ibid*, 169.
[64] *Ibid*, 188.
[65] *Ibid*, 180.
[66] *Ibid*, 192.
[67] *Ibid*, 181.
[68] *Ibid*, 195. Also taking into consideration that the office of the interim president may be exercised for an unlimited period of time. See *ibid*, 190.

Finally, the Court came to the conclusion that the challenged provisions were unconstitutional based on the unbalanced position of constitutional institutions,[69] which includes potential conflicts between them, and can also be described by the 'imperfect system of government' generated by the constitutional reform. In this regard, the requirement of functionality and the systemic coherence of the Constitution was also mentioned.[70] In the final part of the reasoning the Court referred to the argument of the petitioners by stating that the challenged provisions were 'inconsistent with the constitutional principle of the rule of law', as 'there cannot exist in a Constitution any gaps or internal contradictions'.[71]

D. Comparative Perspective

The Court made short references to the jurisdictions of other constitutional courts.[72] The Federal Constitutional Court of Germany (*Bundesverfassungsgericht*) was mentioned regarding its position on the constitutional review of constitutional amendments based on the eternity clauses of the Basic Law (*Grundgesetz*), expressed and mentioned in many decisions. Austria, Bulgaria and South Africa were mentioned as examples on the invalidation of constitutional amendments, while from the extended practice of the Constitutional Court of Turkey relating to this topic it referred to the case from 2008 relating to the headscarves debate at universities (based on a constitutional amendment).[73] It is worth mentioning that – despite the dominant influence of the Indian basic structure doctrine on other jurisdictions – the Court did not refer to relevant cases from India.[74]

A 2014 judgment of the Constitutional Court of Lithuania was also mentioned: this decision has special relevance when analysing the central question of this case, the time perspective. The Lithuanian judgment invalidated a constitutional amendment adopted in 2006, eight years after its enactment, on procedural grounds.[75]

[69] *Ibid*, 197.

[70] *Ibid*, 200.

[71] *Ibid*, 210, 211.

[72] For some comparative studies, see Gary Jeffrey Jacobsohn, 'An Unconstitutional Constitution? A Comparative Perspective' (2006) 4(3) *International Journal of Constitutional Law* 460; Kemal Gözler, *Judicial Review of Constitutional Amendments: A Comparative Study* (Ekin Press, 2008); Gábor Halmai, 'Unconstitutional Constitutional Amendments: Constitutional Courts as Guardians of the Constitution?' (2012) 19(2) *Constellations* 182.

[73] Decision 7/2016 (4 March 2016) CC 100, 101, 103.

[74] On the Indian doctrine of limits to constitutional amendments, see Sudhir Krishnaswamy, *Democracy and Constitutionalism in India: A Study of the Basic Structure Doctrine* (Oxford University Press, 2010).

[75] Decision 7/2016 (4 March 2016) CC 102. In 2006, Parliament abolished Article 125 of the Constitution that provided for the Bank of Lithuania's 'exclusive right to issue bank notes'. This was not in line with the rights of the European Central Bank. In its decision of 24 January 2014, the Constitutional Court declared that the Law amending Article 125 of the Constitution was adopted in breach of the essential procedural requirements of Art 147 of the Constitution and annulled it. See Irmantas Jarukaitis and Gintaras Švedas, 'The Constitutional Experience of Lithuania in the Context of European and Global Governance Challenges' in Anneli Albi Samo Bardutzky (eds), *National Constitutions*

The Court also referred to the regulation of other constitutions regarding the modality of electing the President, as well as rules relating to the dissolution of parliaments – the central parts of the examined constitutional amendment. The position of the Venice Commission was also cited – mostly in relation to the question of stability of the Constitution.

IV. General Comments

The establishment of the Court's competence to posterior review of constitutional amendments is admittedly a novel development in the constitutional culture of the Republic of Moldova – however, it does not offer new lessons from a comparative constitutional law scholarship perspective, as this practice is intensively spreading around the world. Accordingly, in the analysis below we do not pay special attention to the question of competence. Similarly, the review of constitutional amendments on procedural grounds is relatively simple to justify, therefore, it is not the subject of intensive criticism. In the present case the Constitutional Court based its formal review on relatively unambiguous procedural requirements. Therefore, in this regard, the reasoning can be considered convincing. Even the procedure and substance are intertwined, in the present analysis we focus our attention on the substantive review – as the substantive standards used and the contextual elements which influence these raise the most interesting questions and offer the most beneficial lessons.

As presented above, the Constitutional Court did not base its reasoning on a particular substantive standard, rather a mixture of considerations can be mentioned in this regard: fundamental values of the Constitution, the Constitution as a harmonious system, the unity of the constitutional text and the logical connections of the constitutional provisions, the requirement of the balanced position of constitutional institutions and the constitutional principle of the rule of law. Even though the petitioners based their argumentation on the rule of law and the Court recalled this principle in the final part of its argumentation, the Court did not elaborate a particular 'rule of law standard'. The logical connections between the rule of law and the other considerations are not elaborated on in detail in the reasoning; therefore, the reference to the principle is rather symbolic.

The fundamental values mentioned in the Court's reasoning and their relations as forming a harmonious system are relatively close to the most influencing doctrine which can be considered as a basis for review of constitutional amendments, the basic structure doctrine.[76] In this regard, these fundamental values fit

in European and Global Governance: Democracy, Rights, the Rule of Law (Springer, 2019) 997. For a summary of the case, see: www.lrkt.lt/en/court-acts/search/170/ta850/summary.

[76] The approach of the Moldovan Constitutional Court can be considered as part of an international trend that is 'moving towards accepting' the doctrine. See Roznai (n 14).

the standard of forming 'a certain constitutional core, a set of basic constitutional principles which form the constitutional identity and which cannot be abrogated through the constitutional amendment process'.[77] It is easy to demonstrate that the independence of the state, democracy, the republic as a form of government and the intrinsic character of human rights and fundamental freedoms are the core values of the Moldovan Constitution, therefore, an amendment that disrupts the harmony of these is not acceptable. However, the Court did not make an explicit statement in this regard.[78]

The arguments relating to the *unity of the norms of the Constitution* and the logical connections between the provisions raise two interesting theoretical questions. (a) First, every written constitution is intended to be a comprehensive set of norms which are set in coherent and logical order. As the constitution is the supreme law in a given legal system, potential contradictions or conflicts between the norms included in the constitution cannot be solved by taking into consideration other legal acts – that is why the stakes are much higher in the case of constitutional interpretation compared to ordinary legal interpretation. (b) Second, from a theoretical point of view it can be stated that there may be a hierarchy of norms within the constitution, as certain provisions (typically principles) formulate the basis for other constitutional provisions; or seek stronger protection compared to other constitutional norms by way of entrenchment or eternity clauses.

In both considerations presented above, contextual interpretation is a method of constitutional interpretation which supports simultaneously taking into consideration more provisions of the constitution (regardless of their hierarchical status), as well as the relationship between them. However, there can be even conflicts between the norms of a given constitution which cannot be solved by way of constitutional interpretation. In such cases the amendment to the constitution (rather formal or informal) or the annulment of one of the conflicting provisions can put an end to the conflict.

According to our assessment, in the given case the unity of the constitutional norms was not challenged based on the above considerations: there was no explicit contradiction between the provisions of the constitution and the question of different hierarchical status of constitutional norms was also not raised. The constitutional conflict was rather caused by the fact that certain provisions were proved to be not functioning in practice.

This leads us to the examination of facts. In its reasoning, the Court refers more often to the lack of capacity of the Parliament to elect compromise presidential

[77] Roznai (n 14) at 70.

[78] The values mentioned in the reasoning are rather universal than particular, therefore, are closer to a concept of constitutional identity that accords more attention to the generally accepted values in constitutional democracies. This is in contrast with the approach which focuses on more particularistic values of the national identity. In this regard Michel Rosenfeld can be mentioned. See: Michel Rosenfeld, 'Constitutional Identity' in Rosenfeld and Sajó (eds) (n 32) 758. On constitutional identity as a mixture of particularistic values with universal ones, see Gary J Jacobsohn, *Constitutional Identity* (Harvard University Press, 2010).

candidates, and as a result, to the periods of interim presidency, repeated dissolutions of the Parliament and early elections. In this regard one may ask: what is the role of the Constitutional Court in examining facts, and what are the limits of such examination? Facts may be mentioned as the contextual elements of the legal (constitutional) question at hand, as well as substantive elements of the chain arguments. The first case takes place in almost every constitutional dispute (by specifying the role of the petitioner and the circumstances in which she or he identifies a constitutional question), while the second is a relatively rare situation with a limited scope. Classic examples on the examination of facts as substantive parts of the chain of arguments are fundamental rights disputes examined based on the proportionality test. The examination of suitability, as a subtest of proportionality, is a limited, fact-based examination of the relation between the legitimate purpose of the given limitation of fundamental rights and the means chosen by (typically) the legislator. In this case the examination of facts (the lack of capacity of state organs to function in accordance with the norms prescribed in the Constitution) gets even more weight in the chain arguments: this fact-analysis brings the Court to conclude that the unity of constitutional provisions had deteriorated.

According to our consideration, fact-evaluation with such a strong effect is possible in the practice of constitutional courts, however, more transparent grounds are needed. The doctrine of 'the living constitution', which requires taking into consideration the realisation and effects of the constitutional provisions in practice,[79] could have been a more sophisticated and justifiable frame for such examination. In the absence of such reference, the Court could easily face criticism relating to its activity on qualifying political actors on questionable grounds – a strong argument for those who claim less judicial activity in cases affected by the counter-majoritarian dilemma.[80]

The arguments relating to the requirement of the balanced position of constitutional institutions is close to the principle of the separation of powers – however, the Court did not elaborate on this relationship in detail, rather referring to the classic principle as an additional element to take into consideration. The lack of the structured reasoning relating to the separation of powers can be criticised from two perspectives. (a) First, the very basis for the possibility of reviewing unconstitutional constitutional amendments is the difference between the primary (constitutive) power and secondary (constituted) powers, as well as their

[79] One of the very first explanations of the notion of living constitution is given by Mary Barbara McCarthy when focusing on the US Constitution as a 'living letter governing actions and determining positive relations among men' compared to the 'dead letter' it could be without the active role of court (referring to Tocqueville). Mary Barbara McCarthy, 'The Living Constitution 1787:1937' (1938) XIV *Notre Dame Lawyer* 1, 2, 9. However, the above consideration is much closer to the 'living tree doctrine' known from the practice of the Supreme Court of Canada. According to this powerful metaphor, the constitution 'planted in Canada a living tree capable of growth and development within its natural limits'. See *Edwards v Attorney-General for Canada* [1930] AC 124, 136.

[80] See Alexander Bickel, *The Least Dangerous Branch: The Supreme Court at the Bar of Politics*, 2nd edn (Yale University Press, 1986).

relationship based on the vertical dimension of the separation of powers.[81] In this regard, the Constitutional Court and the constitution-amending power are both constituted, therefore, when examining the constitutionality of constitutional amendments, the Court acts in order to preserve the vertical dimension of the separation of powers. This consideration could be a strong basis for the legitimacy of the decision from the formal point of view. (b) Second, by taking into consideration the horizontal dimension of the separation of powers, in the recent case, by way of detailed analysis of the relations between the President, the Government and the Parliament, the Court could have been presented a set of persuasive arguments. As it was mentioned above, the form of government of the Republic of Moldova is not univocal, that is why it would have been of key importance for the Court to interpret the provisions of the Constitution in this regard and to analyse the relations of state institutions based on these requirements.

It is worth mentioning that when expressing its substantive arguments, the Court did not support these genuinely on a textual basis. The eternity clause of the Constitution prohibits amendments that infringe fundamental rights and their safeguards. Even at the core of the case before the Court there were no conflicts relating to the exercise of fundamental rights, the modality of the election of the President, as well as the balanced functioning of the elected branches of government have a direct effect on the exercise of political rights. Therefore, a reference to the eternity clause would offer a strong textual support for the Court's arguments. In addition, the form and system of government is one of the typical groups of norms explicitly protected by eternal clauses[82] – a question at the core of this case. However, in this case the protection of these norms was only implicitly taken into consideration by the Constitutional Court.

V. The Time Perspective as Context

The most specific peculiarity of the case is that the decision on the unconstitutionality of the constitutional amendment in question was taken after 16 years. However, the Court did not assess in its reasoning the *question of time* at all. In the absence of convincing arguments, the timing of the decision could cause serious

[81] Yaniv Roznai, 'Towards a Theory of Constitutional Unamendability: On the Nature and Scope of the Constitutional Amendment Powers' (2017) 18 *Jus Politicum* 5.

[82] One reason for the silence of the Court in respect of eternity clauses could be the particularity of transitional constitutions: in most cases the eternity clauses of these reflect the results of political compromises and settlements. See Silvia Suteu, 'Eternity Clauses in Post-Conflict and Post-Authoritarian Constitution-Making: Promise and Limits' (2017) 6(1) *Global Constitutionalism* 65. In this case the form of government was not covered by the eternity clause – therefore, no political consensus was behind that, which could otherwise support the Court's interpretation. On eternity and transitional constitutions see also Gábor Halmai, 'Transitional Constitutional Unamendability?' (2019) 3 *European Journal of Law Reform* 259.

doubts relating to the legitimacy of the decision in the light of judicial activism.[83] One can also mention the challenges relating to unconstitutional constitutional amendments in general – the arguments highlighting this challenge unambiguously have stronger bite in the case of judicial decisions taken after a longer period of time.

Constitutional review of legal acts, constitutional amendments or even original constitutions after an unusually long period is not novel in constitutional adjudication. The Supreme Court of Canada invalidated an act more than 100 years after its enactment.[84] As mentioned earlier, the Constitutional Court of Lithuania invalidated a constitutional amendment eight years after its enactment[85] – a case mentioned in the reasoning of the respective decision of the Constitutional Court of Moldova, and there is also a unique example from Honduras on the invalidation of certain provisions of the original text of the Constitution, which took place after 33 years.[86]

The Moldovan case offers a recent cause and a perfect example to examine those circumstances which shall be taken into consideration when assessing the time factor in constitutional review cases. Although these considerations might be relevant when reviewing other types of norms, for the purposes of this analysis we will focus on unconstitutional constitutional amendments. First, in the light of the Moldovan case we will highlight those theoretical and practical difficulties that time-sensitive decisions could cause. Based on these, we will present appropriate methodological tools to take into consideration when examining the time-factor of such cases. In our view, these considerations serve the purpose of maintaining the democratic legitimacy of the judicial decisions on a substantive basis, therefore, can prevent potential conflicts that originate in the 'super-counter-majoritarian difficulty'.[87]

We suggest the use of a three-tier test in the case of time-sensitive disputes on unconstitutional constitutional amendments. The three-tiered 'time sensitivity test' focuses: (a) on the effect of the decision on the system of separation of powers, especially on the relationship between the constitutional (supreme) court and the constitution-amending power; (b) the question of incorporation, whether the substance of the challenged amendment can be considered as part of the set of provisions of the original constitutional text; and (c) the 'constitutional-political question', whether the challenged provisions of the amendment can be considered as questions that are the subject of political considerations of the constitution-amending power.

[83] Mihai Popşoi, 'Controversial Ruling by Moldova's Constitutional Court Reintroduces Direct Presidential Elections' (2016) 13(46) *Eurasia Daily Monitor*, https://jamestown.org/program/controversial-ruling-by-moldovas-constitutional-court-reintroduces-direct-presidential-elections/.

[84] *Re Manitoba Language Rights* [1985] 1 SCR 721.

[85] See n 75.

[86] See n 8.

[87] Landau, Dixon and Roznai (n 9), 41.

Although these subtests refer to the substance of constitutional amendments, these can be relevant both in the case of review of such amendments on procedural grounds, as well as on a substantive basis – as these considerations are often intertwined.

A. Theoretical and Practical Difficulties Relating to Time-Sensitive Decisions

Every legal dispute has at least one aspect of time consideration. The identification of the time effect of the piece of legislation under consideration (*ex tunc*, *ex nunc or pro futuro effect*) is a simple task, however, which sometimes requires careful examination of all aspects of the case.[88] From a closer look, in the case of judicial review the time consideration has a different aspect, depending on the procedure which is initiated before the constitutional court or the supreme court. In *ex ante* review cases the court has to take into account the time left until the proposed entering into force of the examined piece of legislation, while in cases of *posterior* norm control the most important question is that of the temporal effect of the decision, whether the court intends to invalidate a piece of legislation with retroactive, present or future effect.[89]

If a piece of legislation under judicial review was enacted relatively long time ago (as in the present, Moldovan case), the court faces particular difficulties.

From the theoretical point of view, one can argue that challenges relating to the counter-majoritarian dilemma have a new aspect in such cases. It can be argued that the decisions of the elected representatives of the past are overruled by the judges of the present who do not have direct authorisation for their activity from the people – neither from the past, nor from the present generations. From the other perspective, one can say that the 'dead hand problem', relating to the originalist method of constitutional interpretation[90] can be mentioned here as well: regulations of the past in many cases are not suitable to guide social relations of the present, therefore, it is not rational to follow these. If one considers one classic answer to the counter-majoritarian dilemma, eg taking into consideration the function of judges to protect long-term, basic values of the constitution against the actual interests of the political majority, one can argue that judicial review after a longer period supports the same purpose. As none of the critical approaches (that of criticising the present judicial decision of past regulations as well as the one criticising the long-term effect of the old regulations) is imperative, it seems

[88] Take, for example, a case in which a former regulation, which is already not in force, has to be applied because of the timing of the facts relating to the case, which constitute the legal basis for the particular legal claims.

[89] See generally Patricia Popelier, Sarah Verstraelen, Dirk Vanheule, Beatrix Vanlerberghe (eds), *The Effects of Judicial Decisions in Time* (Intersentia, 2013).

[90] See Craig R Ducat, *Constitutional Interpretation* (Wadsworth Cengage Learning, 2009) 79.

reasonable to consider judicial review in these cases no more problematic than in the basic form of judicial review.

Judicial review itself and judicial review after a longer period also poses the question of separation of powers. The classic concept has many interpretations.[91] However, preventing undue concentration of powers is a key element of all these. Arguably, judicial invalidation of a constitutional amendment after a long period of time is crucial for the principle of separation of powers. It creates a situation to which the legislative branch must react. At first glance this situation makes no difference compared to cases where the time dimension is not relevant – the legislator has to correct the regulation in both cases, which can be considered as a technical issue. However, if the challenged piece of legislation is a symbolic element of a political programme, the invalidation of it after a long period of time could cause a much more problematic situation for the political organs. Consider, for example, a hypothetical case in which the court invalidates a government policy on compulsory military service or tax regulation years after its enactment. In that case, the political organs are forced not only to correct the regulation, but to rewrite their political programme or agenda. As a result, the decision of the court limits extensively the margin of action of the legislator in a field which otherwise belongs exclusively to political powers. From the separation of powers point of view, this situation could cause an undue concentration of powers on the side of the court. Other theoretical difficulties can also be mentioned in this regard: questioning the existence or effects of legal relations based on the invalidated piece of legislation can be in contradiction with the principle of legal certainty.

One can add that the above arguments are also closely related to the political question doctrine. According to this classic doctrine, a court may decide only in terms of legal questions and not in the case of regulations which reflect the political considerations of political organs. In a very similar fashion to the political question doctrine, in cases of judicial review after a longer period of time the court has to consider the political implications of the case. The only difference is that in 'classic' cases in which the doctrine is applied it is satisfying for the court to focus on the subject of the case (eg the content of the challenged piece of legislation), while in cases of judicial review after a longer period of time the effects of the decisions are also worth taking into consideration – in close relation to the context of the separation of powers, as described above.

As to other practical difficulties relating to time-sensitive judicial decisions it is worth mentioning one aspect. If the challenged piece of legislation was enforced in practice for a longer period of time, it is likely that numerous administrative and judicial decisions were based on that, and accordingly, numerous legal relations

[91] MJC Vile's attempt to describe the particularities of the 'pure doctrine' received the most attention in contemporary constitutional scholarship. According to him, the doctrine has three elements: the 'separation of agencies', the distinction between the three specific functions of government and the separation of persons. See MJC Vile, *Constitutionalism and the Separation of Powers* (Oxford University Press, 1967).

were established. Sustaining these legal relations is a practical challenge for law-enforcement agencies, while in some cases the decision of the court could cause instability in the particular legal field. Instability grows proportionately with the number of cases based on, and the time which has passed after the enactment of the challenged piece of legislation.

Moreover, practical difficulties are intertwined with theoretical ones: in the case of amendments to the existing legal acts, one may ask whether the challenged (new) piece of legislation can be considered as an integral part of the original document, or if it was fully incorporated. One can argue that after a longer period of time the incorporation arguments have stronger bite. The stable, long-time practice of state organs and other legal entities based on the challenged piece of legislation can also be considered as a strong argument supporting incorporation. In this regard, contrary to the opinion of the Moldovan Constitutional Court presented in the recent case, we consider that the question whether the challenged piece of legislation (the original amendment) was later amended, has no relevance. Needless to say, the new amendments to the amended piece of legislation or the original document can later raise all the questions detailed above.

B. The 'Time Sensitivity Test' and its Justification

Taking into consideration the above explained difficulties, in the following section of our analysis we will explain the 'time sensitivity test' formulated on a normative ground. We suggest the application of the test for constitutional courts and supreme courts examining the constitutionality of constitutional amendments in order to base their decisions on arguments which react on the above described challenges. The subtests of the time sensitivity test will be explained based on the above discussed case of the Constitutional Court of Moldova.

The time sensitivity test consists of three subtests: (a) examining the effect of the decision on the system of the separated powers; (b) examining the extent of incorporation of the amendment to the original text of the constitution; and (c) examining the relation between the challenged amendment and political considerations of the constitution-amending power.

The *first subtest* focuses on the effect of the decision on the system of separation of powers, especially on the relation between the constitutional (supreme) court and the constitution-amending power. If one focuses on the function, or more precisely, the purpose of the doctrine of the separation of powers, the classic (Montesquieuan) and simplified understanding of the function of the doctrine can be summarised as 'a bulwark against the abuse of state power and the threat of tyranny'.[92] If one considers the constitutional values to protect when preventing

[92] Aileen Kavanagh, 'The Constitutional Separation of Powers' in David Dyzenhaus and Malcolm Thorburn (eds), *Philosophical Foundations of Constitutional Law* (Oxford University Press, 2016) 221–39.

tyranny, in classic interpretations two interrelated principles come to light, 'limited government'[93] and 'individual liberty'.[94] Both principles are the very theoretical foundations of judicial review, and – taking into consideration the basic function of liberal constitutions in protecting these principles – judicial review of constitutional amendments. In other words, in the separation of powers context the court's duty is to protect the foundations of the constitutional system created by the primary (constituent) power against the actions of the secondary (constitution-amending) power. These actions often have political motivations – that is why this relation can be described as a tension between the judicial and political powers.[95]

The time perspective sharpens this tension: if the decision of the court is taken after a longer period of time, its effects are much stronger on the political (constitution-amending) power. In that case, not only the amendment itself, but the whole legal and political practice relating to it is judged. Therefore, the task of reparation (both in legal and political terms) of political powers is much more complicated. Accordingly, one can argue that in these cases there is a risk of undue concentration of power on the court's side compared to the constitution-amending power – in the absence of compelling arguments in favour of the later review of the constitutional amendment.

In the Moldovan case, there existed such compelling arguments. The institutional deadlocks and governmental crises which occurred repeatedly proved that the activity of governmental branches and related institutions were insufficient to prevent and solve these. Therefore, the effective functioning of the separated powers was at constant risk. Invalidating the amendment to the Constitution by the Constitutional Court after 16 years is without doubt an excessive use of power, however, it was used in a situation where the state of separated powers could not be considered balanced. The purpose of the Court's decision was to push back the governmental branches in that balanced situation – therefore, from the separations of powers perspective the decision can be justified.

The *second subtest* focuses on the question of incorporation, whether the substance of the challenged amendment can be considered as part of the set of provisions of the original constitutional text. Incorporation is an abstract term. From the formal point of view, an amendment which fits the procedural

[93] Vile (n 91) 12.

[94] Vile (n 91) 15.

[95] This interpretation fits into current scientific approaches to the separation of powers: Eoin Carolan claims that in reality there is a separation between the judiciary and the 'political organs' [Eoin Carolan, 'The Problems with the Theory of the Separation of Powers' (2009) SSRN: <https://ssrn.com/abstract=1889304>]; Kavanagh (n 92) 234 describes the separation of powers principle as an instrument of constituting and restraining political power; while Jeremy Waldron is more straightforward when stating that the division of powers (as an element of the set of principles which constitute the separation of powers) is needed to avoid 'excessive concentrations of political power', see Jeremy Waldron, 'Separation of Powers in Thought and Practice' (2013) 54 *Boston College Law Review* 433, 466.

requirements after entering into force becomes part of the amended legal act, in this case, the constitution. Incorporation as a substantive relation between the original and new parts of the constitution can take place only after these previous steps. This substantive relation can be described based on more particularities. The most significant of these could be the 'dialogue' between the norms of the constitution being of different ages. If the new provisions can be interpreted in the light of the basic principles of the original document, and the examination of certain legal relations require taking into consideration both the 'old' and the 'new' provisions (contextual interpretation), then incorporation can be considered successful. On the contrary, if the new provisions cannot be interpreted in the light of the basic principles of the constitution or are even in explicit contradiction with its original norms, the incorporation is unsuccessful, from a substantive point of view.[96] Individual cases can underline both of the above considerations.

Incorporation as a subtest can be used in other ways: if the new provisions are in explicit contradiction with the original ones, the court has to present convincing arguments in order to not invalidate these. In the case incorporation can be supported by the existing 'dialogue' of the different set of norms of the constitution; the court needs compelling arguments for invalidating the new provisions of the constitution. In the case this 'dialogue' is not that intense, less compelling but legitimate arguments could also be acceptable.

In the Moldovan case there was no contradiction, rather a dialogue between the new and original provisions of the Constitution. This dialogue can be described as a somewhat loose relation based on the broad concept of the rule of law and the unspecified (parliamentary) form of government of the country – two of the basic principles which are connected to the challenged provisions of the Constitution. Therefore, the Court did not have to present arguments of special normative force in order to support the declaration of invalidity of the challenged provisions of the Constitution. In this regard we recall that one of the most interesting arguments expressed by the Court, namely, the challenged provisions were not 'validated' by successive modifications, are not relevant. Related practice of state organs is much more important when considering the effects of a challenged piece of legislation than the possible future amendments to it.[97]

The *third subtest* considers the presence of a 'constitutional-political question', whether the challenged provisions of the amendment can be considered

[96] See Roznai (n 14) 220: 'every interpretive effort must be made to reconcile amendments with protected rights and basic constitutional principles. This is the German method of … "interpretation in conformity with the Constitution". Using such an interpretive method, courts may choose the interpretation that is most compatible with the constitution without annulling the entire amendment. This doctrine is compatible with the foundational structuralist interpretation … This is, thus, another "line of defence" of the constitution that avoids the remedy of annulment. Annulment of a constitutional amendment by the courts should be the means of last resort.'

[97] One can add that future amendments could be relevant from a different aspect: if these were unsuccessful attempts to eliminate the conflicts caused by the original amendment.

as questions that are the subject of political considerations of the constitution-amending power. The political question doctrine has an impressive tradition in the jurisdiction of the courts of the United States. At the very basis of the doctrine there is the distinction between political questions and legal questions. Even individual cases could have different aspects; based on the Constitution, courts are entitled to decide only in terms of legal questions, the political ones belonging to the political branches of government. Some of the particularities of political questions can be even specified, eg the textual basis of the Constitution which accords competence to one of the political branches, lack of judicial standards, the need for taking into consideration an initial policy, the connection of the question before the Court with a former policy decision.[98] It is worth mentioning that the doctrine is not exclusively taken into consideration in the practice of US courts: not taking decisions in pure political questions by courts is a fundamental requirement also following from the principle of the separation of powers. A significant example is the margin of appreciation doctrine in the practice of the European Court of Human Rights which accords a great scope of action for states if there is no European consensus in the case. One explanation for this could be that the question at hand is a sensitive one from a political point of view.[99]

Political questions in the case of constitutional amendments can be handled based on a slightly different perspective. From a theoretical point of view, the constituent power is free in establishing the content (the substance) of the constitution if it takes into consideration the basic values and aspirations (including constitutional principles) shared in the given society. After the foundations of the constitutional system are fixed by the constituent power, the constitution-amending power is free in setting modifications to it as long as it respects the procedural and substantive norms related to amendments, set in the constitution. Within these boundaries, the constitution-amending power is free to take into consideration both legal and political purposes. When measuring such amending provisions, the standard for the court is the same: the explicit or implicit limitation clauses or the explicit or implicit norms of the basic structure of the constitution. If certain provisions of the amendment are in contradiction with these, the court can declare them invalid. However, in the case of amending provisions which are closer to political considerations, a more careful investigation is needed. These considerations in most of the cases have weaker connections with (explicit or implicit) eternity clauses and norms belonging to the basic structure of the constitution. That is why the constitution-amending power could allow wider leeway in these cases.[100] Accordingly, the court may reject such considerations after a longer

[98] For a detailed approach, see *Baker v Carr*, 369 US 186 (1962).

[99] See George Letsas, *A Theory of Interpretation of the European Convention on Human Rights* (Oxford University Press, 2007) 91.

[100] We can also mention here cases where political considerations of amendments contradict the basic norms of the constitution. Consider, for example, an amendment to the French Constitution which would alter the secular nature and unitary structure of the state.

period of time only in exceptional cases, with an explicit reference to the political nature of the examined provisions and their relation with the affected norms of the constitution.

The Moldovan case is a good example of a case where the Court failed to do so. One can argue that the decision on the form of government and political structure of the country is a purely political question. However, the Moldovan Constitution does not contain any provision which could acquire special normative force (normative priority) to the provisions establishing these structures. Consequently, the Court would have been required to explain explicitly and in a convincing manner the reasons for alternating the political decision of the constitution-amending power. The Court failed to do so. The argument expressed in the decision, namely, the political branches were not able to amend the problematic elements of the Constitution, is even more problematic from the perspective of 'political-constitutional' questions.

VI. Conclusion

Time matters. And it should matter in judicial review. And, moreover, the time aspect would prima facie seem especially important when the law under review is of a constitutional status and one that has been applied by governmental authorities and citizens – and relied upon – for many years. Of course, the lapse of time between an enactment of a constitutional amendment and its review by a court, even a long period of time, need not result in an absolute bar from judicial review. Especially when matters of great constitutional importance are involved, there is a great public interest in the judicial review, just as the doctrine of laches would not apply or be relaxed when there is an important public interest to be protected.[101]

How can time be incorporated during the process of judicial review of amendments? We suggest a three-tiered 'time sensitivity test' that centres on: (a) the effect of the decision on the separation of powers; (b) how the amendment has been incorporated into the constitutional text; and (c) the extent to which the challenged provisions can be considered as a 'constitutional-political question'. This test would bridge the gap between the time that has passed from the enactment of the amendment to the time it is reviewed by the court. Other judicial mechanisms may accompany this test, especially in the realm of judicial remedies. For instance, as the effect of invalidating a constitutional provision that was in place and applied for a long period, we agree with the Venice Commission that 'it is … important for such a decision to include unambiguous transitory provisions and set a precise time-limit for bringing lower-order norms and the functioning of state institutions

[101] Higgins (n 5) 189.

into harmony with the Constitution in force'.[102] This would manage to mitigate the effect of invalidating an 'old' constitutional provision.

Of course, this chapter does not attempt to provide a full theoretical and practical account of the time aspect of judicial review of amendments, but solely to open a debate on this important question which, in our opinion, has not been given sufficient attention in the literature.

[102] Venice Commission (n 3) para 39.

7

The Effect of Specialised Courts over Time

YIFAT ARAN AND MORAN OFIR

I. Introduction

Specialisation of the judiciary is often perceived as inevitable due to the ever-growing complexity of different fields of human and governmental action. It is also a prevalent direction to take for governments seeking to enhance a justice system that is inefficient and slow. Thus, it has been widely adopted, both in the United States and internationally. As noted by several scholars, however, support for specialisation is typically based on commonsense and intuition rather than actual data.[1]

As part of this general trend, organisations that promote economic growth, such as the Organisation for Economic Co-operation and Development (OECD) and the World Bank, have advocated for specialisation to improve the court system and to stimulate the rule of law. In 2010, Israel joined this movement when it established an Economic Division within the Tel Aviv District Court (hereinafter, EDTA). The EDTA is a relative pioneer outside the United States in

[1] See Lawrence Baum, *Specializing the Courts* (University of Chicago Press, 2011) 4, see also *ibid* at 210 ('To the extent that participants in the policy-making process think explicitly about how specialization might affect court outputs, they tend to act on the basis of folk theories that rest on common-sense notions of causality rather than on extensive and systematic analysis.'); Lawrence Baum, 'Probing the Effects of Judicial Specialization' (2009) 58 *Duke Law Journal* 1667, 1681 ('Debates over specific proposals for specialized courts and the general movement toward greater specialization in the courts reflect a belief that such effects exist. The empirical evidence on the impact of specialization, however, is limited.'); Nora Freeman Engstrom, 'A Dose of Reality for Specialized Courts: Lessons from the VICP' (2015) 163 *University of Pennsylvania Law Review* 1631, 1640 ('though the contention that judicial specialization expedites case resolution, elevates the quality of judicial decisionmaking, and breeds consistency is surly logical and intuitively appealing, it has, too often in the past, been accepted on faith.'); Chad M Oldfather, 'Judging, Expertise, and the Rule of Law' (2012) 89 *Washington University Law Review* 847; Douglas H Ginsburg and Joshua D Wright, 'Antitrust Courts: Specialists Versus Generalists' (2013) 36 *Fordham International Law Journal* 788, 794 ('there is still no empirical foundation for the proposition that specialist judges are more efficient than generalists in the production of judgments').

adopting a special-court model for corporate law as a means to further strengthen a developed market, and as such, the specialised division's success or failure might provide important guidance for creating policy in other countries. It also offers a particularly appealing case in which we provide a rare empirical description to assess the actual impact of a specialised economic court.

In this chapter, we develop a research design that reveals empirical evidence of whether or not – and in what ways – a specialised court offers litigants increased efficiency. We also examine the question of whether and how specialised courts stimulate private enforcement of corporate law.

Direct comparisons between generalist and specialised courts provide the best evidence about the effects of giving jurisdiction to specialised judges, and only a limited number of such studies exist.[2] Most of the empirical work on this field has been done on 'problem-solving courts' (eg, community courts, juvenile courts, domestic violence and so on), which have goals and characteristics distinct from those of the civil and commercial courts.[3] Therefore, those studies offer little relevant insight to the understanding of specialised economic courts.

This research aims to make a first step in filling the empirical gap in the literature regarding specialised economic courts, offering the first comprehensive empirical description of civil enforcement of corporate and securities law through class and derivative actions. The Israeli reform provides a remarkably apt and rich experimental situation because of the unusual comparisons it enables. As we shall discuss in more detail, the specialised court was established in only one district, while in others, general courts continued to hear economic matters. This permits subtle comparisons over time and between jurisdictions while controlling for substantive applicable law.[4] Our analysis makes two principal contributions.

First, we develop methodology to evaluate the outcomes of court specialisation. In this study, we use a quasi-experimental research design to evaluate the results of a specific court reform, the establishment of the EDTA. We compare court decisions under two time frames (before and after the reform) and in two judicial frameworks (the specialised court versus general district courts). The research was

[2] See Baum, 'Probing the Effects' (n 1) 1681; Carolina Arlota and Nuno M Garoupa, 'Do Specialized Courts Make a Difference? Evidence from Brazilian State Supreme Courts' (2016) *European Business Law Review.* Arlota and Garoupa used a criterion group research design to evaluate differences between decisions made by a non-specialised court *en banc* and those of a specialised court panel in constitutional review cases across Brazilian state supreme courts.

[3] Baum, Specializing the Courts' (n 1) 27.

[4] See Baum, 'Probing the Effects' (n 1) 1681–83. ('To a degree, the dearth of comparative studies reflects the difficulty of comparing generalist and specialist judges. Sometimes, a good point of reference for comparison does not exist … Still, meaningful comparisons are possible. In those instances in which jurisdiction is transferred from generalist to specialized courts, scholars can use records of both case outcomes and doctrinal positions to estimate the effects of the jurisdictional change, so long as they take into account possible changes in the composition of the cases that go to court …. Further, with appropriate controls for relevant differences, scholars can systematically compare generalist and specialized courts that hear the same kinds of cases.')

designed with high attention to potential threats to validity, including self-selection bias and censored data situations, and it incorporated a few novel techniques to overcome those challenges.

Second, by implementing this methodology, we provide evidence and insights about the differences between specialised and general courts. This analysis sheds light on the reform's results regarding the enhancement of civil enforcement and the improvement of efficiency of judicial decision-making.

The chapter proceeds as follows: first, we review the historical background of specialised courts. Then, we account for the elements of the research project: (1) discussing the Israeli case and its usefulness for this project in more detail; (2) describing data collection and coding; and (3) showing how the research design was created. Based on this unique design, we present our findings in terms of whether the specialised court, in this instance, delivered the results that specialisation advocates argue are the virtues of such reform: (a) encouraging civil collective actions to enforce corporate and securities laws, and (b) improving court efficiency. Finally, we conclude by outlining the lessons learned from the Israeli experience with implementing a specialised corporate court.

II. Historical Background

The good reputation of court specialisation rests in large part on the records of the Delaware Court of the Chancery, which has its historical roots in the principles of equity developed by the King's Chapel in feudal England. The Chapel evolved into a Court of Chancery, which provided judicial relief to those left remediless because of the procedural rigidity, corruption, and inadequate enforcement machinery of the common law courts. When the Delaware Court was established in 1792, it was remarkable only because it ran against the tide of judicial practice toward general courts.[5] Its successes came with the rise of modern corporations during the nineteenth century, as it evolved into a forum that deals almost exclusively with disputes regarding business organisations. Through its equitable doctrines and remedies, the Delaware Court of Chancery has become a superb forum for resolution of internal corporate controversies. Gradually, it has developed a large body of case law that significantly reduces the uncertainty of doing business as a Delaware corporation rather than being incorporated elsewhere. Delaware's dominance in the corporate world – and the success of its model – is reflected by the fact that more companies are incorporated in Delaware than any other state.[6]

[5] William T Quillen and Michael Hanrahan, 'A Short History of the Delaware Court of Chancery–1792–1992' in *Court of Chancery of the State of Delaware – 1792–1992*, a publication of the Bicentennial Commemoration Committee of the Historical Society for the Court of Chancery of the State of Delaware.

[6] There has been a long and often heated debate in the academic literature regarding the reasons for Delaware's dominance. The two basic contradictory theories are known as 'race to the top' and 'race to the bottom'. Race to the top proponents argue that Delaware offers a body of law that maximises

The idea that a business might look favourably on a city, region or state with courts that engender confidence in the judicial process contributed to a national and international movement toward specialised business courts.[7] Taking Delaware's Chancery Court as a role model, specialisation supporters argued that separation of the business docket from the rest of the docket would improve courts' efficiency, the quality of decision-making and the overall appeal of the court system.[8] In the United States, the movement for specialised courts arose in the 1990s[9] and led to the creation of business and commercial dockets within state trial courts in approximately half of the 50 states.[10] These 'business courts' assign specialist judges to manage and decide commercial and business cases.

The movement towards specialised commercial courts has since become global. Consequently, the World Bank's 2012 annual 'Doing Business' report stated: 'No fewer than 23 economies made it easier to enforce contracts by introducing or expanding specialized courts to deal with commercial cases.'[11] In 2010, Israel established the EDTA, distinctive from the other specialised business courts because it adjudicates only corporate and securities cases and not general commercial disputes.

III. The Israeli Case

The World Bank has recognised Israel as providing one of the highest levels of investor protection through the corporate governance mechanisms and disclosure

shareholder interests, therefore market forces push firms to incorporate there. Race to the bottom proponents claim that Delaware law provides directors and managers the most leeway to exploit firms at shareholders' expense, therefore managers prefer to incorporate the firm there. Lastly, others argue that Delaware is dominant simply because it got there first; Once Delaware took the lead, it made sense for new companies to incorporate there.

[7] This approach was partly affected by Roberta Romano's influential book *The Genius of American Corporate Law* in which she claimed that allowing Delaware to dominate national corporate law is not a problematic feature but rather an important virtue of American corporate law; See Roberta Romano, *The Genius of American Corporate Law* (AEI Studies in Regulation and Federalis, 1993); Roberta Romano, 'Law as a Product: Some Pieces of the Incorporation Puzzle' (1985) 1 *Journal of Law, Economics, & Organization* 225; See also Rafael La Porta, Florencio Lopez-de-Silanes, Cristian Pop-Eleches and Andrei Shleifer, 'Judicial Checks and Balances' (2004) 112(2) *Journal of Political Economy* 445, 470.

[8] See, eg, Rochelle C Dreyfuss, 'Forums of the Future: The Role of Specialized Courts in Resolving Business Disputes' (1995) 61 *Brooklyn Law Review* 1; and for a critical stance see Jeffrey W Stempel, 'Two Cheers for Specialization' (1995) 61 *Brooklyn Law Review* 67, 72.

[9] Rochelle C Dreyfuss, 'Specialized Adjudication' (1990) 9(1) *Brigham Young University Law Review* 377; Baum, Specializing the Courts (n 1); Ellen R Jordan, 'Specialized Courts: A Choice' (1981) 76 *Northwestern University Law Review* 745; Diane P Wood, 'Generalist Judges in a Specialized World' (1996) 50 *SMU Law Review* 1755; Lee Applebaum, 'Some Observations on Modern Business Courts and the Bar's Role in Their Development' (Business, Law Today, 2008 March/April).

[10] DJ Block et al, 'Business Courts: Towards a More Efficient Judiciary' (1997) 52 *Business Law* 947; Christopher R Drahozal, 'Business Courts and the Future of Arbitration' (2009) 10 *Cardozo Journal of Conflict Resolution* 491; See also Oldfather (n 1) (claiming that 'recent years have witnessed the beginning of a backlash against the increasing division of intellectual labor, such that it is appropriate to question the merits of judicial specialization').

[11] World Bank, 'Doing Business (Enforcing Contracts)' (2012) 3.

requirements set in its laws.[12] Like many less developed capital markets, however, Israel was experiencing challenges associated with protecting minority shareholders in an environment of concentrated ownership. Majority shareholders could and did act against the interests of minority shareholders (both small private investors and institutional investors). In 2010 approximately 20 business groups, nearly all of them family owned, controlled 160 publicly traded companies – with a 40 per cent share of the market.[13] Using pyramidal structures, those families were able to invest relatively small amounts of equity in a public company to buy control not only of the target but also of the many other public companies it controls. Through such conglomerates, ten privately owned business groups effectively controlled about 30 per cent of Israel's economy in 2010, which put Israel at the top of the OECD's chart of countries with the most concentrated markets.[14]

Empirical evidence documenting this profound problem was found in the high level of control premium in Israel. Control premium is the difference between the market value of shares and how much a buyer is willing to pay for those shares if they confer control over a company. According to empirical studies, the average control premium in Israel stood at about 27 per cent,[15] compared with four per cent in the United States and less than 10 per cent in most of the countries of Western Europe, North America and Southeast Asia. The high estimated premium in Israel presumably indicated the relative ease with which majority shareholders could take advantage of the companies under their control for the extraction of private benefits at the expense of public shareholders.[16]

To address the issue, the Israeli Security Authority already had its system of regulatory and criminal actions, which resulted in wrongdoers facing a potentially wide array of sanctions. Seeking a complementary mechanism to these administrative tools, it appointed a special committee to suggest additional remedies. Following the Delaware Court of the Chancery Model, the committee recommended establishing a specialised economic court in the hope that it would promote private enforcement in the form of investors' class actions and derivative suits.[17]

[12] As can be seen in Israel's ranking in the 'Doing Business' annual reports by the World Bank. The report examines investor protection laws in 183 economies, specifically focusing on publicly traded manufacturing companies. Israel was ranked among the top 10 every year between 2006 and 2014. In 2015, after the introduction of a new methodology, Israel was ranked 11.

[13] OECD, 'Corporate Governance in Israel' (2011) 15–16 (relying on data from the Bank of Israel).

[14] OECD, 'Israel: Review of the Financial System' (2011); Lucian A Bebchuk, 'Corporate Pyramids in the Israeli Economy: Problems and Policies: A Report Prepared for the Committee on Increasing Competitiveness in the Economy' (March 2012), available at: http://mof.gov.il/Lists/Competitiveness-Committee_4/Attachments/3/Opinion_2.pdf.

[15] Alexander Dyck and Luigi Zingales, 'Private Benefits of Control: An International Comparison' (2004) 59 *The Journal of Finance* 537; Ronen Barak and Beni Lauterbach, 'Estimating the Private Benefits of Control from Block Trades: Methodology and Evidence' (2011) 2 *International Journal of Corporate Governance* 59.

[16] Shmuel Cohen and Beni Lauterbach, 'Differences in Pay between Owner and Non-Owner CEOs: Evidence from Israel' (2008) 18 *Journal of Multinational Financial Management* 4.

[17] The other key recommendations were improvement of director independence, particularly regarding approval of related party transactions; improvement in the workings of auditing (or balance)

At the time, investors in public companies were rarely using class or derivative lawsuits to address violations of corporate and securities law.[18] By 2010, only one class action suit regarding corporate and securities law violations had reached a final decision in a judgment.[19] More than ten years elapsed between a claim's submission and the final result in appeal.[20] Sometimes, by the time the Supreme Court had reached its decision, economic circumstances had made it impossible to collect the damages awarded. In addition, with relatively few precedents set by the Supreme Court, lower court judges and corporate legal advisers had little case law on which to rely.

In recommending the creation of a specialised economic court, the committee was inspired by the US model,[21] which views private enforcement of corporate governance norms through litigation as an integral component of the legal system.[22] Thus, the committee emphasised private plaintiffs' relative advantages in comparison with the regulator in terms of access to information and the ability to petition for effective remedies ex-ante (such as injunction against merger) and ex-post (damages). Moreover, the committee referred to limited empirical evidence suggesting that private enforcement is more effective than public forms of enforcement in the field of corporate and securities laws.[23]

committees, pertaining to approval of company's financial statements; and improvement of corporate disclosure and accountability. See Goshen Committee Report, 'Corporate Governance Code in Israel' (Israel Securities Authority (ISA), 2006).

[18] In addition to the ability to bring a personal suit in any instance of a violation of rights, the Class Action Law – 2006 permits the filing of a class action suit in the name of a group of injured parties under certain circumstances. Moreover, the Companies Law allows a shareholder and a director to file a derivative suit in the name of the company (see section 194 of the Companies Law) (since 2011, a creditor can also file a derivative suit in respect of a prohibited distribution effected by the company, see section 204 of the Companies Law). Detailed information regarding the number of class actions and derivative suits between 2006 and 2010 will be presented in the following chapters.

[19] Application for Civil Appeal 8332/96 *Shemesh v Reichert*, Piskei Din 55 (5) 276, 289.

[20] *Ibid* (15 years); Civil Appeal 2718/09 *Gadish v Elscint Ltd* (05/28/12) (15 years).

[21] Review of the committee's rationales for establishing the EDTA are available in Hebrew: Zohar Goshen, 'Economic Court is a National Interest' (2009) 5 *The Attorney* 36 [in Hebrew] (2009) 36 5 זוהר גושן "בית משפט כלכלי הוא אינטרס לאומי" **עורך הדין**; Daniel Rimon and Tal Amir, 'Economic Court – From Delaware to Israel' (2010) 4/7 Corporations 3 [in Hebrew] (2010 נובמבר) 61, 3 4/ דניאל רימון וטל אמיר "בית-משפט כלכלי – מדלוור לישראל" **תאגידים** ז.

[22] J Maria Glover, 'The Structural Role of Private Enforcement Mechanism in Public Law' (2011) 53 *William & Mary Law Review* 1137 ('[American] system often relies heavily and explicitly on enforcement by private parties to achieve public regulatory objectives. Whereas European nations regulate the conduct of their citizens largely using ex ante regulations promulgated by a centralized bureaucracy.'); See also Deborah R Hensler et al, 'Class Action Dilemmas: Pursuing Public Goals for Private Gain' (RAND Corporation, 2000).

[23] Rafael La Porta, Florencio Lopez-de-Silanes and Andrei Shleifer cooperated with attorneys from 49 countries to assemble a database of rules and regulations governing securities issuance. Their aim was to detect the most important aspect of these laws. Their conclusion was provocative, arguing that capital markets benefit first and foremost from regulation based on private enforcement. In other words, a growth in the capital market requires a legal system in which private plaintiffs are easy to sue for violating corporate and securities laws. Moreover, the research indicates that such private enforcement is preferable to law enforcement by public authorities. Immediately after this paper was published, various reactions, including various attempts to show counter empirical evidences, emerged. The discussion of this question is still in progress, and it provides food for thought for policy crafters

In the same line of argument, relying, inter alia, on Isaac Ehrlich and Richard Posner's 'Economic Analysis of Legal Rulemaking',[24] the reform supporters maintained that governmental enforcement agencies, such as the ISA, are suitable to enforce clear prohibitions ('rules') rather than general criteria ('standards'). They argued that disclosure requirements under securities law are embedded with standards – primarily, the duty to discover all information that a *reasonable* shareholder would consider important in making an investment decision. Defining just what that content might be is a job more suited to a court (in an ex-post enforcement) than to the regulator (in an ex-ante regulation). Responding to the allegation that Israeli regulation sets overly restrictive rules and consequently disrupts economic activity, advocates argued that the regulator could not adopt less strict standards unless a competent court existed to shape the standards' content. By establishing a specialised court for corporate and securities law, advocates believed, these issues would be handled more effectively than in the general judicial system. A better division of labour between the court and the regulatory authority would lead to more principle-based regulations (standards), which by definition would be less restrictive and more flexible.

The reformers expected that by providing remedies to minority shareholders, the specialised court would enhance the quality of public company management, boost the Israeli capital market, and ultimately improve the economy. In the absence of such a court, they reasoned, a market characterised by diffused ownership would not develop. Controlling shareholders simply would lack an incentive to sell control in the market as long as they were able to extract private benefits from the company at the minority shareholders' expense.

Following approval of the committee's recommendation by the Knesset,[25] the Economic Division was established in the Tel Aviv District Court, composed of

worldwide. See, Rafael La Porta, Florencio Lopez-de-Silanes and Andrei Shleifer, 'What Works in Securities Laws?' (2006) 61 *The Journal of Finance* 1; Mathias M Siems, 'What does Not Work in Comparing Securities Laws: A Critique on La Porta Et Al.'s Methodology' (2005) 16(7) *International Company and Commercial Law Review* 300; Howell E Jackson and Mark J Roe, 'Public and Private Enforcement of Securities Laws: Resource-Based Evidence' (2009) 93 *Journal of Financial Economics* 207.

[24] Isaac Ehrlich and Richard A Posner, 'An Economic Analysis of Legal Rulemaking' (1974) 3(1) *The Journal of Legal Studies* 257. Ehrlich and Posner developed a theory of legal precision that distinguishes rules and standards and examines, in turn, the relative responsibility given to the legislature versus the judiciary. They argue that less precise laws (standards) would promote litigation and discourage pre-trial settlements. Therefore, one perquisite of the successful implementation of legal standards is an effective court that could shape the content of broad legal definitions. Precise laws reduce the uncertainty surrounding litigation and allow parties to better predict the litigation outcome and therefore raise the likelihood of pre-trial settlement. However, more precision comes at a cost, as it is difficult to agree on specific circumstances to be resolved under the rule. In particular, so they argued, as the debated issue's complexity rises, the legislature may increasingly delegate decision-making to the judiciary. For broader analytical discussion of the difference between 'rules' and 'standards', see Ronald M Dworkin, 'The Model of Rules' (1967) 35(1) *The University of Chicago Law Review* 14; Duncan Kennedy, 'Legal Formality' (1973) 2(2) *The Journal of Legal Studies* 351; Julius Stone, 'From Principles to Principles' (1981) 97 *Law Quarterly Review* 224; Anthony d'Amato, 'Legal Uncertainty' (1983) 71 *California Law Review* 1; Avinash K Dixit, Lawlessness and Economics: Alternative Modes of Governance (Princeton University Press, 2007).

[25] The recommendations were integrated into a government bill proposal (Courts Law [Amendment 59], 2010) that was brought before the Knesset Constitution, Law, and Justice Committee for

three tenured judges from its bench: Judge Khaled Kabub, Judge Ruth Ronen and Judge Danya Keret-Meir. These judges adjudicate only civil cases pertaining to securities and corporate law, as well as criminal cases pertaining to violations of securities legislation.[26] Elsewhere in the Israeli judicial system, 'general' district judges continue to adjudicate in corporate and securities cases along with the rest of the docket.[27] The reform was limited to the Tel Aviv district, the economic capital of Israel and the place where most of the corporate litigation takes place.

The Israeli case has the makings of a good testing ground for evaluating specialised economic courts using empirical research methods, in particular, because the reform was limited to a single jurisdiction. Empirical comparisons could be drawn not only between the EDTA and other courts, but between all courts before and after the reform. In addition, the EDTA was in effect a deliberate test of the idea that institutional modification of the court system could lead to better enforcement of corporate and securities law by the public. Therefore, it offers an interesting natural setting for this type of research.

IV. Empirical Analysis

A. Data Collection and Coding

To generate our initial list of cases, we used the largest and most comprehensive commercial legal database in Israel: Nevo,[28] supplementing it and checking it against a variety of other sources.[29] We searched for class actions regarding alleged violations of the Companies Act and the Securities Act,[30] as well as alleged violations of other laws that are in the jurisdiction of the EDTA, such as the Joint Investment Trust Act. The search's results were limited to the period of eight years between 15 December 2006 and 15 December 2014, four years before and four years after the reform.

deliberation and approval and then voted on by the Knesset. The Amendment was published on 27 July 2010, and went into effect on 15 December 2010.

[26] The Jurisdiction of the specialised division include all claims derived from Companies Law, Securities Law, Mutual Fund Law, Investment Advisors and Portfolio Managers Law, including Criminal Files, Civil Files, Class Actions, Derivatives Suites and Administrative Files.

[27] The general practice is that specific judges familiar with commercial adjudication address such matters; however, there is no guarantee regarding the presiding judge's identity.

[28] The database's online address is www.nevo.co.il.

[29] Four complementary sources of information were: (1) 'The class actions book' held by the administrator of the judicial system; (2) Lists of class and derivative actions assembled by the Civil Enforcement Department at the Israeli Securities Authority; (3) a census study regarding class actions in Israel between April 2006 and August 2012 that relied on computerised court records; and (4) citations of prior court decisions in the decisions that are already on our list. We also used two additional legal databases, Takdin and Pador. Their online addresses are www.takdin.co.il and www.pador.co.il/PadorGSA/PadorGoogleSearch/SearchResults.aspx.

[30] Those class actions are submitted in accordance to item five of the Second Schedule of the Class Action Law, 5766–2006.

Our extended list of cases was manually reviewed case by case, and irrelevant results were deleted (ie, cases that were not within the subject matter jurisdiction of the EDTA or that were not class or derivative actions). The final list was then manually coded for the following variables: court, judge, nature of the decision (whether procedural or substantive), dates (of submission, of certification, of settlement agreement, of final decision, of decision in appeal), parties' identity, attorneys' identity, cause of action, final outcome and so on.

In sum, we manually coded 242 cases composed of 146 cases adjudicated by the EDTA[31] and the remainder by general district courts before and after the reform. The groups of cases are composed of both class and derivative actions. **Table 1** summarises the generated data and key features distribution across the different groups.

Table 1 Summary of the Generated Data and Key Features Distribution across the Different Groups

	Court	EDTA After	Tel-Aviv Before	General After	General Before	Total
	All cases	146	37	32	27	**242**
Type	**Class actions**	82	30	8	14	**134**
	Derivatives	64	7	24	13	**108**
	Certified	13	2	8	1	**24**
	Settlements	15	0	11	2	**28**
	Appeals	23	3	16	3	**45**
Corporation	**Private**	21	7	15	8	**51**
	Public	118	25	14	16	**173**
Decision	**Procedural**	54	25	14	19	**112**
	Substantive	47	6	10	4	**67**
	Self-Selected Jurisdiction	82	10	10	16	**118**

We compared cases adjudicated by the EDTA with cases adjudicated by general district courts and with cases over time. Both groups' performances are evaluated in comparison with a baseline that was created by cases adjudicated four years prior to the reform (15 December 2006, until 15 December 2010). This comparison of the EDTA's decisions to both the general district courts' decisions from the same period and the general courts' decisions from prior to the reform, offers some unusual opportunities for our research design.

[31] Two cases that were submitted to the Tel Aviv District Court prior to the reform were assigned to judges that later became part of the economic division. The final decisions in those cases were given after the reform. Including those two cases, the number of court decisions given by the economic division is 148.

B. Research Design

We use *quasi-experimental research designs* to test the effects of specialised courts over time. Moreover, we use a before-and-after design to compare a pre-test (pre-reform) measurement with a post-test (post-reform) measurement to test the effectiveness of the intervention. In addition, since the Israeli case involved the creation of a specialised court in just one district while others remained the same, it provides a particularly good experimental situation which allows us to use a significant control group.

Two main threats must be addressed when designing the research: (1) Self-selection by subjects into treatment and control groups might cause the treatment group to consist of a unique composition of cases, different from those in the control group; (2) The treatment and control groups may be un-equivalent in their composition of cases due to other reasons, for example, if cases in the treatment group (ie, the specialised court) tend to involve different parties than those who litigate in other courts. These two sorts of differences between the treatment and control groups may lead to confounding outcomes and consequently to an erroneous interpretation that the treatment caused the effect witnessed in the results.

Given the local jurisdiction rules of civil procedure in Israel, it is reasonable to assume that some plaintiffs' attorneys were able to choose whether to file claims in the EDTA or elsewhere. According to section 3(a) of the Civil Procedure Regulations 5748–1984, jurisdiction over corporate and securities lawsuits is determined primarily by: (1) the defendant's place of residence or of business, and (2) the place of the act or omission that is the cause of the action.[32] Therefore, a problem of self-selection arises. To address this issue, we isolated a subgroup of cases in which the plaintiff's attorney did not have an opportunity to forum-shop, as only one district court held local jurisdiction over the case. While many plaintiffs can choose between two (and sometimes even more) courts that have acquired local jurisdiction, plaintiffs in a substantial number of cases have no choice but to bring the lawsuit before a specific court. Those cases, once identified, create a *random selection group* that enables us to estimate the degree of influence that self-selection on behalf of plaintiffs has over the court's performance.[33] In addition,

[32] Section 3(a) of the Civil Procedure Regulations 5748–1984, states as follows: 'Action not on real estate 3. (a) An action that is not entirely of a real estate character shall be submitted a Court, in the area of whose jurisdiction one of the following is located: (1) the defendant's place of residence or of business; (2) the place where the obligation was created; (3) the place where the obligation was meant or intended to be fulfilled; (4) the place where the property was delivered; (5) the place of the act or omission, which is the cause of the action. (a1) Notwithstanding the provisions of sub-Regulation (a), if the business of the plaintiff has several branches, and one of them is located in the jurisdiction in which the defendant's place of residence or of business is located, the action will be brought to the court in that jurisdiction. (b) If there are several defendants, the action may be brought in any court in which it can be brought against one of the defendants.'

[33] We evaluated the possibility of self-selection on behalf of the plaintiffs by collecting data on the address of the defendants' offices and residences. When the address was not stated within the statement

it should be noted that self-selection in this research is not just a confounder; it is also one of the dependent variables we measure as an indicator of the court's attractiveness. Therefore, the distribution of the self-selecting group among courts was also measured.

To control for the possibility that some other historical event or reform affected the change in the dependent variable, we coded possible historical causes for each case, such as influences of recent amendments of the Companies Act (amendment 16-20), the 2008 global financial crisis, recent changes to the regulatory environment affecting the equity market in Israel, and the effect of the US Supreme Court decision in *Morrison v National Australia Bank*.[34] To control for the possibility that changes in the dependent variable were due to normal developmental processes (maturation), we compared our results to data regarding class action patterns in Israel in roughly the same time frame (April 2006 to August 2012). We also used statistical regression to control for variations among groups in terms of type of action (class vs derivative) and type of defendant corporation (privately vs publicly listed).

C. Research Hypotheses

Several hypotheses have provided the foundation for the move toward specialised economic courts in Israel, the United States and elsewhere. One of them is the notion that specialised courts will encourage judicial enforcement of corporate and securities law through civil class and derivative actions as an alternative or supplemental means to administrative enforcement. Researchers have also argued that expert judges are more efficient because they are familiar with the relevant issues and can adjust the procedure to the peculiarities of financial cases. Thus, specialisation expedites case resolution. Based on the Israeli data, we test these hypotheses using a variety of methods to provide nuanced results.

i. Hypothesis 1: Specialised Courts Encourage Civil Enforcement throughout Class Actions and Derivative Suits

Using the Israeli case as our database, we conducted empirical analysis to analyse the impact of specialisation in two areas: (1) The effect on litigation rates; and

of claim, we searched the registered address of the defendant corporation in the Israeli Registrar of Companies website (http://havarot.justice.gov.il/). To decide to which jurisdiction district each address belongs to, we used the 'map of local jurisdiction' available at the legal commercial database 'Nevo' website (www.nevo.co.il/Nafa.aspx). To validate our assessment regarding self-selection, a sample of ~10% (24 court decisions) was coded by a civil procedure expert, and his results were concordant with ours in 83% of the cases.

[34] *Morrison v National Australia Bank*, 561 US 247 (2010).

(2) when plaintiffs were able to choose where to submit the claim, did they prefer the specialised division to the general courts?

a. The Effect on Litigation Rates

Prior to reform, there were 16 class actions and derivative suits per year nationwide, and 57.8 per cent of them were adjudicated in the Tel Aviv District Court. After the reform, class and derivative actions nationwide increased to 44.5 cases annually, and the EDTA received 82 per cent of those cases. Notably, there was no apparent change in the number of cases adjudicated outside Tel Aviv before and after the reform, suggesting that the existence of a specialised court had an impact.

However, deeper analysis of litigation rates by year demonstrates a consistent linear growth in the number of cases after December 2006. **Figure 1** shows the increase in nationwide corporate and securities litigation rates, from eight cases filed in 2007 to 46 cases filed in 2014,[35] and the strong linear pattern of growth in the number of cases ($R^2 = 0.87$).

Figure 1 Nationwide Corporate and Securities Litigation Rates by Year

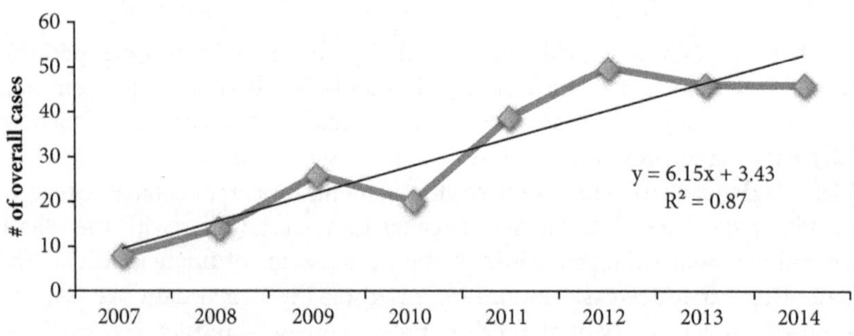

Moreover, as demonstrated in **Figure 2**, a comparison of the pattern of growth in the number of corporate and securities class actions and derivative suits with the overall class action cases (including consumer, antitrust, environmental, employment, etc) submitted in Israel by year[36] demonstrates that the increase we observed in class actions and derivative suits was in consonance with the overall growth in collective litigation in Israel in recent years. We conclude that the increase in litigation rates after the reform is likely a product of a maturation process in the field of private collective action, and we cannot relate it to the creation of the specialised

[35] Note that because we rely on public data and not court records, it is possible that a few of the lawsuits that were submitted in 2014 have not yet been published (at the time of writing).

[36] The number of total class actions in Israel are drawn from Alon Klement et al, *Class Actions in Israel – an Empirical Perspective* (Israeli Courts Research Division, 2014) [in Hebrew] available at http://elyon1.court.gov.il/heb/Research%20Division/doc/25122014.pdf.

court. Notably, this result serves as a reminder of the importance of time series and 'around the threshold' analysis. If we were to look only at the aggregated number of cases before and after the reform, our conclusion could have wrongly asserted a causal relation between the reform and the increase in the number of cases.

Figure 2 Comparison in Litigation Rates between Corporate and Securities Litigation (Left Y-axis) and Overall Class Actions (Right Y-axis) between 2007 and 2012

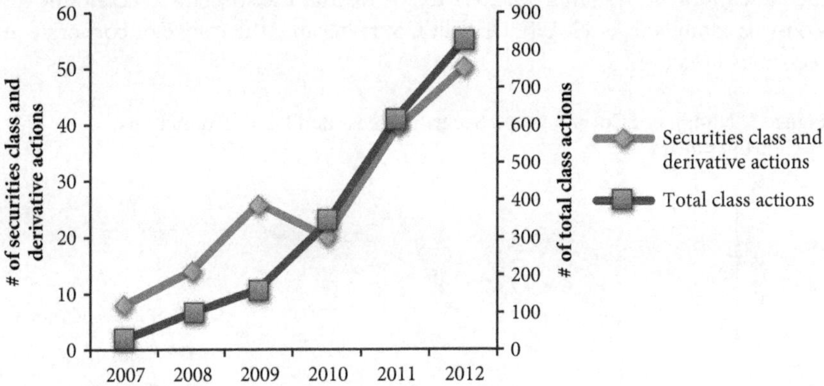

It could also be argued that this increased number of cases is related to historical events such as the world financial crisis of 2008, decentralisation of the cellular and food industry markets and recent amendments to the Companies Law of 1999, among others. However, these historical explanations explained only 27 per cent of cases submitted after the reform, whereas the increase of litigation after the reform is substantially higher.[37] Therefore, maturation seems to provide a better explanation to the pattern of increase in litigation rates.

[37] To account for the possibility that other reasons provide better explanations for the growth in the aggregated number of securities litigations, for each case in our list, alternative explanations were coded as fitting or not. The alternative explanations that were tested are as follows: (1) The world financial crisis of 2008 led to a domino effect that resulted in a liquidity crisis and eventually to an insolvency problem in some Israeli corporate conglomerates that were highly leveraged. This insolvency crisis led, in turn, to litigation regarding alleged liability of the controlling shareholders and the officers in those conglomerates. We found eight cases related to this explanation prior to the reform and 18 cases afterwards; (2) Opening of markets to competition – in recent years, changes in the regulatory environment of the cellular market and a growing competition in the food industry, led some Israeli conglomerates to a liquidity crisis that resulted in litigation. We found four cases related to this explanation prior to the reform and 14 afterwards; and (3) Recent amendments to the Companies Law that have given greater power to public shareholders and others and therefore created more causes of action. We found 11 cases related to this explanation all after the reform; (4) US Supreme Court decision in *Morrison v National Australia Bank* (n 34) reduced the extraterritorial reach of the anti-fraud provisions of the US Securities Laws. Consequently, foreign class members are more likely to be excluded in US global securities class actions, and therefore Israeli shareholders of dual companies that are listed in both the United States and Israel are more likely to seek remedy in Israeli courts. We found four cases related to this explanation, all after the reform.

b. Preference for the Specialised Court

While our results fail to support a link between specialisation and encouragement of civil actions, they do suggest that plaintiffs prefer the specialised forum. We note a major gap in litigation rates between the EDTA and all other district courts after the reform, which suggests that the reform changed the distribution of cases, if not their number. **Figure 3** presents securities litigation rates in the various courts of the Israeli judicial system from 2007 to 2014. After the specialised economic division was established, Tel Aviv District Court became the centre of corporate and securities litigation.[38]

Figure 3 Number of Corporate and Securities Class and Derivative Actions by Court per Year

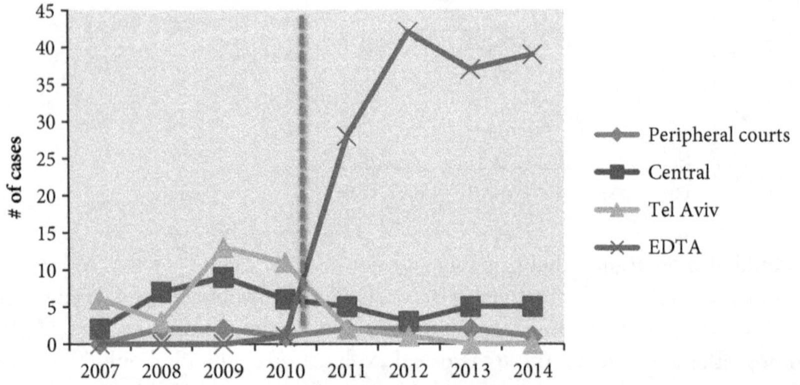

We found support for this observation in the control group of self-selected plaintiffs, those who enjoyed the opportunity to select a forum. As shown in **Figure 4A**, the most popular court prior to the reform was the Central District Court. After the reform, this rush from Tel Aviv to the Central District Court ended. Strikingly, **Figure 4B** shows that after the reform, 90 per cent of the 92 self-selected plaintiffs chose to submit their claims to the EDTA. This result supports the hypothesis that the specialised court is the preferable tribunal in the plaintiffs' eyes.

What could explain this shift in forum selection? We suggest three competing explanations. First, the specialisation of the EDTA makes it more attractive for filing corporate and securities claims – just as one who suffers from cardiovascular

[38] The results may also suggest that the minor decrease in the number of cases that were submitted in both the Tel Aviv District Court and the Central District Court in 2010 is related to plaintiffs' expectation to the establishment of the economic division and their willingness to delay the submission of their claims until then. The Courts Law Proposal (amendment no 59) Jurisdiction for Economic Affairs (5770-2010) was enacted in July 2010 declaring that the economic division would start to operate on 15 December 2010. In accordance, the decrease in the number of cases that were submitted in 2010 is a result of an exceptionally low last quarter in 2010.

Figure 4 Pie Charts of the Selected Forum by Plaintiffs with Ability to Forum Shop A) Prior to the Reform and B) After the Reform

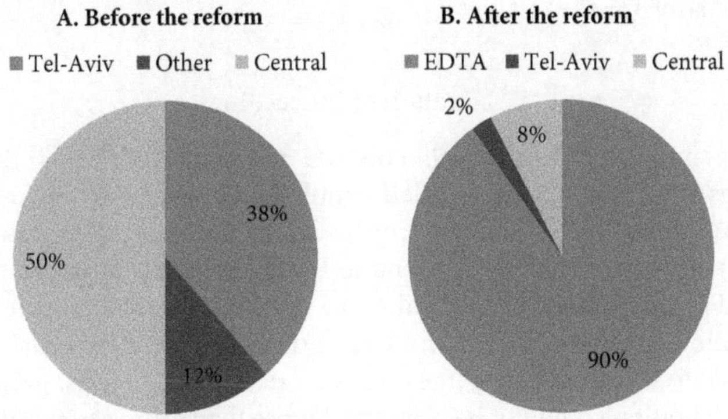

disease would visit a cardiologist and not a primary care physician. Second, the Tel Aviv District Court is the default option for plaintiffs' attorneys, as it is the closest district court to the offices of most of these attorneys. Prior to the reform, many attorneys were willing to make the journey to the Central District Court because it offered quicker proceedings. Following the reform, this advantage was matched by the EDTA (as we shall see below). Third, the EDTA provides greater certainty in terms of the identity of the adjudicating judge. While in submitting an action to a general court there is no, or at most limited, certainty with regard to the judges who will hear the case, when submitting it to the EDTA the plaintiff enjoys the benefit of having only one of three possibilities.

We conclude that while the specialised court did not lead to significantly greater litigation rates, the Knesset's goal of concentrating most of the securities litigation in one court[39] has been achieved. Further research is needed to clarify the role of specialisation in forum selection and to help determine what attracts the litigants: the judges' proficiency, the rapid proceedings, or certainty regarding the judge's identity.

ii. Hypothesis II: Specialised Courts Improve Efficiency

A key rationale for the increasing US and international move towards specialised courts has been the expectation that familiarity with the types of cases and legal issues will shorten the duration of proceedings. Another expectation is that the specialised court will encourage settlements (especially at the early stages of the litigation) due to superior judges' authoritativeness, case management and the certainty as to the applicable case law. Using the EDTA example, we employed

[39] The Courts Law Proposal (amendment no 59) Jurisdiction for Economic Affairs (5770-2010) states: 'The logic of the arrangements suggested in this law proposal is to create a permanent, tight infrastructure of an efficient and professional *judicial system to discuss most of the matters set in economic law*.'

several subtle quantitative tools, including the novel application of hazard functions, more typically used to examine lifespan of patients or machinery, to put these expectations to the test.

a. Duration of Proceedings

An overly simplistic analysis will compare the distributions of duration of proceedings – from submitting the claim until final decision – in the four comparison groups (specialised and general courts before and after the reform, as shown in **Table 2**). This type of analysis is bound to fail, however, as the cases after the reform had less time to be heard, and many of them have not yet reached a final decision. In other words, cases in the 'after' groups did not have equal chances to reach a final decision. Moreover, the cases that reached final decision in the 'after' groups by definition could not be prolonged more than four years.

Table 2 Differences Among Groups in Percentage of Open Cases and Duration to Final Decision

		Duration to Final Decision (days)		**% of Open Cases**
		Average	**SD**	
General (outside of Tel Aviv)	**Before**	673.4	238.6	29.63%
General (outside of EDTA)	**After**	404.3	771.1	53.12%
Tel Aviv	**Before**	1189.4	265.8	29.73%
EDTA	**After**	374.1	529.3	57.53%

To better handle these types of data, we employed a censored time series analysis by assessing hazard functions for each of the four groups. In our analysis, better efficiency is manifested by shorter lifetime of the lawsuit, whereas in other usages of hazard functions such as in clinical trials and machinery testing, researchers typically hope for longer lifetime. **Figure 5** presents the hazard function for the four groups.

The survival analysis presents a complex picture regarding the relationship between court specialisation and court efficiency. Thus, it is clear from **Figure 5** that cases adjudicated in the Tel Aviv District Court prior to the reform had a longer survival rate, in other words, longer duration of proceedings, compared with that of all other groups. Significance analysis[40] determined that this longer lifetime is significant compared with all other groups. **Figure 5** also shows a

[40] P-values were calculated using log-rank test to compare the survival distributions of two samples.

minor but not significant improvement in the efficiency of general courts after the reform compared with the prior condition. Finally, this figure shows that both the specialised division and general courts after the reform have roughly the same case lifetime distribution. Whereas the median lifetime of cases in the Tel Aviv District Court before the reform was 1,601 days (4.39 years), after the establishment of the EDTA, the median was reduced to 634 days (1.74 years). In the general courts before the reform, the median lifetime was 800 days (2.19 years), reduced to 589 days (1.62 years) after the reform's implementation.

Figure 5 Kaplan-Meier Plot for the Duration of Proceedings in the Different Groups

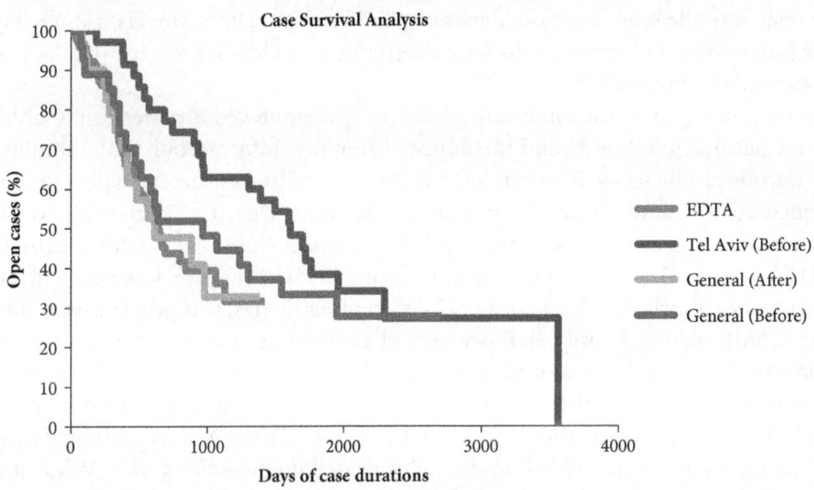

Thus, prima facie, our analysis does not support the hypothesis that specialised courts are necessarily more efficient than general courts. The analysis shows no significant difference in the duration of cases between the EDTA and other district courts following the reform (p-value = 0.9148). Furthermore, our analysis revealed that cases in the EDTA that belonged to the self-selection group (ie, when plaintiffs were able to choose where to file the claim) demonstrated significantly shorter durations (p-value = 0.0120). This finding suggests that improved efficiency could be affected more by lawyers' efforts to expedite the resolution of cases and less by the institutional design of the court itself (specialised or general). Hence, as the theory predicts, self-selection by plaintiffs turns out to be a significant factor in the evaluation of court reform.

Does this imply that specialisation brings no advantages in terms of efficiency? Not entirely. Our analysis also suggests that before the reform, Tel Aviv District Court suffered from poor efficiency, and the reform has resolved this situation. However, other courts have fared better in recent years, too. This systematic improvement can be explained in two ways: (1) Streamlining processes in the court system improved the performance of the system as a whole; and (2) judiciary 'race to the top' causes general courts to try to match the EDTA's pace of proceedings.[41] Although officials in the court system deny such a competition exists, lawyers believe otherwise.[42] Hence, as the theory regarding Delaware Court of Chancery alludes, amelioration of one jurisdiction might lead to a ripple effect that positively transforms the entire system.

Continuing our empirical analysis, we looked for possible confounding factors in the composition of the courts' dockets. Our empirical analysis shows that the lifetime of cases regarding private corporations, which are usually less complicated, is significantly shorter than is the equivalent regarding public corporations (p-value = 0.0013), as presented in **Figure 6A**. Another confounding factor may be related to the type of action: derivative suit or class action. Our data show that the proceedings in derivative suits are shorter than in class actions (p = 0.0050), as presented in **Figure 6B**.[43]

To control for these confounding factors, we employed a regression analysis fit for hazard functions,[44] and the adjusted results reinforced our initial findings of improved efficiency in terms of the cases' duration in the specialised court compared to the duration of cases in the Tel Aviv District Court prior to the reform (p-value of cases' lifetime of Tel Aviv District Court before the reform vs. EDTA = 0.0146). More importantly, while in Tel Aviv, only 15.3 per cent of the cases concern privately held corporations, in other courts, this proportion is 39.0 per cent. In addition, only 38.8 per cent of cases in Tel Aviv are derivative suits compared with 62.7 per cent in other courts. This difference in docket composition may explain the better performance observed in general courts both before and after the reform. In other words, the *EDTA adjudicates more complicated cases at the same pace as the general courts do while handling simpler docket*. While it is easy to miss this nuance, we conclude that specialisation allows the court to cope more efficiently with complex cases, and in turn, promotes the overall efficiency of the bench.

[41] For the American origin of the 'race to the top' theory see nn 6 and 7 above. See also Daniel M Klerman and Greg Reilly, 'Forum Selling' (2016) 89(2) *South California Law Review* 241. (Discussing incentives for district courts to attract litigation and the potential effect of these incentives on procedural and substantive decisions.)

[42] This debate on possible competition is reflected, for example, in an interview held with the former president of the Tel Aviv District Court, Judge Devora Berliner, and published in the Bar Association Journal issue from November 2010. Berliner was asked: 'In the eyes of the lawyers, the Central District Court created a healthy competition in a sense that "envy should multiply wisdom." Today a lawyer knows that when he or she comes to the Central District Court, within two years, the case is likely to end, perhaps even within a year and a half. Can you say that today the Tel Aviv District Court can dispense with a civil case in the same amount of time?' Judge Berliner's reply was: 'The very establishment of the Central District Court reflects the understanding of the court system that the District Court of Tel Aviv has worked over the years under an impossible workload … But I do not see a competition; we are not a factory. I hope that cases that begin today will be resolved within two years. I will not say that it is 100% guaranteed. I'm cautious, but we are really moving toward it …' (*Hapraklit*, November 2010, interview held by Adv Ron Dror).

[43] To reduce variability, the analysis for both type of corporation and type of action is based only on cases adjudicated after the reform. The same analyses for all cases returns highly significant results (Type of corporation p = 1.6e-5; Type of action p = 2.6e-6).

[44] Utilising Cox proportional hazards regression method.

Figure 6 A. Kaplan Meier Plot of the Duration of Cases for Privately vs. Publicly Listed corporations. B. Kaplan Meier Plot of the Duration of Cases for Class Action vs. Derivative Action

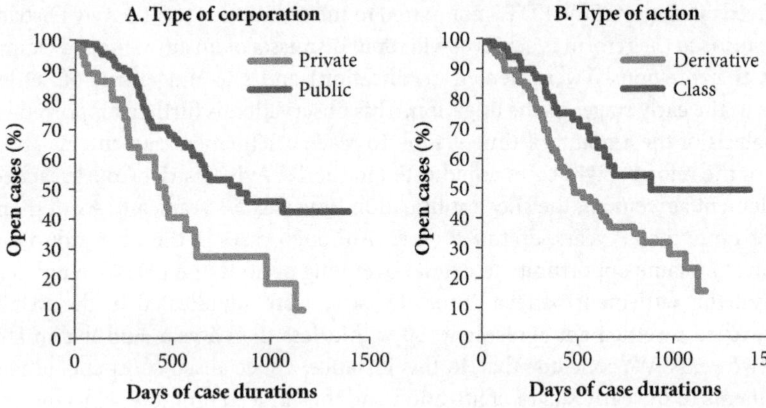

b. Rates of Dismissals and Settlements

A court that sets a high bar for the plaintiffs in the early stages of litigation will likely be more efficient than a court that allows easier access to claimants (however this result might raise concerns as to the justice system's role in offering remedies). Another outcome of specialisation might be a greater percentage of settlements. We therefore compared the end results of cases adjudicated by the different courts. **Table 3** presents the final dispositions of cases in the different groups.

Table 3 Comparison of the Final Disposition of Cases Adjudicated by the Different Courts

	General		Tel Aviv	
Result	**Before**	**After**	**Before**	**EDTA**
Dismissed without prejudice	26.1%	26.7%	26.9%	25.7%
Certification denied	17.4%	20.0%	30.8%	22.9%
Parties settled prior certification	21.7%	33.3%	**15.4%**	**22.9%**
Dismissed on summary judgment	4.3%	6.7%	**3.8%**	**14.3%**
Certification was granted (pending final decision)	8.7%	0.0%	0.0%	8.6%
Parties settled post certification	4.3%	0.0%	11.5%	2.9%
Stay of proceedings	8.7%	6.7%	3.8%	1.4%
Parties settled on appeal	8.7%	0.0%	3.8%	1.4%
Claim was upheld	0.0%	6.7%	3.8%	0.0%

Our data show that, compared to the pre-reform Tel Aviv District Court, there was a rise in the number of settlements prior to certification under the EDTA, alongside an increase in dismissals on summary judgment. This result suggests that the improved efficiency of the EDTA compared to the situation in the Tel Aviv District Court prior to the reform is achieved via rapid dismissal of unsubstantiated claims (cases that previously were denied certification) and encouragement of settlements in the early stages of the litigation. This observation is further supported by an analysis of the amount of time it took to reach a settlement agreement. Thus, prior to the reform, eight cases adjudicated in the Tel Aviv District Court reached a settlement agreement; the shortest litigation time was 2.3 years, and six of them ran for more than 3 years, up to 8.8 years. Although cases in the after-group did not have the same opportunity to extend over long periods, the EDTA completely changed the settlement pattern. Thus, 18 cases were adjudicated in the EDTA and reached a settlement agreement, 10 within less than a year, and five in less than two years. We conclude that, in this instance, a specialised court encourages settlements in the early stages of litigation and thus indeed promotes efficiency.

Did this efficiency come at the expense of justice? In other words, does the impact of efficiency work against plaintiffs by throwing out more of their lawsuits? As the data in Table 3 shows, about the same percentage of cases were dismissed without prejudice (26.9 per cent v. 25.7 per cent), and more cases were certified (uncertified dropped from 30.8 per cent to 22.9 per cent) after the reform. Therefore, we assess that cases that used to be settled only after many years of litigation (post-certification) are now being settled relatively quickly (pre-certification), and the rise in the number of summary judgments is coming mostly from the cases that used to be uncertified: the same result after more litigation time.

In summary, the empirical analysis revealed the complexity of the relationships between specialisation and court performance. Thus, we established that the specialised division was as efficient as general courts even though it adjudicated the more complex cases typical of the economic capital of Israel. However, this improved efficiency is not completely independent of self-selection effect (ie, plaintiffs who choose to go to the specialised division are also more likely to resolve their dispute early on). The efficiency attribute of the specialised division is achieved by both prompt dismissal of unsubstantiated claims and the encouragement of settlements in the early stages of litigation. Thus, efficiency does not come at the expense of access to justice, with roughly the same percentage (65.3 per cent vs. 64.3 per cent) ending in a negative outcome for plaintiffs.

V. Conclusion

In our research on Tel Aviv's Economic Division, we provide a rare empirical account of the outcomes of a court specialisation reform and of the interaction between court specialisation and private enforcement. The results of this research shed light not only on the implications of specialisation, but also, and by no means less importantly, on the challenges and potential stumbling blocks of empirical evaluation in this field. The Israeli case offers an unusual opportunity to perform meaningful comparisons of both transformation within the same jurisdiction from generalist to specialised court and across jurisdictions between generalist and specialised courts that hear similar cases within the same time frame.

As we have described in some detail throughout the previous section, this methodology provided a nuanced and sometimes complex look at this Court's first four years. By utilising quasi-experiment research design, we revealed the transformation that occurred in civil enforcement of corporate and securities laws throughout the specialised division's first term.

Although the reform did not lead to the initiation of a greater number of lawsuits, it did cause a major shift in forum selection, suggesting that plaintiffs prefer the specialised forum. Moreover, although increased efficiency was not exclusively related to specialisation, further analysis suggests that the specialised division is more capable of managing a particularly time-consuming docket. This efficiency is likely to be the result of two factors: (1) better case management which leads to prompt dismissal of unsubstantiated claims and the facilitation of settlements in the early stages of litigation; and (2) self-selection effect (ie plaintiffs who choose to go to the specialised division are also more likely to resolve their dispute early on). Notably, the efficiency does not worsen the outcomes for plaintiffs.

The establishment of the Economic Division has served as a major step forward in bolstering the enforcement of corporate governance in Israel. In their initial term, the division's judges formulated important precedents in substantive cases involving the relationship between majority and minority shareholders in companies and fashioned unique procedural rules that help manage these cases. No doubt this has been an exciting transformative period of great import for the future and strength of the Israeli financial market. Our hope is that the insights from the Israeli experience will provide guidance for constructing judicial policy in other countries and will contribute to the global discussion on the relationships between legal institutions, investor protection and financial growth.

PART III

Legislators and Time

8

Temporary Legislation as a Mechanism for Reaching Consensus. A Critical Analysis in the Absence of *Ex Post* Evaluation

ENRICO ALBANESI

I. Introduction

Several recent scholarly works have analysed temporary legislation,[1] among its other functions, as a mechanism for reaching consensus.

The aim of this chapter is to test the validity of such an argument (which has been tested by scholars especially with regards to the US,[2] the UK,[3] the Netherlands[4] and Israel[5]) with reference to the Italian legal system, by examining the main piece of temporary legislation which is used in the Italian legal order, namely the *decreto-legge*.

[1] See, eg: Ittai Bar-Siman-Tov, 'The Lives and Times of Temporary Legislation and Sunset Clauses' (2018) 66 *The American Journal of Comparative Law* 453; Ittai Bar-Siman-Tov, 'Temporary Legislation, Better Regulation, and Experimentalist Governance: An Empirical Study' (2018) 12 *Regulation & Governance* 192; Antonios E Kouroutakis, *The Constitutional Value of Sunset Clauses. An Historical and Normative Analysis* (Routledge, 2017); Tom Ginsburg, Jonathan S Masur and Richard H McAdams, 'Libertarian Paternalism, Path Dependence and Temporary Law' (2014) 81 *The University of Chicago Law Review* 291; Sofia Ranchordás, *Constitutional Sunsets and Experimental Legislation. A Comparative Perspective* (Edward Elgar, 2014); Frank Fagan, *Law and the Limits of Government. Temporary versus Permanent Legislation* (Edward Elgar, 2013); Jacob E Gersen, 'Temporary Legislation' (2007) 74 *University of Chicago Law Review* 247. See also the special issue (2015) 3 *The Theory and Practice of Legislation*, guest-edited by Sofia Ranchordás. From a drafter's perspective, see also Helen Xanthaki, *Drafting Legislation. Art and Technology of Rules for Regulation* (Hart Publishing, 2014) 188–190.

[2] See Forrest Maltzman and Charles R Shipan, 'Continuity, and the Evolution of the Law' (2008) 52 *American Journal of Political Science* 263; Thomas W Merrill, 'Showcase Panel IV: A Federal Sunset Law. The Federalist Society 2011 National Lawyers Convention' (2012) 16 *Texas Review of Law and Politics* 342–343 and Ginsburg, Masur and McAdams (n 1) 297–98.

[3] With regard to the UK (in a comparative perspective with the US), see Kouroutakis (n 1) 92–93.

[4] With regard to the Netherlands (in a comparative perspective with the US), see Ranchordás (n 1) 66–68.

[5] With regard to Israel, see Bar-Siman-Tov, 'Temporary Legislation' (n 1) 205–206.

This analysis is particularly interesting because the Italian *decreto-legge* deeply differs from the main temporary tool which has been primarily studied by scholars so far: sunset clauses. The key difference is that the *decreto-legge* as such, under the Italian Constitution, is a *provisional* (thus, not only temporary), *short-lasting* (it only lasts 60 days) and *emergency* (thus, not routine) source of law. However, it is *de facto* used routinely nowadays.

The first hypothesis which I will try to demonstrate here is that the large number of *decreti-legge*, which are used in Italy as a *routine* source of law nowadays, derives from the fact that it is a temporary piece of legislation. This type of legislation has an important role in the Italian system. The demonstration of this hypothesis will prove, with regards to the Italian case, the validity of the argument tested by scholars with respect to other jurisdictions, ie, temporary legislation is a mechanism which can be used to reach consensus.

However, a critical analysis of the (mis)use of the *decreto-legge* will follow, by highlighting in terms of the quality of legislation the negative outcomes which result from the use of a *provisional*, *short-lasting* and *emergency* decree as a tool to reach consensus on *routine* legislation in Italy.

In fact, the second hypothesis which will be tested in the chapter is that the use, ie to reach consensus, of such a tool as the *decreto-legge*, has led to negative outcomes in terms of the quality of legislation in Italy. Moreover, the fact that the *decreto-legge* lacks (inevitably, due to its short time duration) those elements which should characterise sunset clauses (namely, an *ex post* evaluation) deprives the *decreto-legge* of such added value, once again in terms of quality of legislation. In other words, although it is temporary legislation – and is used to reach consensus for this reason – the Italian *decreto-legge*, due to its short time duration, cannot be considered, as such, experimental legislation, viz. legislation enacted for a limited period of time in order to examine if a particular legislative measure will effectively achieve certain goals.[6]

The chapter proceeds as follows. First, I will outline the main features of sunset clauses, namely their temporary character and *ex post* evaluation (section II.A) as it is the piece of temporary legislation largely studied by scholars. Following this, I will give an overview of the role that temporary legislation plays as a tool to reach consensus (section II.B), both as outlined by scholars.

Second, the main legal and factual features of the Italian *decreto-legge* will be described – on the one hand, the principles of the Italian Constitution which establish the *decreto-legge* as a provisional emergency decree (section III.A), and on the other hand, its transformation as a decree which is used routinely nowadays (section III.B).

[6] See the distinction between temporary legislation and experimental legislation (the latter being a subgroup or a certain type of the former) established by Bar-Siman-Tov, 'Temporary Legislation' (n 1) 194.

Third, it will be explained why, in the Italian context, the *decreto-legge*, as a provisional emergency decree, has been used as an invaluable tool to reach consensus (section IV).

Fourth, some controversial issues concerning the provisional nature of the *decreto-legge* will be analysed, especially its negative outcomes in regards to the quality of legislation. The (partial) solutions given by the *Corte costituzionale* (the Italian Constitutional Court) to these issues will also be described. Further, it will be underlined that the *decreto-legge* as such (inevitably, due to its short time duration) lacks *ex post* evaluation, which is the added value of sunset clauses from the perspective of quality of legislation (section V).

Finally, the hypotheses outlined above will be tested in light of the analysis carried out in the previous sections (Conclusions).

II. The Main Features of Sunset Clauses and the Use of Temporary Legislation as a Tool to Reach Consensus

A. The Main Features of Sunset Clauses

The main features of sunset clauses, as they have been outlined by scholars, are their temporary character and the presence of *ex post* evaluation. According to Sofia Ranchordás, temporary character means that sunset clauses do not aim at continuity; rather they determine the expiration of a law after a fixed period, unless there are substantial reasons to believe that the former should be extended for a determined period. As for the duration period, Ranchordás does not set a term. There is no limit for this temporality. However, the sunset clause examples from the Netherlands and the US given in her work last between one and six years[7] – a significant length in order to examine whether a particular legislative measure will effectively achieve certain goals.

The presence of *ex post* evaluation means that the effects of the sunset disposition should be assessed in order to verify whether the objective for which it was enacted has been achieved. In other words, the renewal of sunset clauses should be decided on the grounds of an evaluation report.[8] In practice, this does not always happen.

This second feature has been especially outlined by those authors who analysed sunset clauses from the perspective of better regulation, such as Helen Xanthaki[9] and Ittai Bar-Siman-Tov,[10] who both point out that sunset clauses are of little use

[7] See Ranchordás (n 1) 32–33.
[8] Ranchordás (n 1) 33.
[9] See Xanthaki (n 1) 189.
[10] See Bar-Siman-Tov, 'Temporary Legislation' (n 1) 196.

if they are not paired with evaluation or learning mechanisms that enable more informed decisions to be taken at the end of the initial period.

B. The Use of Temporary Legislation as a Tool to Reach Consensus

The key to understanding the pivotal role of temporary legislation as a tool to reach consensus lies especially in its information effects. As Jacob Gersen has pointed out, temporary legislation involves multiple stages of legislative action to sustain a particular public policy; thus, this form of 'staged decision-making' produces informational effects. It facilitates the integration of new information into the policy process, it may allow short-term cognitive biases to diminish and it helps compensate for some forms of asymmetric information in politics.[11] The role of temporary legislation as a mechanism for reaching consensus has been tested mainly with regards to the United States. Forrest Maltzan and Charles Shipan define sunset clauses as 'important vehicles for encouraging a law to be revisited' and 'a mechanism for building coalitions'.[12] Merrill, in outlining a rationale for sunset clauses (viz. experimentalism), underlines that they can serve as 'a mechanism for convincing those who are sceptical about the merits of a proposed law or programme to try it out on a temporary basis'.[13] Ginsburg, Masur and McAdams call temporary legislation a 'form of political compromise', because 'proponents of regulation will accomplish their goal but will, by accepting an expiration date, bear the costs of extension. Opponents of regulation will be less opposed to temporary than permanent ones'.[14]

An example of this perspective can be found in the research conducted by Antonios Kouroutakis more generally with regards to consensus models of democracy.[15] In this context, Kouroutakis sees sunset clauses as a mechanism for reaching consensus between different institutions or between Majority and Opposition, and notes that

> where there is a disagreement, the branches may be willing to compromise and accept an agreement when the legislation is not permanent. This provisional character of legislation permits the branches to review their stance at the expiration of the law.[16]

Focusing his analysis particularly on the UK, Kouroutakis recognises that the use of sunset clauses is expected to be more frequent in a separation of powers system where the legislative function is shared among different institutions, such

[11] See Gersen (n 1) 266–272.
[12] See Maltzman and Shipan (n 2) 263.
[13] See Merrill (n 2) 342–343.
[14] See Ginsburg, Masur and McAdams (n 1) 297–298.
[15] Arend Lijphart, *Patterns of Democracy. Governmentt Forms and Performance in Thirty-Six Countries* (Yale University Press, 2012) 30–45.
[16] See Kouroutakis (n 1) 92.

as the US. However, Kouroutakis also points out that sunset clauses are a useful tool to reach consensus in a majoritarian separation of powers regime, such as the UK, when consensus is an essential condition, viz. when there is a coalition agreement or in the event of strong opposition from backbenchers.[17]

Ranchordás has analysed some examples from the Netherlands in which the use of temporary legislation played a role as 'an adequate instrument to reach a compromise or to convince the opponents or sceptics of a certain rule as to its positive effects or absence of side-effects'.[18]

Additionally, Ittai Bar-Siman-Tov notes that during his examination of temporary legislation within his jurisdiction (Israel), he uncovered some tentative anecdotal evidence suggesting that 'temporary legislation may indeed be used as a tool for softening parliamentary opposition (and perhaps also parliamentary scrutiny)'.[19]

III. The Italian *Decreto-legge* – A Provisional Emergency Decree which is Used Routinely Nowadays

In order to better understand the main hypotheses which will be tested in the following sections, it is first necessary to describe the principles of the Italian Constitution which establish the *decreto-legge* as a provisional emergency decree and its transformation into a decree which is now used routinely.

Then, it will be discussed why, in the Italian context, the *decreto-legge*, as a provisional emergency decree, has been used as an invaluable tool to reach consensus (section IV), and why the misuse (in order to reach consensus) of such a tool has led to negative outcomes in terms of the quality of legislation (section V).

A. Constitutional Principles on the *Decreto-legge* as a Provisional Emergency Decree

The *decreto-legge*[20] is a provisional emergency decree issued by the Government. The Government is allowed to issue a *decreto-legge* only in 'extraordinary cases of necessity and urgency'.[21] When the Italian Founding Fathers drafted this section,

[17] Kouroutakis (n 1) 93.

[18] See Ranchordás (n 1) 68.

[19] See Bar-Siman-Tov, 'Temporary Legislation' (n 1) 196.

[20] On the *decreto-legge* in general, see the three monographs written by: Andrea Simoncini, *Le funzioni del decreto-legge: la decretazione d'urgenza dopo la sentenza n. 360/1996 della Corte costituzionale* (Giuffrè, 2003); Alfonso Celotto, *L'abuso del decreto-legge. Vol. I. Profili teorici, evoluzione storica e analisi morfologica* (Cedam, 1997); and Giovanni Pitruzzella, *La legge di conversione del decreto-legge* (Cedam, 1989).

[21] Italian Constitution, s 77(2).

they were expecting a tool to be used for *real* emergencies, such as earthquakes or floods. One should bear in mind that the *decreto-legge* is a piece of primary legislation. It can repeal primary legislation, but it cannot derogate provisions of a constitutional nature.

As it is an emergency decree, the *decreto-legge* comes into force immediately, ie, once it has been published in the Italian Official Gazette. As it is a provisional emergency decree, the *decreto-legge* expires 60 days after its publication in the Official Gazette and this takes effect from the beginning (*ex tunc*). This means that the *decreto-legge* no longer has effect from the moment of publication itself. Thus, all the effects produced on the basis of that decree during this period are deprived of their legal grounds.[22]

The *decreto-legge* expires 60 days after its publication, but it can be converted into an Act of Parliament via a 'fast-track' procedure within these 60 days.[23] This is the reason why the Government introduces to Parliament a bill of conversion immediately after issuing the *decreto-legge*.[24]

When the Parliament does not convert the *decreto-legge*, it is able to pass an Act to legitimise the legal effects produced on the basis of that decree; however, the Parliament is not compelled to do this.[25]

When passing the converting Act, the Parliament can choose:

- not to amend any provision of the *decreto-legge*; this takes effect *ex tunc* (ie from the publication of the *decreto-legge*). This means that the *decreto-legge* expires 60 days after the publication and the converting Act retrospectively takes the place of the *decreto-legge* for the 60 days during which the *decreto-legge* was valid, without changing the substance of its provisions (*novazione*);
- to fully supersede or not to convert one or more provisions only of the *decreto-legge* (and convert the other provisions without changing their substance); this takes effect *ex tunc* (ie from the publication of the *decreto-legge*). This means that the specific provisions that are fully superseded or not converted expire after 60 days and, in the event of it being superseded, the provisions of the converting Act retrospectively take the place of the superseded provisions for the 60 days during which the *decreto-legge* was valid;

[22] *Ibid.*

[23] *Ibid*, s 77(3).

[24] *Ibid*, s 77(2). The Italian *decreto-legge* is quite different from the emergency laws in other jurisdictions. For example, as for the differences with the emergency laws in Israel, the Italian *decreto-legge* is a piece of primary legislation (and not merely a regulation to which the Parliament potentially gives a stamp of approval in the form of primary legislation); no declaration of a state of emergency by the Parliament is required as a condition for the Executive to issue the *decreto-legge*; its validity is limited to 60 days (and not three months); it is a temporary decree, because it loses its validity if the converting Act is not passed within 60 days; the Parliament is not entitled to extend its validity but it is only entitled to convert the *decreto-legge* into an Act and potentially to amend it for the future. On the emergency laws in Israel, see Suzie Navot, *Constitutional Law of Israel* (Kluwer Law International, 2007) 293–301.

[25] Italian Constitution, s 77(3).

- to add a new provision (within the scope of the *decreto-legge*); this provision takes effect from now on (*ex nunc*). This means that the new provision takes effect from the moment in which the converting Act comes into force;
- to amend one or more provisions of the *decreto-legge*; this provision takes effect *ex nunc* (ie from that time onwards). This means that such provision takes effect from the moment in which the converting Act comes into force, unless the converting Act explicitly lays down that the provision takes effect *ex tunc*, that is, from the moment of the publication of the *decreto-legge*.

B. The Transformation of the *Decreto-legge*: A Decree which is Now Used Routinely

As mentioned above, the *decreto-legge* is constitutionally limited to 'extraordinary cases of necessity and urgency', but in practice it is now being used routinely – far exceeding the rules set out by the Italian Founding Fathers.

Some recent statistics demonstrate this tendency. The percentage of the *leggi* (Acts of Parliament) enacted by the Parliament between 2013 and 2015 is 47.97 per cent of all main Italian legislation. Other percentages are as follows: 15.67 per cent *decreti-legge*, 31.72 per cent other primary legislation decrees issued by the Government (*decreti legislativi*) and 4.64 per cent secondary legislation decrees issued by the Government (*regolamenti di delegificazione*).[26] However, among the aforementioned Acts of Parliament, 27.42 per cent are converting Acts of *decreti-legge*, thus parliamentary legislation which does not substantively come from the Parliament but from the Government.[27] Moreover, if the letters of the provisions of the converting Acts of *decreti-legge* are counted, one would realise that among the 4,148,885 letters of the whole Acts of Parliament, 3,820,873 are included in converting Acts.[28]

This does not necessarily mean that the Parliament is able to regain its full role in legislation when passing converting Acts; a lot of far-reaching amendments to the converting bills are often introduced to the Parliament by the Government itself, sometimes under the pressure of a vote of confidence.

At the end of the day, almost 75 per cent of Italian legislation is enacted through *decreti-legge* and converting Acts, and it is clear to see from these statistics that the *decreto-legge* has turned into a tool to carry out routine legislation.[29]

[26] See Camera dei deputati-Osservatorio sulla legislazione, *Rapporto 2015-2016 sulla legislazione tra Stato, Regioni e Unione europea. XVII legislatura, Volume secondo* (Camera dei deputati, 2017) 334.

[27] *Ibid*, 335.

[28] *Ibid*, 345.

[29] On the role of the *decreto-legge* among the other sources of law nowadays, see the books edited by: Lino Duilio (ed), *Politica della legislazione, oltre la crisi* (Il Mulino, 2013); Roberto Zaccaria (ed), *Fuga dalla legge? Seminari sulla qualità della legislazione* (Grafo, 2011); Marta Cartabia and others (ed), *Gli atti normativi del Governo tra Corte costituzionale e giudici. Atti del Convegno annuale dell'Associazione*

A slight decrease in the number of *decreti-legge* issued by the Government has occurred since the end of the most dramatic stage of the recent economic crisis,[30] but the tendency of the Parliament (or better, by the Government) to introduce far-reaching amendments to the converting bills under the pressure of a vote of confidence is still present over recent years.[31]

From an institutional perspective, this means that nowadays most of the activities of the Parliament are just to amend and/or confirm legislation issued by the Government by *decreto-legge*. Moreover, as mentioned, one should also bear in mind that the Government tends to request confidence votes on converting bills quite often, as it needs to close ranks within its Majority. As two Italian scholars stated, this is a symptom of weakness of an arrogant government.[32]

A negotiation of some amendments within the Majority and sometimes with the Opposition is carried out by the Government during the converting proceedings, as will be seen later.

In 1995, the *Corte costituzionale* apparently discouraged this misuse of the *decreto-legge*. It stated that the existence of 'extraordinary cases of necessity and urgency' underlying a *decreto-legge* was a condition of validity of the *decreto* itself, as established under Italian Constitution, s 77.

However, one should bear in mind that there are two kinds of scrutiny on the potential lack of such a condition.

First, there is the *political* scrutiny carried out by the Parliament when it examines the converting bill. If the Parliament states that no 'extraordinary cases of necessity and urgency' underlie the *decreto*, it has the power to not pass the converting bill.

The other type is the *judicial* scrutiny carried out by the *Corte costituzionale* itself when a point of law regarding the validity of the *decreto-legge* is brought before it. If the *Corte costituzionale* states that no 'extraordinary cases of necessity and urgency' underlie the *decreto*, it has the power to quash it. However, such a judicial review, as the *Corte costituzionale* stated in 1995, can lead the *Corte costituzionale* itself to declare the *decreto-legge* void only when the lack of those conditions is *evident*.[33]

At the end of the day, the *Corte costituzionale* exercised a great level of self-restraint in that judgment. From a legal perspective, such a self-restraint of the *Corte costituzionale* has legitimised the current routine use of the *decreto-legge*.

'Gruppo di Pisa' Università degli studi di Milano-Bicocca, 10-11 giugno 2011 (Giappichelli, 2011); and Andrea Simoncini (ed), *L'emergenza infinita. La decretazione d'urgenza in Italia* (Eum, 2006).

[30] See Valerio Di Porto, 'I decreti-legge tra crisi economica e affievolimento delle patologie' (2018) 2 *Osservatoriosullefonti.it* 1–8.

[31] See Giuseppe Mobilio, 'Il procedimento legislativo nella XVII legislatura: spunti ricostruttivi e distanze dal modello costituzionale' (2018) 2 *Osservatoriosullefonti.it* 25–33.

[32] See Vincenzo Lippolis and Giovanni Pitruzzella, *Il bipolarismo conflittuale. Il regime politico della Seconda Repubblica* (Rubbettino Editore, 2007) 47–53.

[33] See *Corte costituzionale* [1995] No 29.

The *Corte*'s statement actually means that it is nowadays up to the Parliament to state what 'extraordinary cases of necessity and urgency' are from a political perspective; that is to say that almost everything can be seen as extraordinary cases of *political* necessity and urgency. Only when the lack of such conditions is evident (ie when it becomes obvious from some clear indications, such as the lack of the homogeneous contents of the *decreto-legge*), is the *Corte costituzionale* tasked with quashing the *decreto-legge*. This has been done by the *Corte* very few times.[34]

In 2012, the *Corte costituzionale* also stated that the Parliament is not allowed to use the converting act to regulate matters which are not homogenous, because under s 77 the task of the converting act is only to convert/amend a *decreto-legge*. Since then, the *Corte* declared provisions of *decreti-legge* void several times.[35]

IV. The Use of the Provisional *Decreto-legge* as a Tool to Reach Consensus

As mentioned earlier, the *decreto-legge*, established under the Constitution as a provisional emergency decree, has been used as an invaluable tool to reach consensus on routine legislation, ie, also on the grounds of extraordinary cases of *political* necessity and urgency.

The misuse of the *decreto-legge* is the consequence of the weakness of the Italian governments in using legislative process to pass ordinary bills and carrying out their political programmes *via* the ordinary legislative process. This problem results, on one hand, from a unique constitutional feature of the Italian constitutional system (namely, the Italian equal bicameralism) and, on the other hand, from the peculiar political context.

The Italian equal bicameralism is a unique feature of the Italian constitutional system among other parliamentary bicameral systems. This system includes both (*i*) direct legitimation and (*ii*) exactly the same tasks and features of both the *Camera dei deputati* and the *Senato della Repubblica*. It seems to put a brake on the efforts of the governments to carry out their political programmes because they have to gain a vote of confidence from both Chambers and engage in a cumbersome legislative process with each Chamber during the 'ping pong' stage (in fact, both Chambers have the power of veto over the legislative process). Moreover, in this context, governments have to deal with a very peculiar political system which makes it even harder for the Government to get its bills passed in the Parliament.

Let us start with the political system during the so-called *Prima Repubblica* (the historical period between 1948 and 1994).

[34] See *Corte costituzionale* [2007] No 171; [2008] No 128; [2013] No 220.
[35] See *Corte costituzionale* [2012] No 22; [2014] No 32; [2015] No 154; [2016] No 94.

In that period, Italy had a proportional-representation electoral system and a very fragmented multiparty system. Moreover, the *conventio ad excludendum*[36] made any alternative government impossible. There was some form of a tacit agreement between the main parties and the *Democrazia Cristiana*, viz. the main centre party, that they would not make any alliance with the *Partito Comunista Italiano*, viz. the communist party, to create a government. This took place during the Cold War period. For this reason, Giorgio Galli defined the Italian political system at that time as characterised by an imperfect bipartidism.[37]

In the so-called *Prima Repubblica*, it would have not been conceivable to exclude such a relevant political force, as that of the *Partito Comunista Italiano* (with 30 per cent of the votes), from managing the state at all. The *Democrazia Cristiana* sought the *Partito Comunista Italiano*'s approval on the main bills, and, among other things, this led to a consensus democracy instead of a majoritarian one. This was useful for the *Partito Comunista* not to be excluded from managing the state and for the *Democrazia Cristiana* to carry out its political programme easily.

During the so-called *Seconda Repubblica* (the historical period from 1994 onwards), Italy chose electoral majority systems. Beside this, the fall of the USSR and the transformation of the communist party into a social-democratic party led to different parties forming subsequent governments and this has strengthened the role of the Government. Such a transformation put Italy on the path towards a majoritarian democracy, but, as has been mentioned above, there remain important hurdles to overcome before a coherent institutionalisation of such a model[38] is accomplished.

Among other difficulties, the chosen electoral majority system has led to fragmented all-encompassing electoral coalitions being formed – an obstacle for governments unable to carry out their political programme *via* the ordinary legislative process. More recently, due to the pressure resulting from the economic crisis and some institutional deadlocks, the traditional electoral coalitions which had been the central character of the Italian political system in the last 17 years collapsed in 2013, immediately after the elections. Thus, in 2011, 2013, 2014 and 2016, four Italian governments (Monti's, Letta's, Renzi's and Gentiloni's) were supported by parties coming from opposing electoral coalitions. However, no serious political agreement was signed by them on the programme to be carried out and, once again, this has led them to seek consensus on the political programme on a day-to-day basis. Moreover, in 2013, two different majorities were elected, one in the *Camera dei deputati* and one in the *Senato*. This made it even harder to form

[36] See Leopoldo Elia, 'Governo (forme di)' in *Enciclopedia del diritto* (Giuffrè, 1970) Vol XIX, 639.

[37] See Giorgio Galli, *Il bipartitismo imperfetto: comunisti e democristiani in Italia* (Il Mulino, 1967).

[38] See Sergio Fabbrini, 'The Transformation of Italian Democracy' (2009) 1 *Bulletin of Italian Politics* 29–47. On the transformation of the Italian political and constitutional system in the 1990s, see Gianfranco Pasquino, *Il sistema politico italiano* (Il Mulino, 2002) and Giovanni Pitruzzella, *Forme di governo e trasformazioni della politica* (Laterza, 1996).

a government and, once the Government was formed (which was supported by three parties which had run in the elections in separate coalitions), it was difficult for it to carry out its programme *via* the ordinary legislative process.

At the end of the day, the features of a consensus democracy seem to still be present in the Italian political system. As for legislation, governments usually 'escape' from the cumbersome ordinary legislative process. They tend to carry out most of their political programme not by introducing bills in the Parliament, but by issuing *decreti-legge* to be converted into Acts of Parliament. The *decreto-legge* is of great value in such political contexts, such as the so-called *Prima Repubblica*[39] and *Seconda Repubblica*.[40] On the one hand, it lets the Government issue a decree that comes into force immediately and thus carry out its political programme effectively; on the other hand, as the *decreto-legge* is a provisional source of law (ie it must be converted, otherwise it expires *ex tunc*, and can be amended by the Parliament), it forces the Majority to reach consensus in order to have the converting bill passed by the Parliament *via* the 'fast-track' procedure within 60 days.

Such a consensus is to be reached with the Opposition or among the different parties which belong to the Majority itself. As for the so-called *Prima Repubblica*, Alberto Predieri stated that the converting bill could be considered a bill which is strengthened by the fact that the *decreto-legge* comes into force immediately, with regards to which the Government accepts the risk of negotiating amendments with the Members of Parliament,[41] which are often negotiated by the Majority and the Opposition secretly or publicly ('sotto banco o alla luce del sole').[42] As for the so-called *Seconda Repubblica*, Andrea Simoncini remarked on the same task of the *decreto-legge* and finds the reasons for this tendency in the incomplete transformation of the Italian democracy towards a coherent majoritarian model.[43]

In June 2018, Conte's Government was formed, once again, by two parties which had run the elections separately. However, for the first time in the so-called *Seconda Repubblica*, the Government was formed on the basis of a serious agreement negotiated in depth and signed by those two parties.

However, the abuse of the *decreto-legge* has also been carried out by Conte's Government;[44] once again, notwithstanding the agreement signed by the two Majority parties, in the light of political difficulties in carrying out its programme (due to the deep differences between parties' manifestos), Conte's Government continued to use the *decreto-legge* as a tool to 'escape' the ordinary legislative process and to reach consensus among the two Majority parties.

[39] See the analysis of data carried out by Erik Longo, *La legge precaria. Le trasformazioni della funzione legislativa nell'età dell'accelerazione* (Giappichelli, 2017) 115–122 and 139–145.

[40] See *ibid*, 178–196.

[41] Alberto Predieri, *Il Governo colegislatore* in Franco Cazzola, Alberto Predieri e Grazia Priulla (ed), *Il decreto legge fra governo e parlamento* (Giuffrè, 1975) XX.

[42] Vezio Crisafulli, *Lezioni di diritto costituzionale. Vol. II, 1. L'ordinamento costituzionale italiano (Le fonti normative)* (Giuffrè, 1993) 106.

[43] See Simoncini (n 29) 50–52.

[44] See data from 1st June to 28th February 2019 in Enrico Albanesi, *Teoria e tecnica legislativa nel sistema costituzionale* (Editoriale scientifica, 2019) 282.

V. Time Issues Regarding the *Decreto-legge*, its Negative Outcomes from the Perspective of the Quality of Legislation, the (Partial) Solutions Given by the *Corte costituzionale* and the Inevitable Lack of an *Ex Post* Evaluation

In light of the above, it is now necessary to explore some controversial issues concerning the provisional nature of the *decreto-legge* and especially its negative outcomes in terms of the quality of legislation, which derives from the fact that the *decreto-legge* is a temporary source of law. The (partial) solutions given by the *Corte costituzionale* to some of these issues will also be discussed. Once again, in terms of quality of legislation, it will also be underlined that the *decreto-legge* inevitably lacks an *ex post* evaluation, which is the added value of sunset clauses from that perspective.

A. Time Issues Regarding the Duration of the *Decreto-legge*

One of the temporal issues is in regards to the duration of the *decreto-legge* itself.

In the past, when a *decreto-legge* expired after 60 days, the Government would re-issue a new *decreto-legge* with the same content as the previous one. It was called the reiteration (*reiterazione*) of a *decreto-legge*. In doing this, the Government would also include in the new *decreto-legge* a provision to legitimise the legal effects produced on the basis of the previous decree. A *decreto-legge* would be re-issued even dozens of times before being converted into an Act. Meanwhile, every new *decreto-legge* of the chain would include a provision to legitimise the legal effects produced on the basis of the previous decrees.

The Government would do so when it was difficult to convert the *decreto-legge* into an Act, such as when there was not enough political support from the Parliament or simply because the Parliament's agenda was too busy during those 60 days.

However, clearly, if the Government was allowed to re-issue a new *decreto-legge* after expiration and with the same content as a previous decree, the *decreto-legge* would become *de facto* a permanent tool. Moreover, the role of Parliament would be frustrated – first, because legislative powers (ordinarily conferred to the Parliament) can be exercised only exceptionally and provisionally by the Government, and second, because only the Parliament is tasked with legitimising the legal effects produced on the basis of an expired decree.

In 1996, the *Corte costituzionale* declared the *reiterazione* as an unconstitutional practice because:

- it changes the provisional nature of the *decreto-legge*, by prolonging *de facto* the 60-day time limit of validity laid down by section 77 of the Constitution;

- it distorts the 'extraordinary' nature of the cases of necessity and urgency that shall underpin a *decreto-legge*; if the Government assumes that the same cases of necessity and urgency of a previous *decreto-legge* underpin a new *decreto-legge* with the same contents of a previous *decreto*, those cases become routine (and no longer extraordinary);
- it weakens the sanction of the *ex tunc* loss of validity of the *decreto-legge* when the latter is not converted by the Parliament into an Act; the more the *decreto-legge* is re-issued, the more difficult it is from a political perspective not to convert it into an Act of Parliament, because every new link of the chain has generated a higher expectation that the *decreto-legge* sooner or later will be converted;
- from the perspective of the quality of legislation, it has a deep impact on the principle of certainty of law because it is impossible to foresee the duration of the number of reiterations and the final outcome of the entire process, ie whether the decrees would be in the end converted into an Act of Parliament;
- from a more general perspective, especially when this practice is widespread and deep rooted, it has a deep impact on the features of the parliamentary system and the conferring of the ordinary legislative power of the Parliament.[45]

B. Time Issues Regarding Constitutional Judicial Review on the *Decreto-legge*

The second time-related issue regards constitutional judicial review on the *decreto-legge*.

In Italy, a case about the constitutional validity of a piece of primary legislation (ie Acts of Parliament and some decrees issued by the Government, such as the *decreto-legge*) can be brought before the *Corte costituzionale* in two ways: (a) when a Court needs to apply that piece of legislation and it considers its potential invalidity; there is no limit for this and the case can be brought before the *Corte* even years after it has come into force (concrete constitutional judicial review), and (b) when a region considers the potential invalidity of a piece of legislation enacted by the state (and vice-versa); this can be done within 60 days from its coming into force (abstract constitutional judicial review). Once the piece of legislation is brought before the *Corte costituzionale*, it takes time before they rule on this.

This can be particularly problematic when the *decreto-legge*, which has been brought by a court or a region, expires after 60 days after its publication in the Official Gazette.

[45] See *Corte costituzionale* [1996] No 360.

If the *decreto-legge* is not converted into an Act, the *decreto-legge* loses its validity from the time of its publication (*ex tunc*) and the case brought before the *Corte costituzionale* no longer refers to an existing piece of legislation and it is declared inadmissible. This will not be a problem as no other piece of legislation retrospectively covers the 60 days during which the *decreto-legge* was valid.

On the other hand, if the Parliament passes a converting Act, the *decreto-legge* loses its validity from the time of its publication (*ex tunc*), but the Act (with the same contents of the *decreto-legge* itself or with some amendments, at most) retrospectively covers the 60 days during which the *decreto-legge* was valid. The case brought before the *Corte costituzionale* with regards to the *decreto-legge* no longer refers to a piece of legislation. However, if the *Corte costituzionale* declared the case inadmissible, the court or the region would need to bring before the *Corte costituzionale* a new case regarding the converting Act and this would be an unreasonable waste of time.

In 1995, the *Corte costituzionale* found a solution to this problem and stated that when a case is brought with regard to a *decreto-legge* that is converted into an Act, the *decreto-legge* loses its validity but the case is 'transferred' to the converting Act, which will be scrutinised by the *Corte costituzionale*.[46] Even if the piece of legislation is formally different, its contents are substantially the same, and this is the rationale underpinning this solution.

C. Time Issues Regarding the Scrutiny Carried Out in Limited Time by the Parliament on the *Decreto-legge*

The third temporal issue regards the scrutiny of Parliament.

Parliament can scrutinise the *decreto-legge* within 60 days. As the Italian system is bicameral, each Chamber has to carry out the scrutiny and they have 30 days in which to do so. This might be demanding for the Parliament, especially when its agenda is busy, but it is not impossible.

However, the Government tends to issue far-reaching *decreti-legge* that include no homogeneous provisions. Clearly, if the Government was allowed to issue a non-homogeneous *decreto-legge*, it would be *de facto* impossible for the Parliament to consciously scrutinise it fully within such a short time limit. A lack of scrutiny by the Parliament clearly means also a lack of scrutiny on the quality of its provisions.[47]

Moreover, the Government tends to include provisions that are not immediately applicable in the *decreti-legge*. However, there is no reason to use an emergency and provisional decree (even in the broader political sense) to enact

[46] See *Corte costituzionale* [1995] No 84.

[47] On parliamentary scrutiny of the *decreto-legge* from the perspective of quality of legislation, see Annalisa Ghiribelli, *Decretazione d'urgenza e qualità della produzione normativa* (Giuffrè, 2011).

such provisions because the latter could be enacted by the Parliament through an Act passed *via* an ordinary and not a 'fast-track' procedure.

These are the reasons why, under Act of Parliament 1988 No 400, the Government shall issue homogeneous decrees and the provisions included in the decrees shall be immediately applicable only.

However, Act of Parliament 1988 No 400 is not of a constitutional nature. It cannot be used by the *Corte costituzionale* as a parameter to scrutinise the validity and, thus, quash a *decreto-legge*.

The *Corte costituzionale* very wisely stated that when decrees are not homogenous, this can be used by the *Corte* itself as a hint to reveal the *evident* lack of 'extraordinary cases of necessity and urgency' underpinning that provision.[48] The same has been stated by the *Corte costituzionale* with regard to provisions that are not immediately applicable; this can be used by the *Corte* as an indication of the *evident* lack of 'extraordinary cases of necessity and urgency' underpinning that provision.[49]

A lot of far-reaching amendments are introduced to the Parliament by the Government itself. However, the Government is not the only one to blame. New provisions that are not homogeneous are often enacted by the Parliament itself when it passes the converting Act. As can be imagined, as the converting procedure is that of 'fast-track', such amendments are hastily drafted.

However, the *Corte costituzionale* stated that the specific function of the converting Act, and its 'fast-track' procedure is to convert, and at most amend, the *decreto-legge*. This means that the Parliament itself is not allowed to pass the converting Act with provisions that are not homogeneous.[50]

D. The Inevitable Lack of an *Ex Post* Evaluation

As previously discussed (section II), the main features of sunset clauses, as they have been outlined by scholars, are their temporary character and the presence of *ex post* evaluation.

As for their temporary character, examples of sunset clauses analysed by scholars, show that sunset clauses tend to last between one and six years. It is clear to understand the reason behind such duration; it is a significant length in order to allow the legislator to examine if a particular legislative measure will effectively achieve certain goals.

As for the presence of *ex post* evaluation, clearly such a feature derives from the task of sunset clauses as a virtuous tool only if they allow the legislator to verify whether the objective for which they were enacted has been achieved.

[48] See *Corte costituzionale* [2007] No 171; [2008] No 128.
[49] See *Corte costituzionale* [2013] No 220.
[50] See *Corte costituzionale* [2012] No 22; [2014] No 32; [2015] No 154; [2016] No 94.

The *decreto-legge*, as such, lasts only 60 days and lacks an *ex post* evaluation.

I would say that it *inevitably* lacks an *ex post* evaluation because its duration is only 60 days which is not a significant length in order to let the legislator examine if a particular legislative measure will effectively achieve certain goals.

Such a short duration period derives from the fact that under the Italian Constitution, the *decreto-legge* as such has been established, not as experimental legislation, but as a provisional emergency decree which must be converted into an Act of Parliament as soon as possible.

VI. Conclusion

The aim of this chapter was to test, with regards to the main Italian temporary source of law (*decreto-legge*), the validity of the argument stated by some recent scholarly works under which temporary legislation (and especially sunset clauses), among other functions, is a mechanism to reach consensus.

The use of sunset clauses is not recognisable within the Italian legal system. One might ask if the *decreto-legge* could be seen as the main temporary source of law which plays the same role as sunset clauses, at least in terms of being a tool to reach consensus on every-day legislation.

This hypothesis has been proven within this chapter; the *decreto-legge* has been used in Italy as a tool to reach consensus.

As already seen, Kouroutakis underlines in his book that sunset clauses are used in consensus models as a mechanism for reaching consensus between different institutions or between Majority and Opposition (section II). As mentioned, the so-called Italian *Prima Repubblica* was included by Lijparth among consensus democracies and typical features of a consensus democracy are still present in the so-called *Seconda Repubblica* (section II).

Therefore, the conclusion concerning the first hypothesis can be said to be that the *decreto-legge*, as it is temporary legislation (or better, provisional legislation), plays a pivotal role as a tool to reach consensus in the Italian legal order (section IV). The main difference between the *decreto-legge* and sunset clause concerns consensus. With regards to the *decreto-legge*, the consensus between Majority and Opposition or within Majority is reached at the end of the process (when the Parliament passes the converting bill) and not at its beginning (as happens with sunset clauses, when the Parliament enacts them).

This chapter's analysis has also proven the second hypothesis that such a use of the *decreto-legge* to reach consensus in our system is an abuse of a tool which was established by the Italian Founding Fathers as a short-lasting, emergency and provisional decree, not a tool for experimental legislation (section III). Due to its characteristics, in terms of the quality of legislation, such a tool has produced negative outcomes. Moreover, due to its short-time duration, it inevitably lacks an *ex post* evaluation (section V), an element which should characterise sunset clauses as a virtuous tool (section II).

In other words, this chapter has demonstrated that such a use of the *decreto-legge* to reach consensus is the classical example of a tool used for a purpose that is different from the one for which it was intended, which for this reason leads to negative outcomes.

Therefore, although it is true that the *decreto-legge* has played a pivotal role as a tool to reach consensus in Italy due to its temporary nature, my conclusion concerning the second hypothesis is that this has not occurred due to the *decreto-legge* being experimental legislation (its duration is too short to play such a role) or its information effects (see section II). The reason why the *decreto-legge* has played such a role is, on the one hand, its coming into force immediately, and thus, on the other hand, its capacity to force the Majority to seek an agreement with the Opposition or within the Majority itself to have the converting bill passed by the Parliament.

In a recent analysis of the use and the functions of the *decreto-legge* and the other sources of law in Italy from a wider perspective than the one carried out here, Erik Longo[51] saw the *decreto-legge* as a positive tool which can help the legislator change immediately and rapidly existing legislation, a task which is nowadays essential and normal in the modern age of acceleration.[52] In other words, this wider perspective does not look at the *decreto-legge* as a 60-day temporary measure as such; it looks at the *decreto-legge* and the converting Act as a mechanism to rapidly change permanent existing legislation that empirical evidence has demonstrated requires a change. Not surprisingly, from this perspective, Longo enhances the role played by the *ex post* evaluation to assess whether existing legislation really needs to be changed and how. Once again, unsurprisingly and notwithstanding his favourable analysis concerning the role of the *decreto-legge* (at least, with regards to the profile which I have just mentioned), he wishes for a use of sunset clauses in Italian legislation.[53]

A use of sunset clauses in Italian legislation is obviously desirable due to all the advantages analysed in this chapter.

However, the ongoing misuse of the *decreto-legge* is still producing negative outcomes in our legal order in terms of quality of legislation.

In 2014–2016, Italy tried to amend its Constitution to supersede the equal bicameralism – a reform which, in my view, could have also indirectly superseded the abuse of the *decreto-legge*, by making it easier for the Government to use the ordinary legislative process.[54] However, the constitutional reform was rejected by the people in a referendum on the 4th December 2016. The equal bicameralism is still there, in tandem with the peculiar features of the political system, which

[51] See John O McGinnis, 'Laws for Learning in an Age of Acceleration' (2011) 53 *William & Mary Law Review* 305.

[52] See Longo (n 39).

[53] Longo (n 39) 307.

[54] For an analysis of the contest of the constitutional reform and its connections with the quality of legislation, see Enrico Albanesi, 'The Mechanisms Used to Review Existing Legislation in the Civil Law System' (2016) 18 *European Journal of Law Reform* 292–294.

makes it hard for the Government to carry out its programme and get its bills passed in the Parliament *via* the ordinary legislative process, with the misuse of the *decreto-legge* as a consequence.

After the failure of the constitutional reform, as the Italian novelist Luigi Pirandello would say, the Italian *decreto-legge* seems to still be in search of an author.

Or a good drafter, at least.

9

Sunset Clauses: A Contribution to Legislative Quality

HELEN XANTHAKI

I. Hypothesis and Method

Current analysis on sunset clauses tends to focus on constitutional aspects of their utility, their usage, their advantages and their inherent dangers.[1] Little, if any, analysis looks at their role in legislative drafting, both as carriers of a regulatory message and as technical legislative provisions. This chapter aims to address both sets of issues, by defining sunset clauses, placing them in the context of duration or time provisions, addressing their drafting requirements and discussing their role as contributors and supervisors of legislative effectiveness. This ties in well with the topic of the book: time in legislation, in the form of sunset clauses, can contribute to the creation and monitoring of legislative quality.

Legislative effectiveness is embraced as synonymous to, or a measure of, legislative quality by international and national lawmakers and academic and professional drafters.[2]

The hypothesis of this chapter is that sunset clauses contribute to legislative effectiveness and regulatory efficacy. In order to prove this hypothesis, the chapter begins by defining and analysing sunset clauses as a legislative tool. It proceeds with the analysis of legislative effectiveness, and the identification of the role of sunset clauses as contributors to legislative quality. It continues with the presentation

[1] See Sofia Ranchordás, *Constitutional Sunsets and Experimental Legislation* (Edward Elgar, 2014); also see Antonios Kouroutakis, *Sunset Clauses: A Historical, Positive and Normative Analysis* (Oxford University, 2014, PhD thesis); also see Alli Sutherland, 'Ghosting in Tax Law: Sunset Provisions and Their Unfaithfulness' (2018–2019) 46 *Hastings Constitutional Law Quarterly* 479, 498; John Ip, 'Sunset Clauses and Counterterrorism Legislation' (2013) *Public Law* 74, 75; Dan Price, 'Sunset Legislation in the United States' (1978) 30 *Baylor Law Review* 401.

[2] See Helen Xanthaki, 'The Problem of Quality in EU Legislation: What on Earth is Really Wrong?' (2001) 38 *Common Market Law Review* 651, 660; also see Wim Voermans, 'Concern About the Quality of EU Legislation: What Kind of Problem, by What Kind of Standards?' (2009) 2 *Erasmus Law Review* 59, 62;

of Xanthaki's 'enlightened scrutiny test', which is analysed below, and the identification of the role of sunset clauses as guardians of legislative quality. Finally, it concludes by proposing the extended use of sunset clauses as contributors to legislative quality and guardians of regulatory efficacy even beyond short-term legislative texts.[3]

II. Sunset Clauses: A Legislative Exploration

The start and end of the life of legislation is expressed in duration provisions, which convey to the user when their obligation to comply with the legislation begins, and when it ends. Duration provisions is a term that encompasses both commencement and expiry provisions. They address the issue of time relating to the legislative text itself. Thus, they differ from retroactive and retrospective provisions that address time relating to the regulation expressed by the text. Duration provisions relate to the technical, administrative aspect of time in legislation, whereas retroactive and retrospective provisions relate to the substantive aspect of time in law.[4]

The start of the life of legislation is expressed in commencement provisions, namely provisions that set the date when legislation begins to have effect.[5] They do not set the time of passing of the legislation but their time of coming into force. This is the moment that they draw their first breath as part of the system of law of the jurisdiction where they belong. In the UK, commencement provisions are now known simply as the start date of the legislative text. The need to employ a more user-friendly term for commencement arose in the Good Law survey of the UK's

Claudio Radaelli and Franko de Francesco, *Regulatory Quality in Europe: Concepts, Measures and Policy Processes* (Manchester University Press, 2007) 37; C Veljanovski, 'Economic Approaches to Regulation' in R Baldwin, M Cave and M Lodge, *The Oxford Handbook of Regulation* (Oxford University Press, 2010) 19; European Commission, 'European Governance: Better lawmaking' COM (2002) 275 final; EP, Council, Commission, Interinstitutional Agreement on better law-making [2003] OJ C321/1, point 25, http://eurlex.europa.eu/LexUriServ/LexUriServ.do?uri=CELEX:32003Q1231%2801%29:EN:NOT; UK Office of the Parliamentary Counsel (OPC), *Drafting Guidance*, 2017, www.gov.uk/government/uploads/system/uploads/attachment_data/file/666328/drafting_guidance_Dec_2017.pdf; Australian OPC, *Reducing Complexity in Legislation*, June 2016, www.opc.gov.au/sites/default/files/reducingcomplexity_0.pdf; Government of Canada, *Guide to Making Federal Acts and Regulations*, 2001, www.pco.gc.ca/docs/information/publications/legislation/pdf-eng.pdf.

[3] One of the most prominent authors supporting the use of sunset clauses in emergencies is Professor Ackerman: see Bruce Ackerman, 'The Emergency Constitution' (2004) 113 *Yale Law Journal* 1029.

[4] It is worth noting here, however, that substance and form are not completely disconnected. It is argued that sunset clauses may end the form, the legislation, but not the regulation. Legislation may well continue to produce effects after the expiry of the legislation, because it may affect compliance behaviour that goes beyond statutory enforcement. See Frank Fagan, 'After the Sunset: The Residual Effect of Temporary Legislation' (2013) 36 *European Journal of Law and Economics* 209–26.

[5] For an analysis of commencement, see Recent Developments, 'The Commencement of Statutes – Again' (1981) 2 *Statute Law Review* 94–97; see also Mark Gobbi, 'When to Begin: A Study of New Zealand Commencement Clauses with Regard to those Used in the United Kingdom, Australia, and the United States' [2010] 31 *Statute Law Review* 153–216.

Office of Parliamentary Counsel, where iris tracking of legislative users revealed their immense puzzlement over where to find commencement, what it meant, and how it related to the rest of the legislative text. The term has become accepted in the UK but has not caught up in the rest of the Commonwealth, where commencement is commonly used. This is perhaps a reflection of the continued common law tradition of complex and technical start dates, subject to a notification from a Minister.

The end of life of legislation is expressed in expiry provisions, namely provisions that set the date when legislation ceases to have effect.[6] A relatively recent example of a UK expiry provision is found in the Armed Forces Act 2006:

> This Act expires at the end of one year beginning with the day on which the Armed Forces Act 2016 is passed (but this is subject to subsection (2)).

Research in the UK statute book[7] shows 19 Acts carrying expiry provisions.[8]

From a legislative perspective, sunset clauses are a sub-category of expiry provisions. Sunset clauses declare that the Act, at a set time, 'ceases to have effect unless it is reauthorized'.[9] They are defined as 'A provision in a Bill that gives it an expiry date once it is passed into law. Sunset clauses are included in legislation when it is felt that Parliament should have the chance to decide on its merits again after a fixed period.'[10] Thus, sunset clauses are a complex sub-category of expiry provisions, which introduce two regulatory implications: first, the end of the life of legislation at a set time; and second, conditionality upon a decision of Parliament, hopefully made on the basis of post-legislative scrutiny.[11] In recent literature, sunset clauses are mentioned as a tool to adopt a constitution 'on an interim basis, pending its promulgation after a five-year period of correction and further opportunities for public consultation';[12] or as tools for institutional reform in post-conflict, power-sharing arrangements;[13] or as a mechanism of 'orderly

[6] See Helen Xanthaki, 'Legislative Techniques in Rwanda: Present and Future' (2013) 15 *European Journal of Law Reform* 95, 156.

[7] See www.legislation.gov.uk.

[8] See Armed Forces Act 2016; Debt Relief (Developing Countries) Act 2010; Crofting Reform (Scotland) Act 2010; Holocaust (Return of Cultural Objects) Act 2009; Mental Health Act 2007; Conditions (Scotland) Act 2003; Sex Discrimination (Election Candidates) Act 2002; Trade Marks Act 1994; Crofters (Scotland) Act 1993; Education Act 1993 (repealed); Copyright, Designs and Patents Act 1988; Statute Law (Repeals) Act 1986; Agricultural Holdings Act 1986; Mental Health Act 1983; Patents Act 1977; Land Development Values (Compensation) Act (Northern Ireland) 1965; Registered Designs Act 1949; Patents Act 1949; Trade Marks Act 1938 (repealed 31 October 1994).

[9] See EJ Gouvin, 'Are There Any Checks and Balances on the Government's Power to Check Our Balances? The Fate of Financial Privacy in the War on Terrorism' (2005) 14 *Temp Policy and Civil Rights Law Review* 517, 540.

[10] See 'Glossary of Parliamentary Terms', where www.parliament.uk/site-information/glossary/sunset-clause.

[11] See Sofia Ranchordás, 'Sunset Clauses and Experimental Regulations: Blessing or Curse for Legal Certainty?' (2015) 36 *Statute Law Review* 28, 30.

[12] See J Wallis, *Constitution Making during State Building* (Cambridge University Press, 2014) 108.

[13] See Sabine Kurtenbach, 'Introducing Institutional Reform: The Role of Sunset Clauses in Post-conflict Power-Sharing Arrangements' in Nadine Ansorg and Sabine Kurtenbach (eds), *Institutional*

transition' and consumer protection;[14] or even as an effective tool to threaten the independence of agencies.[15]

The following provisions constitute a sunset clause in that they provide the extension of life of the legislation subject to parliamentary resolution:

> (2) Her Majesty may by Order in Council provide that, instead of expiring at the time it would otherwise expire, this Act shall expire at the end of a period of not more than one year from that time.
>
> (3) Such an Order may not provide for the continuation of this Act beyond the end of the year 2021.
>
> (4) No recommendation may be made to Her Majesty in Council to make an Order under subsection (2) unless a draft of the Order has been laid before, and approved by a resolution of, each House of Parliament.[16]

The example above is quite typical of the UK approach to sunset clauses. They are not headed by that term, and the table of contents of Acts does not include the term sunset clause.[17] In fact, the term 'sunset clause' is only found in the Explanatory Note (not the body of the legislation) of The Immigration and Police (Passenger, Crew, and Service Information) (Amendment) Order 2015. The actual provision (s 2) states:

> (3) After article 1(1), at the end, insert—
>
> '(2) This Order shall cease to have effect at the end of the period of seven years beginning with the day on which the Immigration and Police (Passenger, Crew and Service Information) (Amendment) Order 2015 comes into force.'

Nudges to review an Act that expires, namely sunset clauses in substance but not in form, can also be traced in other UK General Acts, such as the Investigatory Powers Act 2016; the Northern Ireland (Welfare Reform) Act 2015; the Enterprise and Regulatory Reform Act 2013; and the Interpretation Act 1978.

The empirical analysis of these Acts confirms that sunset clauses have been accurately defined as provisions which link the expiry of legislation to a qualitative review of its effectiveness. It also reveals that the term sunset clauses, which prevails in academic analysis, has not creeped its way into the body of legislation, at least not in the UK. Expiry is the prevalent term irrespective of whether the revival of the Act is linked to post-legislative scrutiny or not. Yet, sunset

Reforms and Peacebuilding: Change, Path-dependency and Societal Divisions in Post-war Communities (Studies in Conflict, Development and Peacebuilding, 2017).

[14] See Will Roberts, 'Support for Five-Year "sunset clause" on Legacy Commission' (2011) *Professional Adviser* 1.

[15] See Roberta Romano, 'Does Agency Structure Affect Agency Decision-Making: Implications of the CFPB's Design for Administrative Guidance' (2019) 36 *Yale Journal on Regulation* 273, 302.

[16] See Duration of Armed Forces Act 2006 (UK).

[17] Sir Franklin Berman compared UK legislation to international treaties and remarked on the rarity of sunset clauses in UK legislation: see Sir Franklin Berman, 'International Treaties and British Statutes' (2005) *Statute Law Review* 1, 6.

clauses as drafting tools are used frequently in the EU and, with increasing popularity, the UK.[18]

Having defined the term, attention is due to the placement of sunset clauses in the text. The start and end of the life of legislation is a crucial communication to the users. It offers an understanding of when their compliance with the legislation starts and when it finishes. The placement of start and expiry provisions has changed repeatedly in the last decade, from preliminary provision to final provision, and then back to preliminary again. In traditional drafting, especially in the common law, start and expiry provisions were considered merely technical, carrying a message relevant and intelligible solely to lawyers and judges. They were classified as preliminary provisions, amongst the many technical ones placed at the beginning of the legislative text. In modern drafting, their importance for the users was recognised; however, they were demoted to the bottom of the text as final provisions, where they were joined by most traditionally preliminary provisions in a bid to offer the prime spot in the beginning of the text to the main regulatory messages. The idea was that users have a limited attention span, which must be exploited to convey the substance of the regulatory request expressed in the legislative text. Any technical or secondary issues were transferred to the end of the text, available for the trained users who have interest and capacity to read them. In meta-modern legislative drafting published electronically, the primacy of their regulatory message is recognised, but they remain at the end of the text. Their regulatory message (namely the currency or status as currently in force, of the legislative text) is clearly demonstrated visually, for example, by means of a green colour in the title of legislation to show that it is actually in effect.

It is not unusual for simple brief commencement provisions, and, rather more infrequently, even expiry provisions to be placed in preliminary provisions, as a reflection of the importance of their message to the user. But, unfortunately, the start and end date of legislation is often much more complex. In such cases, it would be counterproductive to upgrade a long list of various start and end dates in the beginning of the text, where they can indeed distract the user and steal the precious few minutes of their limited attention span. Such lists must be placed in final provisions[19] or, even better, in a schedule. A hanging clause in preliminary

[18] See HC Deb I July 2003 Written Answer 217 from D Greenberg (ed), *Craies on Legislation* (Sweet and Maxwell, 2004) 384, para 10.2.4, note 7. Also see HM Government, 'Better Regulation Framework: Guidance', 23 August 2018, https://assets.publishing.service.gov.uk/government/uploads/system/uploads/attachment_data/file/735587/better-regulation-framework-guidance-2018.pdf.

[19] 35 Commencement

(1) Subject as follows, this Act comes into force on such day as the Secretary of State may by order appoint; and different days may be appointed for different purposes.
(2) Section 1(1) so far as it inserts the new section 62B, sections 4, 7, 9, 16, 19, 26, 32 and 33, this section and section 36, and Schedules 2 and 4, come into force on the day on which this Act is passed.
(3) Sections 11, 12, 17, 18, 29 and 30 come into force at the end of two months beginning with the day on which this Act is passed.
(4) Section 21(4) to (6) come into force on such day as the Scottish Ministers may by order appoint; and different days may be appointed for different purposes.

provisions directing the user to the end of the text or to the schedule could be an optimal compromise between the importance of the time message for the user and the technical nature of such provisions. And, this constitutes the optimal placement for them: either briefly at the beginning of the text or at the bottom with a hanging provision declaring the primacy of their message.

Having explored their definition and placement, it is time to turn our attention to their effect. There is a question as to whether Acts including sunset clauses need to be formally repealed after their set expiry date. Constitutionally, there is no doubt that the Act dies as soon as the end date arrives. Thus, from a substantive law perspective, a formal express repeal of the dead legislative text is unnecessary and even superfluous. But there is a question of best drafting practice. Although the repeal is not necessary from a legal perspective, nonetheless, clarity is best served with an express repeal. The added value of this method is that the Act disappears from subsequent reprints of the statute book.[20]

Having discussed the definition, placement and effect of sunset clauses, attention must now be turned to their usage. Even in a more traditional approach to the innovative requirement for end dates by default, sunset clauses are necessary in four distinct types of legislation. First, legislation in stages can be useful in cases of controversial issues; for example, liberalising abortion in a deeply religious society may lead to a useless addition to the statute book without an incremental approach. Liberalisation can start by allowing abortion exclusively where the mother's life is at risk, or where the embryo is a product of a criminal act; it can then expand to abortion up to nine weeks of gestation, and then it can proceed further. This incremental approach is served very well by means of a staggered start and end date of the intermediate regulatory regimes. Second, legislation that is experimental or whose benefits are speculative or unpredictable would require an end date; throwing the legislative text in the statute book for a limited period of time can allow the text the opportunity to produce the desired regulatory results, thus earning its place within the statute book. Third, legislation responding to a short-term need requires an end date. A classical example of this would be legislation offering aid to citizens suffering from extreme natural phenomena, such as an earthquake disaster. And fourth, authorisations to use emergency powers always come with an end date; they are a break from constitutional normality and are tolerated only for a well-defined short period of time.

(5) The Scottish Ministers may by order make such transitional, transitory or saving provision as the Scottish Ministers consider appropriate in connection with the coming into force of section 21(4) to (6).
(6) The Secretary of State may by order make such transitional, transitory or saving provision as the Secretary of State considers appropriate in connection with the coming into force of any other provision of this Act.
(7) Power to make an order under subsection (5) or (6) includes power to make different provision for different purposes.

See Growth and Infrastructure Act 2013, section 35.

[20] See AL Diamond, 'Repeal and Desuetude of Statutes' (1975) 28 *Current Legal Problems* 107, 124.

A common criticism of sunset clauses is that they serve little purpose. After all, legislation knows no end. We legislate for an indefinite period of time, or, at least, this is what we continue to contend, even though in the UK the average lifecycle of legislation sits at only 20 years.

Any response arrived at in the debate of sunset clauses against the longevity of legislation seems academic, since sunset clauses serve as more than murderers of allegedly immortal laws. They can serve as an effective tool for enhancing legislative effectiveness and, ultimately, regulatory efficacy.[21]

III. Sunset Clauses as Contributors to Legislative Effectiveness[22]

Legislation constitutes a tool for regulation and, as such, it should be viewed as a part of the legislative process, which in turn is a part of the policy process. From an instrumental, functional perspective, legislation is one of the tools available to governments in their attempt to regulate behaviours in response to undesirable social phenomena. Viewing legislation as a regulatory tool leads easily into the definition of a good law. A good law is simply one that is capable of contributing to the creation of the desired regulatory results.[23] Let us explore this further.

The ultimate goal for regulation is efficacy, defined as the extent to which regulators achieve their goals. Regulatory efficacy is a factual quality, a response to the factual question whether the regulatory exercise has indeed produced the results envisaged by the policymakers. On that basis, efficacy is a quality that appears post-legislatively. It requires not just the passing of the legislation but also its coming into effect. It also requires adequate post-legislative time to reveal itself, and it is dependent on the holistic and harmonious performance of all actors in the policy process. All of them need to play their part well, in order to allow efficacy to blossom and then to slowly flower. Policy officers need to identify the correct, concrete and precise regulatory goal – one that is not too general to become impossible to achieve but at the same time not too narrow to be uninspiring or unworthy of resources. Legal officers and law reformers need to share understanding of the regulatory aims, translate them into law reform, and identify the effect that this reform might have on the rest of the legal system. Drafting officers need to share

[21] For an analysis of the benefits of sunset clauses, see JE Finn, 'Sunset Clauses and Democratic Deliberation: Assessing the Significance of Sunset Provisions in Antiterrorism Legislation' (2009–2010) 48 *Columbia Journal of Transnational Law* 442, 447.

[22] The analysis of effectiveness has been developed in my recent work but originates mainly from Helen Xanthaki, 'On Transferability of Legislative Solutions' in C Stefanou and H Xanthaki (eds), *Drafting Legislation. A Modern Approach* (Ashgate, 2008) 1–18, 17; Helen Xanthaki, *Drafting Legislation. Art and Technology of Rules for Regulation* (Hart Publishing, 2014) 7.

[23] Helen Xanthaki, 'Drafting Manuals and Quality in Legislation: Positive Contribution towards Certainty in the Law or Impediment to the Necessity for Dynamism of Rules' (2010) *Legisprudence* IV 2 115.

understanding of the regulatory and law reform agenda, and identify the most appropriate legislative tool to make efficacy happen. If any of these three levels is flawed, efficacy can only be achieved haphazardly.

Legislative analysis focuses on legislative effectiveness, namely on the production of a legislative text that can contribute to regulatory efficacy. Effectiveness measures the capacity of the legislative text to express the regulation in a manner that can allow it to achieve efficacy. But successful legislative expression is only relevant in the presence of regulatory aims that actually address the social phenomenon in question and that are achievable by means of the regulatory solutions put forward by policy officers. A wonderfully drafted law cannot possibly contribute to the production of the desired regulatory results, if the choice of regulatory mechanism is haphazard and consequently doomed to miss its target from the word 'go'. Just as efficacy, effectiveness proves itself post-legislatively. But, in view of the inherent difficulty of measuring capacity accurately, its scrutiny is also inherently fluid.[24] Despite its fluidity and relativity, effectiveness is a solid criterion of legislative quality. It applies to all types of legislative instruments. If the purpose of legislation is to serve as a symbol, then effectiveness becomes the measure of achieved inspiration of the users of the symbolic legislation. If legislation is to be used as a ritual, effectiveness signifies persuasion of the users who bow down to its appropriate rituality. If legislation is functional, effectiveness measures the extent to which the desired regulatory results materialise.[25]

Legislative effectiveness is nurtured by clarity, precision and unambiguity of legislative expression. Clarity is the quality of being clear and easily perceived or understood. Precision is the exactness of expression or detail. Unambiguity is certain or exact meaning; semantic unambiguity requires a single meaning for each word used, whereas syntactic unambiguity requires clear sentence structure and correct placement of phrases or clauses.

Clarity is nurtured by simplified language and, where appropriate, gender inclusive language. Simplified language pitches legislative communication to the level of legal and subject-related sophistication of the users of the specific legislative text. Simplification refers to choices of words, syntax, structure and context. Gender inclusive language eliminates referents of gender in legislative expression.

Where do sunset clauses enter this paradigm of legislative quality? Sunset clauses serve as tools of clarity, precision and unambiguity; and as tools for efficacy. Let us explore the two angles separately.

Sunset clauses convey clearly, precisely and unambiguously when the legislation will end. They manage to convey one of the most complex and, until recently, vague regulatory implications, namely that of the precipitated end of the

[24] See Klaus Ziegert, 'The Theory and the Measurement of the Effectiveness of Legislation' [1983] 14 *Berichte und Kritik Rechtstheorie* 204, 240–46.

[25] See John Dickinson, 'Legislation and the Effectiveness of Law' (1931) *American Bar Association Journal* 645, 651.

legislation. This allows users to organise their affairs accordingly. Thus, they serve legal certainty and, as such, the rule of law.[26]

In doing so, sunset clauses also inform the user of the regulatory plan. They offer them the knowledge that this piece of legislation is introduced under the condition that it will be assessed over a set period of time. This contextual knowledge may be instrumental in instigating the behavioural changes sought by the regulators. When legislation is experimental, burdensome to implement, or even seemingly nonsensical, users are encouraged to still comply with it in the comfort of the knowledge that its effectiveness will be reviewed and the legislation terminated, if justified.[27]

Moreover, this invitation to participate in the regulatory game reassures users that legislation is passed after due consideration and only if needed, that its effect is indeed monitored closely, and that action will be taken to remove or reinstate it, as proven appropriate by post-legislative scrutiny exercises.[28] And so, sunset clauses contribute to the reversal of users' mistrust of legislation, the law and authority. This in turn feeds into effectiveness, as it supports an environment of regulatory compliance nurtured by trust.

Finally, sunset clauses enhance legislative quality within the statute book, as it allows for legislation – which has either served its purpose and is no longer needed or missed its regulatory targets – to be removed from the whole body of legislative texts.[29] Sunset clauses can be the optimal answer to the labyrinth of legislative texts that plague the statute book with complexity. In fact, they can be the civil law equivalent to law revision in the common law.[30]

Having identified the role of sunset clauses as contributors to effectiveness and, ultimately, efficacy, let us explore their role as guardians of legislative quality.

IV. Sunset Clauses as Guardians of Legislative Quality[31]

Guarding legislative quality is not a simple or mechanical task. The fluidity and relativity of legislation as a product invites a qualitative assessment of

[26] On a full analysis of time in legislation and the rule of law, see Sofia Ranchordás, 'The International Rule of Law Time After Time: Temporary Institutions between Change and Continuity' (2014) *The Netherlands Yearbook of International Law* 67–91.

[27] And parliamentarians may be more inclined to pass the legislation: see Tom Ginsburg et al, 'Libertarian Paternalism, Path Dependence, and Temporary Law' (2014) 81 *University of Chile Law Review* 291, 337.

[28] 'Fast regulatory solutions (or at least faster than lasting legislation) are offered because sunset clauses and experimental legislation have the potential to gather consensus more easily than permanent legislation': see Sofia Ranchordás, 'Innovation-Friendly Regulation: The Sunset of Regulation, the Sunrise of Innovation' (2015) 55 *Jurimetrics* 201, 217.

[29] See Lewis Davis, 'Review Procedures and Public Accountability in Sunset Legislation: An Analysis and Proposal for Reform' [1981] 33 *Administrative Law Review* 393.

[30] See Wim Voermans, Chris Moll, Nico Florijn and Peter Van Lochem, 'Codification and Consolidation in the European Union: A Means to Untie Red Tape' (2008) 29 *Statute Law Review* 65, 81.

[31] Xanthaki's paradigm of enlightened legislative scrutiny is fully developed in H Xanthaki, 'An Enlightened Approach to Legislative Scrutiny: Focusing on Effectiveness' (2018) *European Journal of Risk Regulation* 431–34.

legislative quality,[32] one that is equally flexible and fluid. But, at the same time, the accurate assessment of legislative quality requires a concretely set test that can block possible manipulation in order to produce 'the right' results.[33]

The prevalent method of evaluating effectiveness, as the criterion for legislative quality, is Mousmouti's effectiveness test. It proposes an assessment of objectives, content, context and results.[34] This innovative approach to legislative quality has value, in that it is original as the first test of effectiveness, and it is innovative as the first concretisation of the effectiveness concept. However, its admittedly useful departmentalisation into objectives, context, clauses and results encapsulates a collateral danger, ie it further sub-divides the already fragmented approach to legislative scrutiny into sub-categories for evaluation by present regulatory actors. However, fragmentation of legislative scrutiny is inherent not only in Mousmouti's effectiveness test but also, perhaps even more so, in current methodology of impact assessments and of cost benefit analysis.[35] These are excellent in assessing fragments of the mosaic that is legislation and regulation.[36] But legislation is not fragmented, and neither is regulation.[37]

In order to holistically reflect the overall patchwork of regulation, the assessment of legislative quality must recognise the interdependence of regulatory efficacy, legislative effectiveness, clarity/precision/unambiguity, simplification and gender inclusivity. Such a holistic assessment invites a comprehensive evaluation

[32] The link between quality and scrutiny is detailed in Organisation for Economic Co-operation and Development (OECD), 'Implementing Administrative Simplification in OECD Countries: Experiences and Challenges', www.oecd.org/dataoecd/0/37/37026688.pdf; and Dirk Meulen, 'The Use of Impact Assessments and the Quality of Legislation' (2013) 1 *Theory and Practice of Legislation* 305.

[33] See Elena Griglio, 'Post-Legislative Scrutiny as a Form of Executive Oversight Tools and Practices in Europe' (2019) 21 *European Journal of Law Reform* 118, 122.

[34] See Maria Mousmouti, 'Operationalising Quality of Legislation through the Effectiveness Test' (2012) 6:2 *Legisprudence* 201; see also Maria Mousmouti, 'Effectiveness as an Aspect of Quality of EU Legislation: Is It Feasible' (2014) 2 *Theory and Practice of Legislation* 309; and Maria Mousmouti, 'Making Legislative Effectiveness an Operational Concept: Unfolding the Effectiveness Test as a Conceptual Tool for Lawmaking' (2018) 9 *European Journal of Risk Regulation* 445.

[35] See, eg, World Bank Group, *Better Regulation for Growth: Governance Frameworks and Tools for Effective Regulatory Reform: Regulatory Quality Indicators* (World Bank, 2010).

[36] Maria Mousmouti lists the sets of indicators of regulatory quality as follows: 'The OECD Product Market Indicators measure the extent to which regulatory environments enable competition; the World Bank Indicators of governance and institutional quality cover the quality of regulatory outcomes within the broader framework of governance; the World Bank Doing Business Indicators focus on business related regulations in different areas based on information from both official sources as well as through the perceptions of government agencies and law firms; the OECD Indicators of government capacity to produce high quality regulations record the content of regulatory policies, quality tools, institutional arrangements, dynamic aspects and performance or outcome indicators. Further, indicators are used by governments either for conduct studies aiming to assess whether established practices were followed, impact studies that focus on the impact on decision making and accuracy studies which assess the quality of ex ante regulatory analysis by comparing predicted versus actual impacts. The OECD develops indicators for regulatory management systems.' See Mousmouti, Operationalising Quality of Legislation' (n 34) 198.

[37] This reflects the holistic description of effectiveness as a concept that brings together policy, law and drafting. See Sir Stephen Laws, 'Legislation and Politics' in D Feldman, *Law in Politics, Politics and Law* (Hart Publishing, 2013) 90.

of legislation as a tool for regulation, bringing together the currently fragmented evaluations of efficacy, effectiveness and cost-efficiency. It requires contribution from all actors in the policy/regulatory process.[38]

Such an evaluation is useful both at a pre-legislative stage, as a means of confirming the possible capacity of legislation to lead to efficacy, and also at a post-legislative stage, as a means of confirming whether the legislation, in fact, led to regulatory efficacy.[39] It will also involve questions relating to the whole policy/regulatory process, as effectiveness is not about words and legislative expression. It encompasses legislative expression (words and syntax), structure of the legislative text, context within the statute book, layout of publication, and explanatory materials. This holistic, overarching concept of effectiveness reflects and is nurtured by the holistic concept of legislative drafting and legislative scrutiny.

Thus, during pre-legislative evaluation, effectiveness demands answers to the following tests, all contributing answers to the question as to whether the legislative text is capable of producing the desired regulatory results.

1. Testing the conceptual effectiveness of the regulatory concept:
   a. What are the desired regulatory results?
   b. How can they be measured?
   c. Within how long are they to be achieved?
   d. Can the selected enforcement mechanisms produce the desired regulatory results within the set times?
   e. Can the text be implemented without stumbling on socio-legal hurdles?

2. Testing the effectiveness of the legislative concept:
   a. What is the legal mischief and how is it connected to the desired regulatory goals?
   b. Is this addressed by non-legislative means in the jurisdiction?
   c. How is the new legislation expected to address the mischief?

3. Testing the effectiveness of the legislative communication:
   a. What are the main groups of possible users of the text?
   b. What level of awareness of the topic do they have?
   c. What is their level of legal awareness?
   d. What questions does each group ask?
   e. Does the current structure provide accessible answers to the specific questions of each of the user groups?
   f. Does the current structure facilitate the unhindered identification of the relevant answers?

[38] On the need for better scrutiny of concrete regulatory results, see Paul Stephenson, 'Why Better Regulation Demands Better Scrutiny of Results' (2017) *European Journal of Law Reform* 97.

[39] M Zamboni, 'Legislative Policy and Effectiveness: A (Small) Contribution from Legal Theory' (2018) *European Journal of Risk Regulation* 6.

4. Testing the effectiveness of legislative expression:
   a. Does the language of each provision communicate its message to each of the intended user groups in a manner that they can understand (topic and law)?
   b. If this is difficult, what tools have been used to address complexity, ie examples, definitions or interpretation.
5. Testing the effectiveness of presentation:
   a. Is the layout simple?
   b. Does the text refer the user to relevant provisions? Perhaps via hyperlinks or explanatory notes.
6. Testing the effectiveness of monitoring:
   a. How often
   b. With which results?

The elements of this effectiveness test function consecutively. It is a concept of 'sudden death'. If the text fails one test, it automatically fails the ones that follow. So, the exercise stops at the moment of failure at any point in the scrutiny process. Failure in one test is evidence of ineffectiveness of the text.

During post-legislative evaluation, effectiveness demands consecutive answers to the following questions:

1. What were the desired regulatory results pursued by means of the legislation?
2. Have the selected regulatory mechanisms led to the desired regulatory results, as evidenced by empirical data?[40]
3. Has the legal mischief been addressed by new legislation, as evidenced by data?[41]
4. Were there any issues arising from the choice of legislative expression and how can these be addressed?
5. Is efficacy achieved?
6. If not, was the legislation effective as evidenced by questions 3 and 4?
7. If not, how can the issues be addressed: repeal, repeal and re-enact, or amend the legislation?
8. Within how many years does the legislation need to be re-evaluated?

This type of scrutiny claims value in that it is original as the first holistic exercise of legislative scrutiny; and it is effective as it contributes greatly to a closure of the current gap between legislative intent and legislative effect.[42] Xanthaki's

[40] OD Oliver-Lalana, 'Due Post-Legislative Process? On the Lawmakers' Constitutional Duties of Monitoring and Revision' in K Messerschimdt and AD Oliver-Lalana (eds), *Rational Lawmaking under Review. Legisprudence According to the German Federal Constitutional Court* (Springer, 2016) 259.

[41] J Rachlinski, 'Evidence-Based Law' (2011) 96 *Cornell Law Review* 901–23, 910.

[42] ER Beerworth, 'The Evaluation of Legislation' in R Tomasic (ed), *Legislation and Society in Australia* (Allen and Unwin, 1980) 66, 68.

scrutiny method forces policy officers to reveal concrete regulatory aims and their preferred method of achieving them. In turn, this forces law reformers to translate the regulatory aims to mechanisms of law reform, namely changes in the law, which forces and allows drafters to express this in an effective manner.

In doing so, Xanthaki's method of legislative scrutiny offers Parliament and its legislators an insight into what the legislation requires to perform, how it is hoped that this will be achieved, and on what basis the assessment of the success of the legislative text will take place at the post-legislative stage. Parliament can then use this data to evaluate the effectiveness of the legislative text both as part of their scrutiny when passing the legislation and as part of their scrutiny of the legislation in legislative reviews.[43]

It is precisely here that sunset clauses can serve as guardians of legislative quality. By providing for the end of the legislation, unless Parliament extends its duration, they empower Parliament to impose a full and meaningful post-legislative exercise[44] after a set number of years,[45] on the basis of which Parliament can either passively allow the legislative text to die or pro-actively extend its life further, either as it originally stood or after amendment. Let us explore this further.[46]

Since legislation is a tool for regulation, its performance is increasingly judged against a concrete set of measurable criteria for post-legislative scrutiny. These can be, and increasingly are, used for periodic review of legislation. Providing them in legislation by means of a new category of purpose or objective clauses – combined with a sunset clause for action resulting from periodic review – would not just reflect but mainly enhance the closure of the life cycle of regulation and legislation. Forcing government departments to conduct a post-legislative scrutiny exercise, as a means of persuading Parliament that the legislative text is effective[47] and can continue its life, is a proactive way of returning regulators to the efficacy and effectiveness agenda. Currently post-legislative review is assigned to government departments, albeit without the persuasive mechanism of a sunset clause. This can result in haphazard post-legislative scrutiny. Using sunset clauses as a tool to supplement and strengthen post-legislative scrutiny would greatly enhance legislative quality.

Sunset clauses enhance legislative quality not just within the legislative text but also at the level of the statute book. They do so by automatically updating the

[43] For an analysis of post-legislative scrutiny in the UK, see Tom Caygill, 'A Tale of Two Houses: Post-Legislative Scrutiny in the UK Parliament' (2019) 21 *European Journal of Law Reform* 87.

[44] See Lydia Clapinska, 'Post-Legislative Scrutiny of Acts of Parliament' (2006) 32 *Commonwealth Law Bulletin* 19, 198.

[45] See contra Nicola McGarrity, Rishi Gulati and George Williams, 'Sunset Clauses in Australian Anti-Terror Laws' (2012) 33 *Adelaide Law Review* 307, 308, who report that sunset clauses are accused of providing a convenient political excuse for shortcutting initial parliamentary debate about controversial legislation, thereby postponing the substantive debate until the legislation comes up for expiry or renewal.

[46] See Lewis Anthony Davis, 'Review Procedures and Public Accountability in Sunset Legislation: An Analysis and Proposal for Reform' (1981) 33 *Administrative Law Review* 393, 395.

[47] See P Suber, 'The Paradox of Self-Amendment: A Study of Law, Logic, Omnipotence, and Change (Peter Lang International Academic Publishers, 1990) 14.

statute book every three or five years, depending on the duration of legislation. It makes perfect sense for end provisions to be inserted in legislation as standard, allowing for exceptions only where the possibility of a lack of legislative regulation if the legislation dies unintentionally may create regulatory chaos or unease. In the EU, sunset clauses are part of the Smart Regulation agenda and are becoming increasingly frequent. But, then again, sunset clauses, on their own, are not effective regulatory review weapons; they need to be supplemented by meaningful post-legislative scrutiny exercises undertaken on the basis of measurable criteria.

The starting point for post-legislative scrutiny is the following question: what were the original desired regulatory results? If the regulatory results have been produced, the capacity of the legislative text to contribute to the regulatory effort is proven. If the results are not there, the social scientists participating in the scrutiny exercise need to identify the error. If the error lies with the choice of the regulatory mechanism, then alternative means of regulation can be sought and the legislation can be repealed, or the legislation can be supported by additional regulatory mechanisms. If the problem lies with the content of the legislation, the legal officers (namely law reformers or legal experts in the relevant substantive field of law) must investigate the effects of the legislation on the legal system, and identify the continuing mischief. The mischief can be a result of a legal or drafting error. If the legal system corrected itself in response to the legislation, the lawyers must identify how this occurred and how to prevent it from happening.

Although sunset clauses cannot by themselves guarantee legislative quality, they do serve as the guardians to the portal that is holistic and effective legislative scrutiny. They get the ball rolling by first ensuring that post-legislative scrutiny will take place for legislation that will continue to be in effect after its expiry date, and second that Parliament has an active role to play in post-legislative scrutiny. Of course, sunset clauses cannot force the executive to extend the legislation, nor interest groups to campaign for or against it,[48] nor the legislature to scrutinise effectively[49] and avoid using them as a snooze button to democracy.[50] And nor should they. They are simple tools that open the gates to a possibility of effective post-legislative assessment. But, combined with concrete purpose clauses, they can be mighty guardians of legislative quality.

[48] See Forrest Maltzman and Charles Shipan, 'Change, Continuity and the Evolution of the Law' (2008) 52 *American Journal of Political Science* 252, 254.

[49] For example, Parliament may lack the capacity to question effectively a Departmental review of legislation: see Editorial, 'Post-Legislative Scrutiny' (2006) 27 *Statute Law Review* 1, 3.

[50] For an exhaustive analysis of sunset clauses and how their use can snooze democracy during critical times, see Antonia Kouroutakis and Sofia Ranchordás, 'Snoozing Democracy: Sunset Clauses, De-Juridification, and Emergencies' (2016) 25 *Minnesota Journal of International Law* 29, 35.

V. Conclusions

This chapter set out to shed new light on sunset clauses, by departing from the analysis of their constitutional role and focusing on their legislative role instead. The hypothesis was that sunset clauses contribute to legislative effectiveness and regulatory efficacy. The methodology chosen involved: 1) defining and analysing sunset clauses as a legislative tool; 2) analysing legislative effectiveness as a criterion for legislative quality, and identifying the role of sunset clauses as contributors to legislative quality; 3) analysing Xanthaki's enlightened scrutiny test, and identifying the role of sunset clauses as guardians of legislative quality; and 4) proving that sunset clauses are both contributors to legislative quality and guardians of regulatory efficacy.

Sunset clauses were defined as a sub-category of expiry provisions, which in turn are a sub-category of duration provisions. Sunset clauses are complex expiry provisions in that they convey the end of the life of legislation at a set time, and they allow an extension of that life subject to parliamentary evaluation of the effectiveness of legislation.

Sunset clauses are placed at the end of the legislative text but combined either with a visual indication of the status of the legislative text as currently in force or not, or a hanging provision at the start of the text. This reflects the balance between the primacy of their regulatory message and the need to keep the prime time of the users' attention for the main regulatory communication.

From a legislative perspective, and contrary to legality or constitutionality perspectives, the express repeal of the expired legislative text is needed to clarify the status of the text as current or expired.

In traditional drafting, sunset clauses are necessary in four distinct types of legislation, namely: 1) controversial laws; 2) experimental legislation; 3) legislation responding to a short-term need; and 4) authorisation of emergency powers. But sunset clauses are equally useful in all types of legislation, because they can serve as an effective tool for enhancing legislative effectiveness and, ultimately, regulatory efficacy.

Sunset clauses enhance legislative quality because they can serve as tools of clarity, precision and unambiguity, and as tools for regulatory efficacy. As sunset clauses convey clearly, precisely and unambiguously when the legislation will end, they promote legal certainty. Consequently, they strengthen the rule of law. In disclosing their end clearly, precisely, and unambiguously, sunset clauses reveal the regulatory plan to the user, thus instigating increased compliance with and trust of legislation, the law, and authority. This feeds back into effectiveness and efficacy. Finally, sunset clauses serve as periodic cleaners of the statute book, taking the role of small-scale, focused law revision.

Sunset clauses also serve as guardians of legislative quality. In Xanthaki's holistic scrutiny method, sunset clauses empower Parliament to impose a full and meaningful post-legislative exercise after a set number of years, on the basis of which Parliament can either passively allow the legislative text to die or proactively extend its life further, either as it originally stood or after amendment. Thus, single-handedly, sunset clauses manage to: force the executive to evaluate the legislation; nudge Parliament to undertake substantive and effective post-legislative scrutiny; and cause the statute book to rid itself of legislation that proved unworthy of its place in the statute book (either by omission to extend or by failure to pass parliamentary scrutiny).

The conclusion of this analysis is that sunset clauses are certainly worth considering as a standard provision in most legislative texts. However, as is the case with all matters concerning drafting, they do carry with them some risk. The obvious problem is the risk of an unintentional regulatory gap. It is possible to imagine a situation where the executive is not aware of the imminent expiry of the legislation and does not move to extend it; the legislation expires, and regulation of this area stops without warning, without provision, and without any replacement.[51] One could argue that the risk of unintended regulatory gaps may, at least in certain cases, outweigh the benefits of a sunset clause. Of course, this is not a risk that can be dismissed generally at an abstract level; the threat is present, and, in certain cases, could lead to regulatory chaos. But, drafting is a series of dilemmas. Thus, the introduction of a sunset clause requires a balancing exercise between the benefits of the clause against the risk of a potential unintended regulatory gap. Criteria for making that choice could be the topic of the regulation, the imperative (or not) need for constant regulation in the field, the possible ability of surrounding legislation or even future self-regulation to cover the area of activity involved, or the agility of executive and legislature in the jurisdiction in question to act quickly and effectively, should the need occur. The decision can only be ad hoc and so the danger of a possible, unintended regulatory gap cannot stand in the way of sunset clauses as a tool for legislative effectiveness and regulatory scrutiny.

An additional weakness of the otherwise mighty sunset clauses is their dependence upon the regulatory sophistication of the jurisdiction that they serve. In order for sunset clauses to perform satisfactorily, they require regulators that are aware of, and able and willing to, express their regulatory aims as policy goals and to share them with the drafters; legal officers that have the capacity to express these policy goals in law reform terms; and drafters or lawmakers that can translate law reform terms into sophisticated, concrete and effective modern purpose/objectives clauses. In other words, for sunset clauses to work effectively, they need to feed and flourish from a sophisticated regulatory environment, which is dedicated to regulatory efficacy, legislative effectiveness and economic cost-efficiency.

This last paragraph seems to negate the praises sung for sunset clauses in this chapter. Indeed, it is questionable whether sophisticated regulatory environments exist, let alone whether they are allowed to function in the ideal scenario described

[51] See Paco Francoli, 'Bills Need Sunset Clauses, Urges Grit Senator Banks: "Some Pieces of Legislation Are Sort of Sitting There Almost Hanging over Our Heads"' (2003) *The Hill Times* 677.

above. One would expect that they tend to bow down to pressures of expediency and political compromise. But political reality cannot take away from the fact that sunset clauses have already been useful tools for legislative effectiveness and regulatory efficacy. And thus, it seems that the current undeveloped regulatory environment against which they are set does not hinder their value. If anything, it actually enhances it by showcasing how sunset clauses can contribute to efficacy and effectiveness. Who knows what sunset clauses can achieve within a developed regulatory environment, strengthened further by equally concrete and holistic purpose/objectives clauses.

This chapter suggests that it is worthwhile giving them, and purpose/objectives clauses, a chance to close the current gap in the cycle of regulation.

10

The Legisprudential and Political Functions of Temporary Legislation

ITTAI BAR-SIMAN-TOV AND GAYA HARARI-HEIT*

I. Introduction

Temporary (or 'sunset') legislation refers to statutory provisions enacted for a limited time and set to expire unless their validity is extended. Temporary legislation reverses the default rules for policy continuation: whereas the traditional default rule in legislation is that 'the statute's legal validity continues in perpetuity, the default rule for temporary legislation is that legal validity terminates at the sunset date'.[1] This unique feature of temporary legislation makes it an attractive legislative tool for various legislative functions and uses.

In this chapter, we explore the different functions of temporary legislation, distinguishing between what we term 'legisprudential' functions and political uses. Concerning its legisprudential functions, temporary legislation is used for professional legislative reasons to further legislative ends. It is employed as a regulative-legislative tool to deal with situations in which the nature of the law, policy or circumstances requires the temporality of the norm. Concerning its legisprudential uses, temporary legislation is chosen when the temporary nature of the law is deemed particularly helpful for increasing the effectiveness of dealing with these types of situation. Concerning its political uses, in contrast, temporary legislation is used for narrow political reasons. It is employed as a political tool for overcoming objections and facilitating the passage of the law (regardless of whether temporality is required to achieve the purposes of the law or to deal with the policy situation).

* We are indebted to our friends at the Knesset Legal Department, Daphna Barnai, Gali Ben-Or and Merav Israeli, for helpful conversations and suggestions. We thank Itay Cohen, Chani Koth and Merav Knafo for excellent research assistance and help with translations. We gratefully acknowledge the financial support of the Minerva Center for the Rule of Law under Extreme Conditions, Faculty of Law and Department of Geography and Environmental Studies, University of Haifa.

[1] Jacob E Gersen, 'Temporary Legislation' (2007) 74 *University of Chicago Law Review* 247, 261.

Temporary legislation has garnered increased attention in the past decade,[2] including in theoretical discussions about the functions or uses of temporary legislation.[3] Yet, there is a scarcity of empirical studies about the actual uses of temporary legislation in practice.[4] This chapter contributes to filling this gap, by empirically exploring the various functions of temporary legislation as employed by the Israeli Parliament (the Knesset). While the common wisdom in many jurisdictions is that temporary legislation is rarely used,[5] Israel provides a great case for studying temporary legislation and how it is used in practice, simply because temporary legislation is relatively widely used.[6]

Section II describes this study's multi-method methodology, combining qualitative and quantitative methods. Subsequently, section III discusses the study's findings about the legisprudential functions revealed in the Knesset's employment of temporary legislation. Section IV is focused on the study's findings about the political uses revealed in the Knesset's use of temporary legislation. Finally, section V concludes with a normative discussion about the appropriate uses of temporary legislation.

II. Methodology

In this study, we used a multi-method approach, in which qualitative content-analysis was complemented by quantitative analysis.[7]

We examined the functions and uses of temporary legislation through in-depth qualitative content analysis of temporary laws, their draft bills, and their legislative processes.[8] The materials we examined included the explanatory notes attached to bills and legislative records from the plenum and committee debates. We focused on the stated purpose of enacting the law as a temporary measure, that is, the reasons that the bill or the legislators gave for choosing temporary legislation.

[2] Ittai Bar-Siman-Tov, 'The Lives and Times of Temporary Legislation and Sunset Clauses' (2018) 66 *The American Journal of Comparative Law* 453.

[3] See, Gersen (n 1) 273–78; Sofia Ranchordás, *Constitutional Sunsets and Experimental Legislation: A Comparative Perspective* (Edward Elgar Publishing, 2014) 74–108.

[4] John E Finn, 'Sunset Clauses and Democratic Deliberation: Assessing the Significance of Sunset Provisions in Antiterrorism Legislation' (2009) 48 *Columbia Journal of Transnational Law* 442, 444, fn 6.

[5] Ranchordás (n 3) 9; Sylvia Veit and Bastian Jantz, 'Sunset Legislation: Theoretical Reflections and International Experiences' in Alberto Alemanno and others (eds), *Better Business Regulation in a Risk Society* (Springer, 2012).

[6] Ittai Bar-Siman-Tov, 'Temporary Legislation, Better Regulation and Experimentalist Governance: An Empirical Study' (2018) 12 *Regulation & Governance* 192.

[7] Laura Beth Nielsen, 'The Need for Multi-Method Approaches in Empirical Legal Research' in Peter Cane and Herbert M Kritzer (eds), *The Oxford Handbook of Empirical Legal Research* (Oxford University Press, 2010).

[8] Klaus Krippendorff, *Content Analysis: An Introduction to Its Methodology*, 3rd edn (Sage, 2013); Jonathan B Slapin and Sven-Oliver Proksch, 'Words as Data: Content Analysis in Legislative Studies' in Shane Martin, Thomas Saalfeld and Kaare W Strøm (eds), *The Oxford Handbook of Legislative Studies* (Oxford University Press, 2014).

We concentrated on revealing the participants' perspectives during the legislative process, and chose to present their express words.

The methodological approach that guided us was that an in-depth qualitative approach was particularly suitable for revealing the full complexity of our researched phenomenon and reaching a deeper and richer understanding of the way the Knesset employs temporary legislation, all the while being aware of different critiques as to the relatability of qualitative research and of the methodological literature which discusses ways to cope with these.[9] We therefore attempted to provide 'thick descriptive data' and present direct and extensive quotes from the texts which we relayed upon (as much as possible, given the text limits).[10]

To reveal the political function of temporary legislation as a means to facilitate the passage of legislation, we complemented the qualitative approach with an extensive quantitative study that measured the passage rate of temporary governmental bills *vis-à-vis* all governmental bills. The study covered all governmental bills considered and enacted between the 12th Knesset and 20th Knesset.[11] This timeframe covers nine Knessets across a period of 30 years (from 1988 and up to 24th June 2018). For each of these nine Knessets, we calculated the passage rate of all governmental bills in comparison to the passage rate of governmental bills that were temporary legislation. Passage rates were calculated by comparing the number of governmental bills considered by a particular Knesset and the number of enacted laws that originated from governmental bills by the same Knesset. Data were collected from the National Legislation Database on the Knesset website.

III. Legisprudential Functions of Temporary Legislation

As noted in the Introduction, the provisional nature of temporary legislation makes it an apt means for various functions or uses. The legisprudential functions entail that temporary legislation is used as a regulative-legislative tool for furthering legislative aims or meeting policy challenges, in situations that call for a provisional solution or when there is a need for the informational benefits of

[9] See, eg, Michael Quinn Patton, *Qualitative Research & Evaluation Methods*, 2nd edn (Sage, 1990) 10, 94–97, 187;Lisa Webley, *Qualitative Approaches to Empirical Legal Studies* in Peter Cane and Herbert Kritzer (eds), *The Oxford Handbook of Empirical Legal Research* (Oxford University Press, 2010) 926.

[10] Yvonna S Lincoln and Egon G Guba, 'But Is It Rigorous? Trustworthiness and Authenticity in Naturalistic Evaluation' (1986) *New Directions for Program Evaluation* 73.

[11] We began with the 12th Knesset, because this Knesset is commonly considered as the turning point, following several important changes and reforms, that shaped the contemporary Knesset. Reuven Y Hazan, 'Executive-Legislative Relations in an Era of Accelerated Reform: Reshaping Government in Israel' (1997) 22 *Legislative Studies Quarterly* 329. We chose to focus on governmental bills (as opposed to private member bills), because a previous study revealed that most temporary legislation was a result of governmental bills. For a more elaborate explanation, see Ittai Bar-Siman-Tov and Gaya Harari-Heit, 'Temporary Legislation's Finest Hour?: The Rise of Temporary Legislation in Israel and Principles for its Improvement' (forthcoming, 2019) 41 *Tel Aviv University Law Review*, https://papers.ssrn.com/sol3/papers.cfm?abstract_id=3064829 (in Hebrew) (last accessed 19 September 2019).

temporary legislation stemming from the provisional sequential nature of temporary legislation.

Our conception of legisprudential uses builds upon Gersen's discussion of 'technocratic applications' of temporary legislation. Gersen notes that while temporary legislation can produce strategic or political benefits for legislators, 'at least some of the time, the motivation for adopting temporary measures appears to be more benign. Indeed, temporary statutes are often an appropriate match to specific policy challenges. These uses of temporary legislation might be termed technocratic.'[12] He identifies three such uses: (1) filling gaps in existing law; (2) responses to policy problems that are perceived to be temporary; and (3) experimental regulation.[13]

Interestingly, our examination of Knesset debates revealed a similar conception of the functions of temporary legislation. For example, in discussions held by a joint committee of the Finance Committee and the Constitution, Law and Justice Committee, the Finance Ministry's Legal Adviser explained:

> In our legislation there are two types of temporary provisions from a conceptual point of view. There are temporary provisions that stem from a temporary need, when I make a temporary provision in view of that need and it provides a response for that temporary need. There are temporary provisions that begin as a trial and a test … in general there is sometimes a need to conduct a trial, we go into something and we don't yet know how it will work out, and you want to test the matter.[14]

We also found a similar conceptualisation of the uses of temporary legislation in the Israeli Supreme Court decisions about constitutionality of various temporary laws.[15]

This study has indeed found examples of these uses of temporary legislation in Israel. Interestingly, however, we have uncovered additional legisprudential uses, beyond the three uses discussed by Gersen. The main uses that have been found are detailed below, including examples taken from relevant temporary laws.

A. Temporarily Filling a Legislative Void (Placeholder Legislation)

An especially illustrative example of this type of use is the Military Service Law (candidates for military service who practise their religious beliefs) (Temporary

[12] Gersen (n 1) 273.

[13] Gersen (n 1) 273.

[14] Yoel Baris, The 18in Knesset Joint Committee of the Finance Committee and the Law and Constitution Committee Minutes (11 May 2010) No 2, Knesset Archives, Jerusalem, Israel.

[15] HCJ 466/07 *Galon v Attorney General* 65(2) PD 44, para 25 of President Beinisch's ruling and para 42 of Justice Meltzer's ruling (2012) (Isr.). HCJ 8260/16 *College of Law & Business v The Knesset)* 6 September 2017), Nevo legal database (by subscription, in Hebrew), para 24 of Vice-President (ret.) Rubinstein's ruling (Isr.).

Order) 5761-2001. The background to the enactment of this law was the Court's ruling in *Rubinstein v Minister of Defence* that the arrangement by virtue of which the Minister of Defence ordered the postponement of the service of Yeshiva students is no longer legal, and that an arrangement on this matter must be established via primary legislation of the Knesset.[16] The Government subsequently set up a Commission in order to formulate recommendations for a proper arrangement for the enlistment of Yeshiva students, and thereafter worked on enacting a law that would enact the committee's recommendations. In the meantime, the Knesset asked the Court time after time to extend the suspension of the outcome of its judgment. Since the Knesset discovered that it was unable to complete the process of enacting said law and the final date of the extension of the Court was about to expire, it requested another postponement from the Court; however, the Court ultimately lost its patience and refused the request. This was the background which led to the enactment of the temporary provision. The purpose of this temporary law was not to establish a new or permanent arrangement (nor even to conduct a temporary experiment), but only to fill a legislative void in order to enable the continuation of the old arrangement – by virtue of which the Minister of Defence ordered the postponement of the service of Yeshiva students – between the expiration date of the suspension of the Court's invalidation and the completion of the enactment of the law that will constitute the recommendations of the Commission. This is aptly described by the Court in the *Ressler* case, which dealt with the constitutionality of the abovementioned temporary law:

> The wording of the temporary law indicates that it is intended to fill the void left by the decision of the Court […] not to suspend the validity of the judgment in the *Rubinstein* case […] and to anchor in legislation for a limited period of time […] the authority of the Minister of Defence to postpone the military service of yeshiva students […] on the basis of the same terms [that existed] before the Court ruled in the *Rubinstein* case that the arrangement […] is unlawful.[17]

Another example is the Detention of Suspected of Security Offenses Law, which was passed (and extended several times) to a large extent in order to allow the Government and the Knesset to complete the enactment of the Anti-Terrorism Law, 5766-2016. Indeed, in the process of enacting the original temporary law, and also during its extensions, the Committee's chairpersons and other legislators explicitly reiterated that the purpose of the law is to fill the void until the completion of the legislative process of the Terrorism Law, which would, inter alia, regulate the authority that has been enshrined in the temporary law.[18]

[16] HCJ 3267/97 *Rubinstein v Minister of Defence* [1998] IsrSC 52(5).

[17] For a more detailed description of the events, see HCJ 24/01 *Ressler v The Knesset* 65(2) PD 699 (2002) (Isr.)

[18] See, eg, Menachem Ben-Sasson, The 17th Knesset Minutes (26 June 2006) No 28 ('What will happen in the coming year and a half, in which the law exists … During this period, a law against terrorism will be enacted …'); Itzhak Levi, The 17th Knesset Minutes (18 December 2007) No 182 ('The law was enacted a year and a half ago as a temporary law, but now we are extending it for another three

B. Temporarily Freezing of the Legal Situation (Moratorium)

A similar, although slightly different, use of temporary legislation is as a means of temporarily freezing the legal situation (moratorium). The purpose behind this use is to prohibit certain activities or legal changes for a specified period, usually in order to enable the legislature to formulate a position on an issue.

An illustration of this type of use of temporary legislation is found in the Cloning Law.[19] Article 1 of the original version of the law explicitly stated that the purpose of this law was to set a fixed period of five years, in which no type of genetic intervention will be carried out on humans, in order to enable an examination of the moral, legal, social and scientific aspects of these types of intervention and their implications for human dignity. The law was enacted in the late nineties, and the explanatory notes of the bill stated, inter alia:

> Since the publications about the cloning of Dolly the sheep in Scotland in February 1997, a heated public debate has arisen around the world and in Israel about the issue of cloning […] for the creation of human beings. This issue has diverse ethical, moral, social, legal and medical implications, whose examination requires a period of time in which the society can profoundly formulate its position […]
>
> Since the public debate is still in its infancy, and a uniform position has yet to be formulated among the participants in the debate, this bill proposes a time-out of five years, during which an exhaustive public debate will be carried out and the implications of the medical developments and their ethical, moral, legal and social implications will be clarified.
>
> During this period, the performance of any genetic intervention will be prohibited […][20]

The law thus established a moratorium (or 'time out' in the bill's words) of five years in order to formulate a position on the new and complex issue of human cloning.

C. Temporary Response to Problems Perceived as Temporary and Responding to Emergency or Crisis Situations

Another category is temporary legislation designed to address current needs or problems perceived as transitory, such as dealing with crises, emergencies and

years, because the Ministry of Justice has not yet prepared a permanent proposal regarding these provisions. These provisions must be included in the Terrorism Law. The terror law is not ready'.); David Rotem, The 18th Knesset Minutes (20 December 2010) No 192 ('We request that the temporary law be extended by two years. At this time, the committee will discuss a new terror law, which will come to it and regulate, among other things, this matter as well').

[19] Prohibition of Genetic Intervention (Human Cloning and Genetic Modification in Reproductive Cells) Law 5759-1999.

[20] Prohibition of Genetic Intervention (Human Cloning and Genetic Modification in Reproductive Cells) Bill 5758-1998.

special security needs that are perceived as temporary. A clear example of a law enacted with such a declared purpose is the Citizenship and Entry into Israel Law on family unification with residents of the Occupied Territories.[21] The law was enacted following the second *Intifada*,[22] and its stated purpose was to provide a temporary solution to the special security needs created by this reality, which was presented as temporary. This was the sole declared purpose found in the explanatory notes to the law, and was also the prominent and major (albeit certainly not the exclusive) purpose presented in the parliamentary debates on the law.

MK Yuri Stern, Chair of the Interior and Environment Committee, said: 'We are in a state of war with the Palestinians, and this legislation is emergency legislation for a state of war.'[23] This was also stated by the Minister of the Interior, Avraham Poraz, at another meeting: '[…] It is very similar to the extension of the emergency regulations, because this law is entirely a law for emergencies.'[24]

Attorney General Menachem Mazuz also noted the following during a discussion two years later at the Knesset Internal Affairs and Environment Committee on a proposed amendment to the Temporary Order:

> The law was enacted against a background of a certain security situation, and in order to address a security problem […] it was an enacted as an arrangement for a limited period, which represents the fact that the background to this arrangement is the special security situation prevailing at the time.[25]

Attorney Danny Geva, legal advisor to the Prime Minister's Office, had similarly said:

> The purpose behind this law is very clear: there is a problem, this is not a normal situation. There are special circumstances on which this bill is based, and as its name indicates, it is a temporary provision. As a temporary provision, it adapts itself to the circumstances […] At present the security situation is still defined, unfortunately, as an armed conflict between the Palestinian Authority and Israel […] I wish that the situation will change for the better and there will be no need for it.[26]

However, it is important to note that we discovered that temporary legislation in this category is not limited to terrorism legislation or legislation designed to deal with security emergencies. It turns out that the concept of emergency or crisis

[21] Citizenship and Entry into Israel Law (Temporary Provision) 5763–2003.

[22] The Palestinian uprising against the Israeli occupation of the West Bank and Gaza Strip.

[23] The 16th Knesset Internal Affairs and Environment Committee Minutes, *General Reservations about the Proposed Citizenship and Entry into Israel Law* (14 July 2003) No 47, Knesset Archives, Jerusalem, Israel.

[24] The 16th Knesset Internal Affairs and Environment Committee minutes, *Voting on the Articles of the Citizenship and Entry into Israel Law* (30 July 2003) No 66, Knesset Archives, Jerusalem, Israel.

[25] The 16th Knesset Internal Affairs and Environment Committee Minutes, *Comments and Reservations about the Proposed Citizenship and Entry into Israel Law (Temporary Order) (Amendment), 2005* (28 June 2005) No 459, Knesset Archives, Jerusalem, Israel.

[26] The 16th Knesset Internal Affairs and Environment Committee minutes, *Comments and Reservations about the Proposed Citizenship and Entry into Israel Law (Temporary Order) (Amendment), 2005* (28 June 2005) No 456, Knesset Archives, Jerusalem, Israel.

embodied in this category is very broad. Thus, for example, we found a series of temporary laws designed to provide a temporary solution to the inflation crisis of the 1980s as well as to other economic crises or unemployment problems. Recently, another series of temporary laws were enacted to deal with the housing crisis in Israel.

This use appears to be one of the most common uses of Israeli temporary legislation. Thus, in a previous study, where we examined in-depth a set of 68 temporary laws passed by the Knesset over the years and the declared purposes of their legislation, we found that 27 of them (39.7 per cent) belonged to this category.[27]

D. Temporary Legislation as a One-Time Arrangement for a One-Time Event

Alongside the use of temporary legislation in response to the needs of the hour in the sense discussed above, we have also revealed a slightly different category in which temporariness is even more focused: temporary laws that can be classified as one-time laws. Thus, for example, we have found several temporary laws, each intended to regulate the funding of elections for only one specific election.

We also found temporary laws designed to regulate one transient event, such as arrangements relating to the events celebrating a decade of Israel's independence. In the abovementioned previous study, which examined the purposes of 68 temporary laws, we found that ten of them (14.7 per cent) belonged to this category.[28]

E. Temporary Legislation as a Deviation from the Norm Without Changing the General Norm ('Temporary De-Juridification')[29]

A distinct use of temporary legislation concerns the desire to establish a specific arrangement that will not change the general norm and will not affect the entire legal system. This use is related to the one discussed in subsection III.C above, because it often stems from the perception that an emergency requires arrangements deviating from the norm, but this should be a temporary deviation that will not change the general and permanent norm.[30] The aim of temporary legislation here is to prevent the normalisation of exceptional arrangements, not only

[27] Bar-Siman-Tov (n 6).
[28] *Bar-Siman-Tov* (n 6).
[29] Antonios Kouroutakis and Sofia Ranchordás, 'Snoozing Democracy: Sunset Clauses, De-Juridification, and Emergencies' (2016) 25 *Minnesota Journal of International Law* 29.
[30] *Ibid*; Bruce Ackerman, 'The Emergency Constitution' (2004) 113 *Yale Law Journal* 1029, 1040.

in the practical sense of the sunset (or duration) clause, which prescribes that the validity of the exceptional arrangement will expire within a defined time, but also in a more symbolic sense, based on the perception that its enactment as temporary legislation marks it out as merely a temporary deviation from the norm.[31]

In this case, too, the discussions surrounding the Citizenship and Entry into Israel Law provide clear examples. For example, MK Moshe Gafni said: '[...] it must be in a temporary order [...] I do not agree that this will be a permanent law. It is a stain on the statute book of the State of Israel.'[32] MK Azmi Bishara also stated: 'This is the reason that you are always extending the temporary order. A civilized or uncivilized country cannot include such a law among its legal system.'[33]

However, this use of temporary legislation does not necessarily concern emergency legislation. This use expresses a broader view that a law that deviates from the constitutional principles of the legal system must be temporary, in order to clarify that the deviation from these principles is temporary and specific only, and does not express a permanent change or a general withdrawal from the constitutional principles. A clear expression of this can be found in the Basic Law: Freedom of Occupation, in the notwithstanding clause entitled 'The validity of a deviant law.' The clause states that the validity of a law which violates the freedom of occupation and does not meet the conditions of the limitation clause, but is enacted under the terms of the notwithstanding clause (ie, in a law passed by a majority of 61 Knesset members, which explicitly states that it is valid despite the provisions of this Basic Law), 'shall expire at the end of four years from the date of its commencement, unless an earlier date has been set forth therein.'[34]

F. Temporality as a Means of Mitigating the Constitutional Violation (Temporality as a Means of Proportionality)

A further use of temporary legislation in Israel is as a means of reducing the constitutional violation, especially in legislation that restricts freedoms or violates constitutional rights. A clear example of this can be found in the Citizenship and Entry into Israel Law. For instance, a legal advisor from the Ministry of the Interior said during one of the discussions on the original bill: 'Because of the severity

[31] John Ip, 'Sunset Clauses and Counterterrorism Legislation' (2013) [2013] *Public Law* 74, 82.

[32] The 17th Knesset Internal Affairs and Environment Committee Minutes, 22–23 (20 March 2007) No 135, Knesset Archives, Jerusalem, Israel.

[33] The 16th Knesset Internal Affairs and Environment Committee Minutes (28 June 2005) No 459, Knesset Archives, Jerusalem, Israel.

[34] Basic Law: Freedom of Occupation, 5752-1992, s 8. *cf* Rivka Weill, 'Juxtaposing Constitution-Making and Constitutional-Infringement Mechanisms in Israel and Canada: On the Interplay Between Common Law Override and Sunset Override' (2016) 49 *Israel Law Review* 103, 106. ('The sunset override ... must be limited in duration. Only its sunset characteristic transforms it into a mechanism of infringement rather than one of amendment.')

of this matter, the legislation is done by way of a temporary order that is valid for one year [...]'[35] This explanation for the use of temporary legislation was also explicitly mentioned in Mazuz's comments on the original bill:

> The bill ... includes certain balances. Contrary to the [original] government decision, which was more general and more sweeping, the bill contains several restrictions or limitations in order to balance the general prohibition imposed by law. The most important limitation is the very fact that the proposed law is a temporary order for a year, not a general provision that is not limited in time.[36]

In a later debate, Mazuz added:

> Everyone was aware that the restrictions imposed by the law are very difficult from a humanitarian and legal point of view. We in the government legal system sought to balance and soften the law in various ways. There were two main components that we introduced into the original law. The first was that the law was enacted as a temporary order and not a permanent arrangement.[37]

Indeed, a recurring theme we have found in parliamentary debates on various and varied laws is the perception of time delimitation to ensure a more proportionate solution.[38] Although we did not examine this quantitatively, given the variety of examples encountered, it seems that this use of temporary order as a means of softening constitutional violations is common in Israel.[39]

G. Temporary Legislation as a Tool for Experimental Regulation

An additional use of temporary laws is as a tool for experimental legislation. The purpose behind this type of temporary law is to try a new policy or method temporarily, to obtain information about the effectiveness of the law and its

[35] Daniel Salomon, The Knesset Internal Affairs and Environment Committee Minutes (14 July 2003) No 47, Knesset Archives, Jerusalem, Israel.

[36] *Ibid.*

[37] Minutes No 459, *ibid*: fn 72.

[38] Due to space limitation, two examples out of many will suffice: The Education, Culture & Sport Committee Minutes No 701 of the 18th Knesset, 14–15 (15 October 2012) ('One of the things that we want to emphasize is the fundamental point that this law, even though it involves some violation of freedom of occupation, is a law ... [whose violation is not] excessive. The central balance of this law, beyond its provisions, is the fact that this law is limited to three years, that is, a temporary order.'); The Internal Affairs and Environment Committee & The Constitution, Law and Justice Committee Minutes No 6 (15 March 2010) ('It was a temporary order because this arrangement also has difficulties. We believe that this is a proportionate, reasonable and balanced provision, but part of the proportionality was that it be a temporary order. In the draft they put before us, the legal advisors dropped the temporary order, but the government wants it to remain in a temporary law.').

[39] Yaniv Roznai, 'Book Review – Sofia Ranchordás, Constitutional Sunsets and Experimental Legislation: A Comparative Perspective' (2016) 64 *American Journal of Comparative Law* 790; Suzie Navot, *The Constitution of Israel: A Contextual Analysis* (Bloomsbury Publishing, 2014) 57–58, 201; Suzie Navot, 'The Constitutional Dialogue: A Debate Through Institutional Mechanisms' (26 February 2019). Mishpatim Online. Available at SSRN: https://ssrn.com/abstract=3342189.

implications, and then, at the end of the time assigned to the temporary law, to draw conclusions from the information accumulated during the trial period. Legislation of this type is required mainly in cases of legislating under conditions of uncertainty and lack of information.[40]

A clear example of this use of temporary legislation is the Law for Protection of Literature and Authors in Israel (Temporary Provision). This law, which sought to deal with a structural problem in the book market, was legislated for three years, with the declared purpose of conducting an experiment and gathering information before the decision on permanent regulation. This was explicitly stated in paragraph 23 of the bill:

> It is proposed to enact the law as a temporary measure for a period of three years, since this law is meant to correct a severe market failure in the book industry in the country in light of its unique structure. Therefore, it is proposed to test the effectiveness of the law over a limited period of three years. Based on the data collected during the period of operation of the law, the minister in charge could […] extend […] the period of validity of the proposed law.

This goal was repeated many times by several of the pivotal participants in the legislative process. The Minister of Culture, for example, remarked:

> The law is temporary […] After three years, its consequences and its results will be examined. Someone said here before, one of the members of the Knesset […] that you hope there will be results. That's just it, that's the point. This is the reason that it is for three years, to examine the consequences.[41]

In a later debate, she added: 'This is a temporary order. It is for three years, and we are constantly monitoring and hoping that this will indeed bring order […] [T]his is the dynamics of […] a parliament that legislates, tests, examines and corrects.'[42]

In a previous study examining the purposes of 68 temporary laws, we found that 12 of them (17.6 per cent) were enacted as temporal because of the experimental purpose. This use of temporary legislation seems to be relatively new: ten of these 12 experimental laws were enacted after 2000, and seven of them after 2006.[43]

H. Temporary Legislation as a Dynamic Mechanism for Regulating Dynamic Areas

A similar use of temporary legislation is as a response to issues or areas, in which developments (such as technological developments) are still expected, and

[40] RAJ Van Gestel and G Van Dijck, 'Better Regulation through Experimental Legislation' (2011) 17 *European Public Law* 539; Sofia Ranchordás, 'The Whys and Woes of Experimental Legislation' (2013) 1 *The Theory and Practice of Legislation* 415; Zachary J Gubler, 'Experimental Rules' (2014) 55 *Boston College Law Review* 129.

[41] Livnat, Plenum first reading, 25 July 2012.

[42] The Education, Culture & Sport Committee Minutes No 58 of the 19th Knesset (17 July 2013).

[43] Bar-Siman-Tov (n 6).

therefore it is not advisable to establish a permanent law; rather, it is preferable to set a flexible mechanism.[44] Despite the similarity with experimental legislation, which is aimed at accumulating knowledge from the experimental period in order to obtain an optimal permanent law, there is an important distinction between the two. The goal here is not to be a first-step towards permanent legislation, but rather to allow the law to remain temporary, in order to create a flexible mechanism that will allow legislation to be updated every few years in light of developments in the field.[45] It should also be clarified that this use is different from the use of a temporary law as a temporary solution to a temporary problem, since here it is not expected that the issue requiring regulation will cease to exist.

A clear example of this type of temporary legislation can be found in the Cloning Law. As noted above in subsection III.B, one central aspect of the law was the establishment of a moratorium. Yet, another consideration was the use of temporary legislation to create a dynamic mechanism for monitoring developments in the field and, consequently, for the periodic examination of the law and its need. This is explained in the law's explanatory notes:

> It is proposed to instruct the Helsinki Committee […] to monitor developments in the field of human genetic testing, to advise the Minister on these matters, as well as the question of the validity of the law, the need to change it, update it, cancel it or extend it in light of the developments […]
>
> For this law not to be an obstacle to the progress of scientific research and developments in the field of genetics and medicine in Israel, it is proposed that the Committee submit to the Minister of Health each year a detailed report on these issues. This mechanism will ensure periodic examination of the law and the need for it.[46]

The Chair of the Science and Technology Committee, MK Uri Maklev said, for instance, in one of the discussions on extending the law:

> Sometimes when it's in a temporary law, you have the opportunity to make changes, whereas when the law is permanent, [the arrangement becomes fixed even when it's not justified. The temporary nature of the legislation] gives it a dynamic of changes and of minimizing gaps or making the necessary adjustments. As I said earlier, if the law […] was not for seven years but was forever, I do not know how long it would take to bring this amendment.[47]

Indeed, various speakers emphasised that although theoretically the legislature can always change any law, including a permanent law, in practice a temporary law is a more flexible and dynamic tool, and therefore more suitable for adapting

[44] Sofia Ranchordás, 'Innovation-Friendly Regulation: The Sunset of Regulation, the Sunrise of Innovation' (2015) 55 *Jurimetrics Journal* 201, 219.

[45] Richard H Pildes, 'The Future of Voting Rights Policy: From Anti-Discrimination to the Right to Vote' (2006) 49 *Howard Law Journal* 741, 772.

[46] Prohibition of Genetic Intervention (Human Cloning and Genetic Modification in Reproductive Cells) Law, 5759-1999, 483.

[47] The Science and Technology Committee Minutes No 55 of the 20th Knesset (22 May 2016) 17.

to the dynamic realm of cloning.[48] In addition, they noted that a sunset clause ensures that the legislature will force itself to convene and discuss the issue of cloning every few years.[49]

***

In sum, we see that the Israeli Parliament employs temporary legislation for a wide array of legisprudential uses. This array is richer than the three 'technocratic applications' identified by Gersen and repeated in discussions about temporary legislation by both the Israeli Supreme Court and Israeli Parliament. We see, moreover, in the Knesset's legislative process, deliberate and explicit discussions on the legisprudential function that justifies using temporary, rather than permanent legislation in various circumstances. Finally, we see that these legisprudential uses are not mutually exclusive, as a certain law could be enacted as temporary for various legisprudential purposes.

IV. Political Uses of Temporary Legislation

In addition to the various legisprudential uses of temporary legislation, one of the main uses of temporary legislation is a tool for garnering agreement and soothing political opposition to a bill.[50] Some have therefore conceptualised temporary legislation as a means to lower enactment costs,[51] or as the 'spoonful of sugar that helps controversial legislation go down'.[52] This is what we term the 'political use' of temporary legislation.

It should be clarified that formally, the legislative process of a temporary bill is identical to the legislative process of a regular bill – they both must pass the same hurdles of three (or four) readings in the plenum, as well as committee debate. Yet, in practice, temporary legislation may overcome these hurdles faster and with greater ease, since the perception that the law is only a temporary arrangement may pacify or at least ameliorate opposition to the bill.

To explore these dynamics of the political use of temporary legislation in Israel, we employed a dual approach: a qualitative approach presented in section IV.A and a quantitative approach presented in section IV.B.

[48] The Science and Technology Committee Minutes No 37 of the 16th Knesset (12 August 2016) 23.

[49] *Ibid*. *Cf* also Tom Ginsburg, Jonathan S Masur and Richard H McAdams, 'Libertarian Paternalism, Path Dependence, and Temporary Law' (2014) 81 University of Chicago Law Review 291, 315–25.

[50] Ariel Porat and Omri Yadlin, 'Promoting Consensus in Society Through Deferred-Implementation AGreements' (2006) 56 *University of Toronto Law Journal* 151; Eyal Zamir, *Law, Psychology, and Morality: The Role of Loss Aversion* (Oxford University Press, 2015); Frank Fagan and Fırat Bilgel, 'Sunsets and Federal Lawmaking: Evidence from the 110th Congress' (2015) 41 *International Review of Law and Economics* 1.

[51] *Cf* Gersen (n 1) (arguing that temporary legislation lowers enactment costs, but does not necessarily lower overall transaction costs, because maintenance costs should also be factored in).

[52] Nicola McGarrity, Rishi Gulati and George Williams, 'Sunset Clauses in Australian Anti-Terror Laws' (2012) 33 *Adelaide Law Review* 307; Finn (n 4).

A. Uses of Temporary Legislation to Overcome Resistance to a Bill

Through extensive and in-depth qualitative exploration of the legislative processes of temporary bills in the Knesset, we uncovered a variety of examples of the ways in which temporary legislation soothes resistance to the bill.

First, we revealed multiple indications that the different participants in the process themselves express the opinion that temporary legislation raises less resistance. For example, we came upon incidents where Members of the Knesset explicitly said that if the bill would have been permanent, they would have opposed it, but they were willing to support it as temporary legislation. Clear examples of this could be seen in the parliamentary debates on the Citizenship and Entry to Israel law. For example, MK Moshe Gafni had said: 'I think it must not be a permanent law. I oppose a permanent law on this issue [...] I am against the idea of this law [...] I do not accept the idea, but I accept the temporary provision.'[53] MK Shai Hermesh similarly stated:

> I want to say just one thing, I am against legislating this law [...] [But] I am definitely in favour of the temporary provision because I am fundamentally optimistic [...] Better days will come and we will not need this law. Let's not create laws that maybe one day we will not need.[54]

We also found statements by Knesset Committee Chairs, in which they explicitly stated that their parliamentary scrutiny is laxer when enacting temporary legislation. For example, in debating the Criminal Procedure Law (Detainees Suspected of Security Offenses) (Temporary Provision), which was eventually found to be partially unconstitutional by the Court,[55] the Committee Chair openly admitted:

> Had this been a law that is not temporary, perhaps it would have been treated differently, my consideration of different sections [of the bill] would also be different. Since this is a temporary provision, and not a temporary provision for several years, I imagine it will be a maximum of one year, then we will definitely be able to examine the issues during this year, and if we will want to make additional legislation – we will do so, and if not – we won't.[56]

This citation may demonstrate a more general theoretical effect of temporary legislation: it may create a 'moral hazard' effect.[57] That is, the knowledge that there will be a future chance to examine the law may cause legislators (and even Committee Chairs) to shirk their responsibility to scrutinise the law when enacting it.

[53] Gafni, Internal Affairs and Environment Committee, 20 March 20077.
[54] Hermesh, Internal Affairs and Environment Committee, 20 March 2007.
[55] CrimApp 8823/07 *Anonymous v State of Israel* (2010).
[56] Levi, Constitution, Law and Justice Committee, 20 June 2006.
[57] On the moral hazard effect (in a different context), see Adrian Vermeule, *Judging Under Uncertainty: An Institutional Theory of Legal Interpretation* (Harvard University Press, 2006).

We also found instances where temporary provisions were used as tools to overcome differences within the Government. An interesting example can be seen in the enactment of the National Authority for Road Safety law. This government bill was not originally proposed as temporary legislation, and in fact, during its discussions the Minister of Transportation at the time, Meir Shitrit, argued that there was no good reason to enact the law as a temporary statute.[58]

Since the legislative process of this bill was not completed during the 16th Knesset, the process continued during the 17th Knesset, this time promoted by the Minister of Transportation Shaul Mofaz. During the third meeting of the Economics Committee, in its preparation of the bill for second and third reading, the matter of temporary legislation rose for the first time. Minister Mofaz opened the debate with the suggestion of promoting the bill as a temporary law in order to establish the National Authority for Road Safety as quickly as possible. Mofaz argued that there was disagreement in the Government as to the question of establishing the Authority as a separate statutory authority, outside of the Ministry of Transportation, and that enacting the law as a temporary statute would help to resolve the disagreement. Mofaz stated: 'We have decided [...] to establish the National Authority of Road Safety as a statutory authority, as a temporary law, in order for it to be established in January 2007[...]'[59] Mofaz then continued to explain that enacting the law as a temporary measure

> solves my problem with the Prime Minister, since a large portion of the Ministers as well as the Prime Minister wanted this matter to be inside the Ministry. I convinced them and got their agreement to the formulation I have just talked about.[60]

A particularly interesting case about the use of temporary legislation to overcome disagreements within the Government was discovered in a bill regarding the electronic supervision of detainees. The bill had a sunset clause and after the committee members expressed their wonder for the need to use temporary legislation in this case, it turned out that this was part of a solution to a disagreement within the Government. As the Chair of the Committee, MK Gafni explained:

> It turned out that there is a disagreement between the Ministry of Finance and the Ministry of Welfare [...] and in this disagreement the compromise was to enact the law as temporary law for 3 years and afterwards to have a discussion. I said: this is out of the question, it disgraces the Knesset, it is like a coalition negotiation between the Ministry of Finance and the Ministry of Welfare in primary legislation, and we absolutely disagreed [...] There cannot be such legislation, which contains a temporary provision because they do not get along.[61]

[58] The Finance Committee Minutes No 534 of the 16th Knesset (24 November 2005) 14–15.

[59] The Finance Committee Minutes No 40 of the 17th Knesset (10 July 2006) 4–5. Also, at p 50, Adv Melly Siton, the legal advisor of the Transport's ministry, says, although in a less direct manner, that a temporary legislation was added to the statute as a means to overcome the state's policy which resists as a rule to the establishment of statutory corporates.

[60] Electronic Control Law on a Detainee and a Conditionally Released Prisoner (Legislative Amendments) 5714-2014, 68.

[61] Knesset Minutes No 187 of the 19th Knesset (following: Minutes No 187).

This is an instructive example of how the Government attempts to use temporary legislation to solve inter-governmental disagreements. It is also fascinating since it demonstrates that Knesset committees may resist such use of temporary legislation when they believe that the sole reason for using temporary legislation is resolving ministerial disputes.

In another matter, MK Gafni claimed that temporary legislation enables the Government to extend statutes in controversial matters not only to overcome disagreement between ministers, but also as a way to avoid public and media criticism. This was what Gafni claimed regarding the Government's request not to repeal the Law on Partial Payment of a Convalescence Allowance in the Public Service in 2009 and 2010 (Temporary Provision), but to leave it in the statute book even though it had expired. He argued that:

> We [the Committee] are proposing to repeal the law, which is the simplest thing. Legally speaking this is what needs to be done, [… whereas …] they [the Ministry of Finance] say, no. We will leave it; we will make a temporary law.[62]

In explaining what he believes to be the Government's motivation, Gafni stated:

> If the law is repealed and they want to do it again, they will have to come to the government with the whole plan and then the Ministers will come and ask why, why decrease convalescence allowance from the employees.

On the other hand, if the expired temporary law had remained on the statute book and had not been officially repealed, they would not have to come up with this new law. Instead,

> they will say no convalescence, nothing, we are not taking anything from the workers, we are only [temporarily] extending the validity of this law that was enacted in 2009 […] When a minister will be interviewed by the media and asked why he voted [for this decrease], he will say that he did not vote [for it] but merely extended the date. That's what they intend.[63]

Finally, we also found statements indicating that Knesset Members perceive temporary legislation not only as a tool to overcome parliamentary and public objections, but also as a means of dealing with future judicial review and reducing the fear of invalidating the law. This was explicitly stated by the Chair of the Interior and Environmental Protection Committee, MK Ophir Paz-Pines, when referring to the Citizenship and Entry into Israel Law: 'If this law is not enacted as temporary legislation, it has no chance in the world of passing the Court.'[64]

These examples reveal the uses of temporary legislation as a means of facilitating legislation, overcoming parliamentary and governmental objections to the

[62] Gafni, Finance Committee, 27 June 2011.

[63] *Ibid.* To be fair, the officials from the Ministry of Finance did provide an alternative explanation for their position and stated that they did not intend to extend the law's validity, but the Chair did not believe their explanation.

[64] Paz-Pines, Internal Affairs and Environment Committee, 20 March 2007.

bill, perhaps even evading effective parliamentary, public, media and judicial scrutiny. These findings clearly illustrate the abovementioned claim in the theoretical literature that sunset provisions are 'the spoonful of sugar that helps controversial legislation go down'.

B. Passage Rates of Temporary Governmental Bills

To complement the qualitative and anecdotal evidence about the political use of temporary legislation, we quantitatively examined whether the passage rates of government bills that included temporary provisions are higher than the passage rates of all government bills in general.

This examination revealed that the passage rates of temporary government bills are indeed significantly higher than the passage rates of all government bills. On average, the gap stands at almost 20 per cent in favour of temporary government bills for all nine Knessets examined: 86.9 per cent of government bills that contained sunset clauses passed the legislative process and were enacted into law, in comparison with a 67.3 per cent passage rate for amongst all government bills.

Figure 1 below shows the gap between the passage rates of all government bills and the passage rates of temporary government bills from the 12th to the 20th Knessets. As we can see, in all nine Knessets, the passage rates for temporary government bills were higher, and in some, the gap is particularly dramatic (reaching a peak difference of almost 37 per cent during the 17th Knesset).

Figure 1 Passage Rates for Government Bills

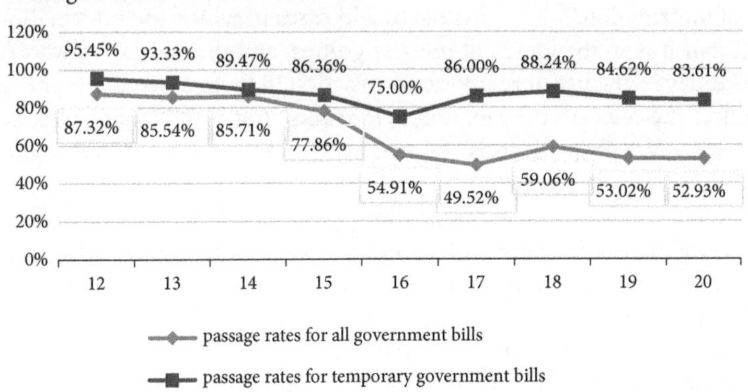

There may obviously be various explanations for our finding that the passage rates of temporary government bills are significantly higher than those of all government bills. Within the scope of this research, we do not attempt to examine and rule out these explanations. Hence, we are cautious not to conclude causality. Yet, it can cautiously be said that this research reveals at least a circumstantial connection between the characteristic of bills as temporary or permanent and their passage rates. Stronger, more concrete conclusions require further research. In the meantime, it is safe to say that the findings are in concert with the findings from the qualitative part of this study. Moreover, they are additionally in line with Fagan and Bilgel's findings that the inclusion of a sunset provision increased the likelihood that a bill becomes law in the US Congress.[65]

V. Discussion: Appropriate Uses of Temporary Legislation

The empirical part of this study has demonstrated that the Knesset uses temporary legislation for a variety of legisprudential functions. At the same time, it revealed that temporary legislation is also employed by the Government for political uses, and apparently quite successfully. Against this backdrop, we conclude with a normative discussion about the appropriate uses of temporary legislation. We begin with the legisprudential uses and then turn to the political use of temporary legislation.

A. Legisprudential Uses

In our opinion, one of the most appropriate uses for temporary legislation is as a tool for experimental legislation.[66] This legislation can be justified primarily to regulate or resolve problems in areas where there is a high degree of uncertainty or lack of information.[67] It is possible to add cases in which the information already exists, but it is in the hands of interest groups, and the legislature wants to verify its accuracy.[68] The use of temporary legislation in these cases allows the legislature to collect the relevant information before adopting a permanent arrangement, to make a more informed decision, and thus to achieve better and higher quality legislation.

It should be noted that almost every piece of legislation has a certain degree of uncertainty; hence, this tool should be used only in cases where it is feasible to obtain the relevant information within a reasonable period. Thus, for example, in the proposed Law to Combat the Phenomenon of Drunkenness, the Ministry of Justice requested that the temporary law would be made permanent, because, according to the Ministry, an additional period of time would not help in dealing

[65] Fagan and Bilgel (n 50).
[66] Ranchordás (n 3).
[67] Helen Xanthaki, *Drafting Legislation: Art and Technology of Rules for Regulation* (Hart Publishing, 2014) 189.
[68] Frank Fagan and Michael Faure, 'The Role of Lawmakers, Lobbyists, and Scholars in the Normative Evaluation of Timing Rules' (2011) 160 *University of Pennsylvania Law Review* PENNumbra 61.

with the uncertainty: 'We will not be wiser in two years than we are now.'[69] Therefore, temporary legislation should be used when there is specific and defined information that the legislature wishes to examine, and when the use of this tool will indeed facilitate its achievement.

This is reinforced when the legislature seeks to apply an innovative or experimental arrangement. One can also see signs of the phenomenon in Israel, in the form of several laws that the Knesset explicitly declared as pilot programmes.[70] Indeed, in the case of such laws, the use of temporary legislation is the most appropriate, as this allows the legislature to apply a new regulation in a careful manner while studying the effects of the legislation and improving it if necessary. However, it is important to emphasise that including a sunset clause in the legislation will not in itself guarantee the improvement of the quality of the legislation; rather, there is a need for additional elements, such as proper *ex post* evaluation mechanisms and extension mechanisms.[71]

Another justified use of temporary legislation is as a flexible tool for dealing with areas that are inherently dynamic. This refers to cases in which there is reasonable assessment that significant changes and developments are expected to occur which are difficult to anticipate; therefore, it is reasonable to assume that the legislature will have to update the policy it is formulating in the law from time to time, in accordance with these developments. These are mainly cases of technological change, but there may also be social or economic changes that justify the use of temporary legislation.[72] As we have seen, the Cloning Law is an excellent example of this. In cases of dealing with areas that are inherently dynamic, as well as in cases of experimental legislation, there is indeed a clear advantage for temporary legislation over permanent legislation.

Other legitimate and justified uses that the Knesset makes of temporary legislation in our opinion are as temporary placeholder legislation, temporarily freezing the legal situation (moratorium) and temporary legislation as a one-time arrangement for a one-time event. In these cases, we believe that temporary legislation is indeed a suitable tool for dealing with the need of the hour or temporary policy challenges, and that it is even preferable to regular legislation. The advantage of using temporary legislation in these cases is that it fills a gap in a quick and easy manner until a more profound and meaningful response by the legislature is found, or until it is simply no longer necessary (in the case of a single transitory event).

Apart from the practical need for such legislation, it seems that there is also a normative justification, as this will ensure that the statute book will not contain

[69] The 19th Knesset Internal Affairs and Environment Committee Minutes No 7 (April 29 2003).

[70] Bar-Siman-Tov (n 6) at 199.

[71] Sarah Tsvavner, *Private Legislation In Parliaments Worldwide* (The Knesset Research and Information Center, 2000); Chen Friedberg, 'The Knesset as a Legislative Body: Diagnosis and Recommendations' in Gideon et al Rahat (ed), *Reforming Israel's Political System* (The Israel Democracy Institute and Am Oved, 2013).

[72] Ranchordás (n 44).

laws that refer to situations that have passed and are no longer relevant.[73] We also believe that in these cases temporary legislation may raise fewer difficulties and risks than other uses (as long as the legislation will truly be temporary, of course, and its actual use will be appropriate). However, because the legislation of such temporary laws is often accelerated and more superficial, it is inappropriate to enact complex laws in this way. Moreover, extending these laws should be exercised with extra caution and their repeated extension without any in-depth examination should certainly be avoided.

More complex difficulties arise around the use of temporary legislation as a temporary solution to problems that are perceived as temporary, especially in response to emergency situations or crises and even more so when the temporary emergency legislation is also seen as a means of deviating from the usual and general principles of the legal system ('temporary de-juridification'). This use has its advantages and there will be cases where it will be justified.[74] However, this also involves added risks, especially the concern that the sunset legislation could facilitate enacting legislation that is unjustified even as a temporary measure, as well as the concern that the temporary law will lead to permanent and irreversible changes.[75] Therefore, it should be emphasised that with this use of temporary legislation, it is increasingly important to consider the way temporary legislation is used in practice. Here there is also a heightened fear of repeated extensions.

Considering the question of when it is appropriate to make use of temporary legislation under this type of use, a critical concern is whether there is a reasonable and honest assessment that the temporary problem or emergency is indeed likely to be temporary. That is, there should be an educated and well-informed assessment of the question regarding the amount of time in which the temporary problem or crisis will come to pass. There is no justification for temporary legislation when it is impossible to establish that the problem is expected to pass within a relatively short period of time. An instructive example of this is counterterrorism legislation. Although emergency legislation to deal with immediate threats is sometimes justified, in practice the end of the war on terror rarely seems apparent or within sight. This is certainly true for Israel.

The clear practical example of these concerns, which has already been discussed, is the Citizenship and Entry into Israel Law, which prohibits family unification with residents of the occupied territories. As previously mentioned, this law was enacted at the height of the second *intifada* and its promoters stressed that it is an emergency piece of legislation meant to provide a temporary solution to a difficult and temporary security situation. We have also seen that participants in the

[73] Xanthaki (n 67).

[74] Kouroutakis and Ranchordás (n 29).

[75] For a more extensive discussion of these concerns, see Bar-Siman-Tov and Harari-Heit (n 11). See also McGarrity, Gulati and Williams (n 52); Ip (n 31).

parliamentary debates stressed that the proportionality of this highly problematic law stems from its temporary nature and some legislators argued that they voted for it only because it is temporary. Unfortunately, although the second *intifada* was a passing event and the waves of terror attacks may pass, the broader threat of terrorism, as well as the Israeli-Palestinian conflict, are issues without solutions in sight.[76] This allows the Government to repeatedly claim, year after year, that the security situation is still difficult; thus, the need for the law has not changed. Consequently, this law, which was enacted as a temporary emergency law for one year in 2003, has already been extended 19 times and (as of September 2019) is still in force.

Another use, which may also be complex and problematic, is the use of the temporality of the law as a means of softening the extent of constitutional violations. As stated above, there is a widespread perception that when restricting individual freedoms, the use of temporary legislation constitutes a more proportionate solution. However, experience also hints at the risks arising from this common perception: the risk that unconstitutional laws, that are not even temporarily justified, will be passed, the risk that constitutionally problematic laws will be passed without sufficient parliamentary scrutiny and oversight and the risk that the enactment of the temporary law would be made only as a means of overcoming parliamentary opposition or as an attempt to immunise the law from judicial review.

Therefore, we believe it is time to break free from the perception that the mere fact that the law is temporary makes it more proportionate or less harmful. The legislature should not belittle this legislation or take it lightly merely because of its temporary nature. Instead, there should be careful assessment of each temporary law that infringes upon rights to examine whether in a concrete case the temporariness of the law does indeed affect the proportionality of the infringement involved. First, it must be kept in mind that there may be severe and disproportionate violations to the extent that their temporality cannot make them proportionate. Second, even if temporality is relevant to the proportionality in a given case, it is necessary to examine how and in what manner it influences proportionality and, moreover, what elements in the assessment of proportionality are at issue. For instance, one must first examine whether the harm is necessary and whether there are any less harmful means available. In extreme cases which justify the temporary adoption of more severe measures that are not justified under other conditions, mechanisms of evaluation and extension must be put in place to ensure that the law is not extended automatically and that efforts would be devoted to the frequent re-examination of the arrangement, including a review of the very need for it, its ramifications and possible alternatives.

Finally, it is necessary to consider, from the perspective of the potential harm bearer of the law, whether the fact that the norm permitting the harm is temporary

[76] Yaacov Bar-Siman-Tov, *Barriers to Peace in the Israeli-Palestinian Conflict* (Jerusalem Institute for Israel Studies, 2010).

affects the intensity of injury to that person. There are cases where the temporary violation of the right reduces the intensity of the harm. Thus, for example, the expropriation of a licence or the freezing of certain economic rights for a short and limited period will usually be more proportionate than permanent denial. On the other hand, sometimes the temporary norm that allows harm does not affect the intensity of the harm involved. To use an extreme example for illustration, from the point of view of a person who has been harmed by a law authorising the state to harm his body or his life, the fact that the law was enacted as a temporary order does not reduce the intensity of the harm. Therefore, when the legislature chooses a tool of temporary legislation, it must consider its possible consequences and ensure that it does not lead to irreversible harm.

B. Political Uses

Some supporters of temporary legislation believe that using temporary legislation as a means of overcoming opposition to the bill is a legitimate and appropriate use.[77] We disagree with this conclusion. In our opinion, when the sole purpose of enacting the law as temporary is to overcome objections to the law, there is no justification for enacting it as temporary law. It is thus inappropriate to use temporary legislation not only to overcome disagreements within the Government, but also as a means of overcoming opposition in the Knesset. This position stems from several arguments.

First, our fundamental position is that principled justification is needed in order to deviate from the basic principles of the certainty, stability and clarity of law. These are fundamental principles of proper lawmaking[78] and temporary legislation deviates from these principles.[79] When a temporary law is enacted to serve one of the legisprudential functions, this justifies deviating from such fundamental principles. On the other hand, when there is no real reason for choosing temporary legislation other than overcoming opposition to the bill, no such justification exists.

Second, allowing and legitimising temporary legislation even when no legisprudential function exists, will considerably increase the extent of the use of temporary legislation. This, in turn, raises the risks of overuse in a manner that will impair the ability to use temporary legislation as a tool for improving lawmaking. Excessive use of temporary laws does not enable a learning process and informed decision-making process. By contrast, it increases the fear that repeated wholesale

[77] See, eg, Antonios Kouroutakis, *The Constitutional Value of Sunset Clauses: An Historical and Normative Analysis* (Routledge, 2017) 130; Porat and Yadlin (n 50); Zamir (n 50).

[78] Lon L Fuller, *The Morality of Law* (Yale University Press, 1977) 79–80; Patricia Popelier, 'Legal Certainty and Principles of Proper Law Making' (2000) 2 *European Journal of Law Reform* 321.

[79] Rebecca M Kysar, 'Lasting Legislation' (2011) 159 *University of Pennsylvania Law Review* 1007.

extensions of temporary legislation would be made without proper examination,[80] or that temporary laws will expire not because of an informed decision, but only because Parliament was busy and did not schedule enough time for an extension.[81] Moreover, excessive use of temporary legislation impairs the legislative process and may overwhelm the legislature with laws that will demand its attention time after time.[82] This concern appears to be particularly relevant to the Israeli legislature, which is already flooded with more bills than most legislatures around the world and provided that MKs are often unable to effectively staff all parliamentary committees and maintain effective parliamentary oversight of the Government and its legislative initiatives.[83]

Third, the question of whether and to what extent facilitating agreement and overcoming parliamentary opposition is a worthy purpose in itself, is derived, inter alia, from the question of whether and to what extent it is difficult to pass legislation in a legal system. In legal systems such as the United States, where the legislative process is much more cumbersome and demanding and, therefore, there are often claims that Congress is paralysed and unable to function properly,[84] this purpose may be used as justification. In Israel, on the other hand, legislative procedures impose only few limitations on the ability to legislate, and, in our view, there are insufficient restraints on the legislature's (and especially the Government's) ability to legislate.[85] Hence, there is no need or justification for further alleviation of the ability to legislate.

Fourth, when the temporality of the law does not serve any proper purpose, but rather the sole purpose is the desire to pass the law more easily, this may be an indication that the law itself does not serve any proper purpose and that its sole purpose is to fulfil coalition needs.[86]

A fifth point, which forms our main argument, is that a tool for reaching agreements and overcoming disputes means, in practice, a means of evading effective parliamentary, public and media scrutiny and perhaps even judicial review, as we saw in section IV.A. This effect is particularly problematic when one takes into consideration that legislation can often be controversial and raise objections since it is constitutionally problematic and violates rights, as we have seen in the case of the Citizenship and Entry into Israel Law.

[80] Bar-Siman-Tov (n 6) at 210. Veit and Jantz (n 5) at 281.

[81] This was the historical experience in England. See Kouroutakis (n 77).

[82] Bar-Siman-Tov (n 6) at 207–208; See also McGarrity, Gulati and Williams (n 52); Kouroutakis (n 77).

[83] Chen Friedberg, 'The Knesset as a Legislative Body: Diagnosis and Recommendations' in Gideon et al Rahat (n 71).

[84] Sarah A Binder, *Stalemate: Causes and Consequences of Legislative Gridlock* (Brookings Institution Press, 2003); Sarah Binder, 'The Dysfunctional Congress' (2015) 18 *Annual Review of Political Science* 85; Josh Chafetz, 'The Phenomenology of Gridlock' (2013) 88 *Notre Dame Law Review* 2065.

[85] Ittai Bar-Siman-Tov, 'The Law of Lawmaking' (2016) 37 *Tel Aviv University Law Review* 645, 691–94.

[86] An example can be seen in the Amendment No 3 to Basic Law: The Government, which in a temporary provision removed the limitation on the size of the government for the duration of the twentieth Knesset.

Moreover, this effect is particularly problematic when it comes to government bills – in Israel, most temporary legislation being governmental. Temporary legislation at the hands of the Government may become a further measure of harming the Knesset's status as the legislative branch. In this sense, one can find a similarity between temporary legislation and another phenomenon in legislation: the Government's use of arrangements laws (omnibus legislation) as a means of overcoming parliamentary resistance. Arrangements laws have been described in the legal literature as 'a unique governmental tool that helps the government speed up legislative processes, overcome parliamentary obstacles, initiate and bring about the completion of legislative activities without systematic discussion, proper supervision and control […]'[87] Justice Beinisch has rightly said that this is a legislative mechanism 'that serves the government as a means to overcome parliamentary obstacles (in other words, to refrain from effective parliamentary scrutiny of the government's legislative initiatives)'.[88] This chapter reveals that similar things can be said about the temporary legislation. In our opinion, there is no justification for providing an additional tool to the Government's toolbox to force its will on the Knesset and, resultantly, weaken its status as the legislative branch.

One of the justifications for temporary legislation in the scholarship from other countries is its ability to empower the legislature vis-à-vis the executive and serve as a tool for effective legislative oversight of the executive.[89] We agree that this could theoretically be true and that with proper design of evaluation clauses and extension clauses, and with proper employment of these mechanisms, this potential benefit could be realized. Yet, our study reveals a different and previously unrealized effect of temporary legislation on the separation of powers, in which temporary legislation could actually be a tool to empower the executive vis-à-vis the legislature. This is at least the concern based on the Israeli experience, in which most temporary legislation results from governmental bills and where the Government uses temporary legislation to facilitate the enactment of legislation over parliamentary opposition.

VI. Conclusion

Temporary legislation can be employed to promote a variety of legisprudential functions, as well as for political purposes. In this study, we revealed how temporary legislation is used in practice in Israel. The rich array of legisprudential uses of

[87] David Nachmias and Eran Klein, 'The Arrangements Law: Between Economics and Politics' (1999) 17 *The Israel Democracy Institute* 7.

[88] HCJ 4885/03 *Israel Poultry Association v Government of Israel* 14(2) PD 36 (2004) (Isr.).

[89] Kouroutakis (n 77); Josh Chafetz, 'Congress's Constitution' (2012) 160 *University of Pennsylvania Law Review* 715; Josh Chafetz, 'A Beautiful Sunset (Provision) for NSA Surveillance' (*The Hill*, 2015) http://thehill.com/blogs/pundits-blog/homeland-security/243169-a-beautiful-sunset-provision-for-nsa-surveillance.

temporary legislation in the legislative practice of the Knesset was demonstrated. Moreover, the political uses of temporary legislation and the ways they operate in practice have been shown.

In turning to a normative discussion about proper use of temporary legislation, it is our opinion that the use of temporary legislation should be prudent and limited to cases in which it serves a legitimate legisprudential purpose. Among these legisprudential functions, the use of temporary legislation is most appropriate when the legislature seeks to collect additional information or to run experimental programmes. Legislation that may severely infringe rights requires special caution in the use of this legislative tool; temporary legislation for the sole purpose of overcoming opposition to bills should be avoided.

By revealing how temporary legislation is used in practice and through our normative prescriptive suggestions about proper use, we hope to contribute both to the bourgeoning global academic debate on temporary legislation and its proper operation in practice.

11

Speeding Up the Legislative Process: To What End and at What Cost?*

WIM VOERMANS

Ours is an epoch in which social and economic processes are undergoing a multi-pronged acceleration that raises many difficult questions for legal scholarship. Social and economic acceleration challenges the noble aspiration to establish fundamental constitutional 'rules of the game' capable of serving as an effective binding force on legal and political actors for a relatively long span of time.[1]

I. Do We Need to Speed Up the Legislative Process?

Anybody involved in modern politics today can tell you that time and timing are essential, especially when it comes down to enacting legislation – timing is an art, while time is a costly resource. However, time and timing are not typically the things we study in-depth or as such in legislative studies, as Borghetto and Giuliani note. They observe that:

> Despite the fact that the temporal dimension of a process should be at the core of its analysis, this rarely occurs in legislative studies. Success or failure of bills – that is, the outcome of that process – attracts much more attention, and is duly examined against all the possible institutional and political causes. Time enters the scene mostly under two different perspectives – either as a wide frame constraining the paths followed by legislators, or as a proxy of something else.[2]

* This article is a revised version of an article that Hans-Martien ten Napel, Reijer Passchier and myself wrote in 2015. Wim Voermans, Hans-Martien ten Napel and Reijer Passchier, 'Combining Efficiency and Transparency in Legislative Processes' (2015) *The Theory and Practice of Legislation* Vol 3(3), 279–94. The contribution reports on a comparative research project we conducted into the speed and performance of legislative processes in different jurisdictions.

[1] William E Scheuerman, 'Constitutionalism in an Age of Speed' (2002) *Constitutional Commentary* Vol 19(2), 353–90, quote from 355.

[2] Enrico Borghetto and Marco Giuliani, 'A Long Way to Tipperary: Time in the Italian Legislative Process 1987–2008' (2012) 17(1) *South European Society and Politics*, 23, 25–26.

Even though time is an essential element in legislation and its relevance is widely recognised, it is seldom directly analysed, Borghetto and Giuliani conclude, probably due to the misguided presumption that legislative processes are a given.[3] While legislative processes and procedures are not static and unchangeable playing fields, as we can see all around us, most legislative processes and procedures nowadays are in a state of flux.

Many governments today are facing two important, potentially clashing, demands. On the one hand, citizens want governmental action to be ever more efficient: expedient government service delivery is growing in importance.[4] Government agents and their apparatus are – for a variety of technological, economic and societal reasons – under increasing pressure to adapt quickly to changing circumstances. The need for speed and adaptability meets (and sometimes collides) with citizens' demands for a more transparent government, as well as an increasingly (inter)active role in the decision-making process. There is an ever-growing demand for openness and transparency in modern societies, as a recent report by the European Parliament observes.[5]

Although efficiency – herein predominantly understood as the optimal balance between careful preparation and speed of the process – and transparency are not mutually exclusive notions, the realisation of both concepts in the fullest sense on the occasion of policy and decision-making remains a complex task. Balances are hard to strike. When more people, or groups of people, are involved in decision-making processes, these processes run the risk of slowing down and becoming less efficient. This could also apply vice versa, whereby informed citizens' participation in decision-making processes (made possible by forms of openness and transparency) can be frustrated by the (perceived) need for governments to take swift action. However, some have argued that information and communication technology (ICT), and the opportunities it offers for participation, can overcome the divide between the need for speed and the demand for public inclusiveness and openness of the legislative process.[6] Different legislatures have picked up this challenge and are indeed implementing ICT systems to enhance transparency and public participation in the legislative process.[7] This development consequently

[3] *Ibid*, 23, 41.

[4] See John Wanna, John Butcher and Benoît Freyens, *Policy in Action: The Challenge of Service Delivery* (UNSW Press, 2010).

[5] European Parliament, Directorate-General for Internal Policies (Policy Department Citizens' Rights and Constitutional Affairs), 'Openness, Transparency and Access to Documents and Information in the European Union', PE 493.035. Brussels 2013, 7.

[6] See for instance Euripidis Loukis, Maria A Wimmer, Yannis Charalabidis, Anna Triantafillou, Rimantas Gatautis, 'Argumentation Systems and Ontologies for Enhancing Public Participation in the Legislation Process' (2007) *Participation and Democracy* 19.

[7] See, for instance, the EU itself: Commission of the European Union, '*e*Participation in the Context of Legislative Processes' (Work Programme 2006, DG Information Society and Media 2006). Or the Dutch Legis-programme: Ministry of Justice, 'Legis samenwerken aan betere wetgeving' (Legis – cooperating for better legislation) (2012) which resulted in a new IT supported legislative framework.

begs the question: does the use of ICT indeed provide the proverbial 'third way' between speed and inclusiveness?

In this contribution, I will try to illustrate the modern-day dynamics of the interplay between the need for expedience and efficiency, on the one hand, and the demand for openness, inclusiveness and transparency on the other hand by looking into one of the main governmental decision-making processes: the legislative process. Particularly in the field of legislation, the balancing of both efficiency and transparency is of the essence for modern legislatures in parliamentary democracies. Laws expressed by acts and legislative instruments can only be truly effective if they rest on broad societal support.[8] As we will argue, a transparent and inclusive legislative process functions as a kind of democratic check on government action: it guarantees sufficient deliberative activity before a government may act. Thus seen, transparency does not so much constitute a threat to, but rather becomes an integral part of efficient and effective lawmaking, as Beetham already noted in 2006.[9]

A 2012 comparative study commissioned by the Dutch Ministry of Security and Justice, and carried out by an interdisciplinary team of researchers from Leiden University empirically confirmed this proposition. The research analysed the performance of legislative processes in three European countries – Finland, Slovenia and the United Kingdom – in terms of efficiency and transparency and compared the results to the performance of the Dutch legislative procedure. With respect to the promising outlook ICT offers in combining the need of speed and inclusiveness, the study further looked into the way legislative processes in these countries use techniques to rise to the challenges. Among other things, it revealed that, indeed, the demand for greater efficiency needs to be dealt with in the context of the simultaneous call by citizens for more transparency and participation. Nonetheless, the speed of the legislative process does not appear to suffer as a consequence of meeting these demands.[10]

[8] Mere majorities of electoral mandates based on elections do not suffice as such – for true democracy (ie the one that commands sustained legitimacy and compliance), inclusiveness, fairness and openness of decision-making processes play important roles as well. They reinforce the underlying 'regulative ideas' of democracy, notably popular control and political equality. These need to be reinforced time and time again – legitimate decision-making is before anything a sustained *belief*. See David Beetham, 'Democracy: Key Principles, Institutions and Problems' in Inter-Parliamentary Union (eds), *Democracy: its Principles and Achievement* (Inter-Parliamentary Union, 1998) 21. Available at www.ipu.org/PDF/publications/DEMOCRACY_PR_E.pdf (last accessed on 20 September 2019).

[9] See David Beetham, *Parliament and Democracy in the Twenty-First Century; A Guide to Good Practice* (Inter-Parliamentary Union, 2006) 115–52. Available at www.ipu.org/PDF/publications/democracy_en.pdf (last accessed on 20 September 2019).

[10] Wim Voermans, Hans-Martien ten Napel, Michal Diamant, Marga Groothuis, Bernard Steunenberg, Reijer Passchier and Stefan Pack (eds), *Legislative Processes in Transition. Comparative Study of the Legislative Processes in Finland, Slovenia and the United Kingdom as a Source of Inspiration for Enhancing the Efficiency of the Dutch Legislative Process* (Leiden/The Hague: WODC, 2012). Available at https://openaccess.leidenuniv.nl/handle/1887/20333 (last accessed on 20 September 2019).

Throughout this contribution, this 2012 study will serve as a guiding rail to illustrate some of the ways in which different jurisdictions in Europe (notably Finland, Slovenia, the United Kingdom and the Netherlands) have managed to combine, or at least balance, the need for legislative efficiency and transparency. This is used to demonstrate how traditional legislative processes presently grapple to translate the will of the citizens into effective legislation, how modern administrations still need democratically underpinned legislative procedures as the basis for the legitimation of (their) decisions, how efficient delivery of decisions and careful (lengthy) scrutiny interact are. On the basis of this material, the concepts of, respectively, efficiency and transparency and, in particular, the way modern legislatures examined in the study use ICT to overcome sometimes opposing demands on their legislative processes are discussed. In so far as possible, a few 'best practices' that show how legislative processes can (and cannot) adapt to new present day demands are highlighted.

II. Modern Society, Traditional Government and Legislative Processes

Tom Tyler has detected a shift in the way we accept authority and – thus – in the way authorities (such as governments or the legislature) can arrive at legitimate decisions. Traditionally, authorities (such as legal authorities) could expect to be obeyed because the public believed – ie, were morally convinced – that these authorities held title (be it institutionally, by power of their expertise, training or title) to decide over issues and/or dictate the law.[11] This seems to have changed. There indeed seems to be some evidence that authorities and institutions are viewed as more legitimate and, therefore, their decisions and rules are more willingly accepted, when they exercise their authority through procedures that people experience as being fair.[12] In modern society, citizens have also become increasingly accustomed to non-hierarchical structures. People, organisations and companies tend to stand on virtually equal footing with one another.[13] In the last 20 years, the Internet, in particular, has given a powerful boost to this emancipation process of citizens in this respect, so much so that modern citizens indeed seem to expect a similar 'horizontal' relationship with public authorities, including their government.

Expectations of horizontal treatment on an equal footing run counter to the setup of government institutions and decision-making processes that are

[11] Tom R Tyler, *Why do People Obey the Law?* (Princeton University Press, 2006).

[12] See Tom R Tyler, 'Public Trust and Confidence in Legal Authorities: What do Majority and Minority Group Members want from the Law and Legal Authorities?' (2001) 19(2) *Behavioral Sciences & the Law* 215.

[13] Raad voor het openbaar bestuur (Dutch Council for Public Administration), *Vertrouwen op democratie* [Trusting Democracy] (Raad voor Openbaar bestuur, 2010) 39.

predominantly vertically organised (top down), especially with respect to legislative processes. Policy-making depends – because of the representative structure of democracies – at best indirectly on the views of citizens. According to the Dutch Council for Public Administration, this state of affairs causes a gap between society and government.[14] In its 2010 report, the Council argued that a horizontal society is defined by dialogue and participation.[15] The decisions of a modern government therefore need to have a more direct basis in society.[16] A greater legitimacy of the legislation can be achieved by paying attention to the legislative *process*. Transparency and a participatory role for citizens in government are thus increasingly important. Accordingly, many citizens nowadays 'do not lean anymore on representative democracy alone, but in essence want to represent themselves (…).'[17]

At the same time, however, citizens (and businesses) also ask for a more efficient government. Societal problems need to be solved quickly and effectively. As Parliamentary Acts are still seen as the proper basis for governmental action, much of this begins with the preparation of legislation. Traditionally, lawmaking is seen as something that not just takes, but also needs, time. That is the rationale for the different stages and steps of legislative procedures worldwide: to allow for careful preparation, consideration, input and scrutiny. Although this view still exists, it is being increasingly eclipsed by a rival view which sees legislation in a more instrumental manner: as a tool that can be used to solve long-term, as well as short-term issues; as a commodity on a time-to-market clock.

III. Studying Legislative Performance

Under the influence of the shorter life cycle of legislation, improved technical possibilities and the role of the media in political and societal debates, the legislative process in many jurisdictions is under pressure to become more efficient, while at the same time guaranteeing the necessary high quality of the outcome. For the Netherlands, in particular, the time pressures and constraints on the legislative process have become more topical than ever. Due to developments in the political landscape, no fewer than five elections took place between 2002 and 2012, with each cabinet having an average life cycle of two years. Getting the results in and achieving the agenda of the successive cabinets proved hard indeed, especially since most of the policies sought for required reforms to be implemented by parliamentary legislation. With an average time span of 12–14 months to pass through the legislative procedure (ministerial preparation, consultation of the Council of

[14] *Ibid*, 40.

[15] Raad voor het openbaar bestuur (Dutch Council for Public Administration), *Gij zult openbaar maken* [Thou shall provide open access] (Raad voor Openbaar bestuur, 2012) 3.

[16] *Ibid*.

[17] Speech of J Wallage, the former Chairman of the Dutch Council for Public Administration at www.youtube.com/watch?v=TteCh2QdtSs and www.raadopenbaarbestuur.nl/documenten/publicaties/2010/02/17/vertrouwen-op-democratie (both last accessed on 20 September 2019).

State and passage through both houses of parliament), it was hard to get any result in on time for the consecutive Dutch cabinets of the last decade.

Against this backdrop, the Dutch Government created a taskforce to work under the tell-tale name, '*for faster legislation*'.[18] The taskforce – working between 2011 and 2012 – was especially interested in possible measures and methods to accelerate the legislative process. This laid the basis for the comparative study we are reporting in here. The main question of interest was: can the performance of the legislative process in the Netherlands be improved? Can it be accelerated and, perhaps, even more importantly, can it become more efficient if we compare it to the legislative process of other jurisdictions?

A. What is Efficiency?

In legislative studies, 'efficiency' is a notoriously difficult concept to grasp and define.[19] One thing that can be noted, though, is that efficiency has to do with 'optimisation'. In an efficient legislative process, the available means to make a law are used in an optimal way: an efficient legislative process is, in other words, one that *performs* well. To measure whether a legislative process performs well is also very difficult, since it will depend on the functions the process serves (legal/constitutional, scientific/professional, political, socio-economic, etc),[20] the demands set on it and the expectations different actors/participants have. This is problematic in light of the lack of objective criteria to measure these considerations. On the other hand, there are certain aspects that might give an indication of performance – and put in the context of efficiency – *in comparing* one legislative process with its equivalent in another jurisdiction. The same can be done with the level of openness and public participation in the process. This 2012 comparative study followed up on the aspects of performance that seemed to interest the Dutch taskforce, namely:

1. Pace/speed of the whole process in terms of time consumption;
2. (types of) Procedure, dedicated (emergency) procedures;
3. Consultation;

[18] The taskforce worked within the framework of the Interdepartmental Commission for Constitutional Affairs with respect to Legislative Policy (ICCW).

[19] For a (partial) attempt at operationalisation, see: Koen J Muylle, 'Improving the Effectiveness of Parliamentary Legislative Procedures' (2003) 24(3) *Statute Law Review* 169, 170–73.

[20] According to Ignace Snellen, government policies and the legislative process are constantly 'pegged' between different rationalities (ie, closed sets of criteria actors use to judge the appropriateness of their actions). To complicate matters these rationalities are mutually exclusive most of the time – they do not go well together. Snellen discerns the legal rationality (conformity with legal principles, coherence of the system), economic rationality (demands efficiency and responsiveness in the means whereby resource-distribution is effected), professional or technical rationality (demands the application of proven theories and models) and the political rationality (process in which unequal distribution is performed in the face of equal claims). See Ignace Snellen, 'Conciliation of Rationalities: The Essence of Public Administration' (2002) 24(2) *Administrative Theory & Praxis* 323.

4. Transparency of the process, openness to the public;
5. Public participation;
6. Planning (including procedures of prioritisation, 'budgets', time limits, discontinuity principle);
7. Coordination between the phases in the process and actors (especially the relationship between government and Parliament, but also in relation to – actors involved in – implementation and enforcement); and
8. Use of techniques (especially ICT).

This set of performance aspects had been elaborated upon, with a focus on ten performance indicators (element six was broken down into different subparts) that were phrased as questions.

The method of research used as part of the 2012 project was quite straightforward. In a preliminary desktop research, it first compared the legislative processes of 26 other Member States of the European Union on each of the ten aspects mentioned above.[21] From these 26, three countries with high scores on more than four of the aspects of interest were selected (to be read from developments related to these aspects reported in these countries) for a further in-depth study, including a site visit and (ten) interviews with actors involved in these processes. On this basis, Slovenia, Finland and the United Kingdom were selected and visited. The key actors involved in the interviews were persons such as Members of Parliament (MPs), civil servants from ministries, implementing agencies and representatives from the media and academia.

IV. Highlights of the Study – Bird's Eye View

As noted before, and as the 2012 study showed, it is very hard to give an appraisal of the performance and – based on that – the overall efficiency and openness/inclusiveness of a given legislative process in more or less objective terms. What could be done is to give an overview on how legislative processes in relative terms rate on different aspects relating to performance indicators. In this paragraph, a bird's eye view of the comparison on the different performance indicators is presented. The overview merely gives an impression of some of the highlights of the study and predominantly compares the three countries involved in the in-depth study. It compares them very briefly on the basis of some of the most interesting performance indicators used as part of the study.

A. Pace, Speed and Planning

The average time it takes for a bill to pass through Parliament is about a year, the study shows (including in the Netherlands and other countries). If we factor in

[21] See Voermans, et al (n 10).

the departmental preparation of a legislative proposal – that is the stage before a proposal becomes a bill – then the whole legislative process takes up to two years. A great deal of time is spent in the examined countries on coordination between governmental departments and ministries.

Dedicated legislative procedures, like emergency procedures, can and do speed up things. The Slovenian legislative process, for instance, stands out in the study because of its speed. On average, the legislative process during the Parliamentary phase takes between two to three months. This speed is mainly due to a combination of a dedicated procedure and the strict work programme concluded between Government and Parliament. There is, however, according to domestic observers, a downside to all this. The speed is so overwhelming that there is limited (too little even) time to reflect on and scrutinise bills carefully. This sometimes damages the quality of Slovenian legislation. For this very reason, quite often 'repair' laws modify recently adopted laws by way of quick-fix amendments.

Although the UK's legislative process takes much longer in comparison to Slovenia's (on average two years in total), the UK seems to face a similar problem. Its legislative process uses a rigorous planning system on the basis of political priorities – using a system of bidding for parliamentary time slots, combined with the discontinuity principle which, in effect, stops the process and passage of a bill at the end of the parliamentary term. On occasion, these restraints and pressures do compromise the required scrutiny according to some of those interviewed. Bills sometimes seem to be more or less 'rushed' through the House of Commons, forcing the House of Lords into a position of 'corrective scrutiny'. This position is not really welcomed by the latter in light of its lack of a democratic mandate. The development in the relative positions of the Houses in the legislative process draws on the discussion of the reform of – in particular – the House of Lords.[22] Time pressure and the finality of deadlines at the end of Parliaments – or parliamentary sessions – are especially felt during the so-called 'wash-up'. This is the name for the practice whereby in the few days between the calling of a general election and the dissolution of Parliament – or just before a new session – outstanding bills are rushed through the Commons and the Lords on the basis of deals made privately between the Government and the opposition party (parties). This practice has been criticised by many authors; however, at the same time, it is more or less endemic to the present system.[23] To solve these kinds of problems, some solutions have already been put into place. So-called 'carry-over motions' make it possible to extend the parliamentary debate on a bill into the next session. A more realistic planning takes away some of the pressure too.

[22] See also Kate Malleson and Richard Moules, *The Legal System*, Core Texts Series 4th edn (Oxford University Press, 2010) 49–56.

[23] Ruth Fox and Matt Korris, 'Reform of the Wash-up: Managing the Legislative Tidal Wave at the End of a Parliament' (2010) 63(3) *Parliamentary Affairs* 558.

These two examples, from the UK and Slovenia, demonstrate that greater speed in the legislative process does not necessarily make the process more efficient. Greater speed can have side effects and may thus, in the end, come at a greater cost.

B. Consultation

Wide consultation of a legislative proposal has different, potentially benign, effects. It engages stakeholders and members of the public, contributes to careful preparation and may help in seeing the problem and solutions from different angles. Thus, it enables the proposers to take on board different perspectives, oversee more aspects and alternatives, prevent blind corners and make for a more enforceable and implementable result. It may also benefit the overall transparency and the legitimacy of the proposed measure. In this respect, investment in preparation might even speed up the ensuing handling and debates on the bill that result from it. Not duly informing and consulting experts, citizens or stakeholders may result in overlooking elements and interests, as well as missing out on the necessary support of the addressees and actors that will ultimately have to implement the law.

In the UK, the Government and Parliament for these reasons are committed to improve the overall engagement of the public in the legislative process. There is – as a rule – wide consultation on policies and proposals that may develop into legislation. A two-step procedure of White papers and Green papers is used to discuss and consult government policy on a step-by-step basis. Stakeholders and interest groups, as well as citizens, are invited to put forward their views throughout the process. ICT and the Internet are used more and more often for the purposes of consultation. In Slovenia, government ministries are even obliged to consult the public very early on in the process through the so-called *e-Democracy Portal*. During the second stage of the bill drafting phase, the consultation process is more 'traditional' in nature. Ministries then tend to seek advice from the so-called 'usual suspects', ie, 'standard' stakeholders that are always consulted.

In Finland, not only is the input of citizens and stakeholders taken on board, but also that of experts who are consulted as a standard element of the legislative process. The expertise of civil servants – internal actors in the legislative process – is valued very highly. When the proposed law is discussed in Parliament, the Parliamentary Committee in question hears the civil servants who drafted the law. Furthermore, the Committee also hears other experts and stakeholders. After discussing the technical features of the law, the Committee considers the political aspects.

Consultation with stakeholders and experts can contribute to the quality of a law; consequently, legislative mistakes can be prevented. The Finnish practice of discussing a law in Parliament with the relevant civil servants, without the media being present, can ensure that the quality and (technical) merits of the law will be discussed first, without political interests taking over.

C. The Use of ICT

Planning the legislative process, disseminating legislative information and consulting experts, citizens and stakeholders (in other words: the efficiency of the legislative process) can all be facilitated and sometimes even improved through the use of ICT.[24] Various kinds of new communication techniques are used (or considered) for a wide range of information exchange and dialogue between political actors, government and citizens. With the exception of the United Kingdom – where the legislative process is still largely print-based and a certain adherence to tradition seems to prevail – the use of ICT has become an important factor in attempts to improve the overall performance of the process in the countries under study (not only the countries involved in the in-depth study).

The use of ICT in the Finnish legislative process, for instance, is fairly advanced, especially in the Parliamentary phase of the legislative process. Document structures have been standardised in order to allow electronic access, transferability and input. Two important benefits of this are the long-term accessibility of information in documents and the strengthening of collaboration between Parliament and the ministries. Internet consultation of citizens in the legislative process has existed for quite some time, with E-democracy applications such as Otakantaa.fi and Kansanvalta.fi first launched more than 15 years ago (1999) and since then having been used by citizens and organisations from many parts of society.

ICT plays an important role throughout the legislative process in Slovenia as well. One example of an important ICT system operating within the legislative process is the IPP system, which functions within the ministries as a virtual environment where different versions of proposals and resultant reactions to them are filed. This system also functions as a tracking system with which deadlines are monitored. Within this tracking system, a multi-stage structure is operated in which hierarchical lower units have the responsibility, within the line ministry, to observe the deadlines within the time frame set for the preparation of the bill. ICT is not only used to speed up the process, but makes it more inclusive as well.

ICT is not only used to assist in the operation of parts and elements of the legislative process. It is also used as a means to reform the legislative process itself. The legislative procedures and processes we currently know are largely technique-dependent (based on physical written documents) and paper-based.[25] In the bulk of Western parliamentary democracies, and indeed in most Commonwealth jurisdictions, the majority of legislative procedures were hatched out in the nineteenth or beginning of the twentieth century. Legislative procedures and processes, as a consequence, are deeply rooted in paper and – in turn – the paper underpinnings

[24] See T Arnold-Moore, 'Public Access to Legislation and the Democratic Process' (2004) 5 *RegelMaat* [Dutch Journal for Legislative Studies] 161.

[25] See W Voermans, 'Free the Legislative Process of its Paper Chains: IT-inspired Redesign of the Legislative Procedure' in Ig Snellen, Marcel Thaens and Wim van de Donk (eds), *Public Administration in the Information Age: Revisited* (IOS Press, 2012) 237–51.

have shaped the rules and the processes. The legislative procedure and process is commonly shaped as a conveyer belt on which a piece of paper (be it a draft or bill) is transported and each actor involved has a sequential input. There is, for instance, no simultaneous editing possible, nor any group work. The strong paper base of legislative processes also shows very clearly in drafts or bills that propose amendments to existing legislation. Since a precise and consolidated print of a single legislative text is the required outcome of most legislative processes, amending bills give detailed instructions to the editor or publisher on the parts and elements that need to be changed. Most of the time, the general idea is that amendments are fitted or inserted into the existing text, creating a new single document. As a rule, most jurisdictions do not repeal old acts as a whole and, rather, tend to replace them with a new one in the event of minor amendments;[26] they simply give instructions to insert new elements into an existing text. This is a common enough and – according to most parties involved – evident practice, which departs from the premise that a people's representatives are themselves best set, situated and suited to write and scrutinise parts of law enshrined in legislation, even if this involves very technical subjects and difficult interest representation. It also departs from the idea that democracy is best served by direct involvement and hands-on control from MPs about the text of legislation. ICT provides opportunities to rethink the way the legislative procedure and its processes are set up: to redraw it as it were. Some jurisdictions have already gone that way and reformed the way government prepares proposals and bills (Estonia for example) and the way Parliament handles legislation.[27] The Legis-project in the Netherlands (2009) is an initiative[28] of the Dutch Government that is aimed at *redesigning* the legislative process by using electronic and ICT services.

D. Transparency and Openness

Citizens and stakeholders' attitudes as regards their involvement in the legislative process have changed in the information age we live in. In general terms, they want to be better informed and to have more opportunities to participate. Traditional legislative processes are set up to make the participation between the legislative actors, that is the decision-makers involved, possible. Hence, the focus is on the internal communication and information transfer between ministries, formal

[26] *The Dutch Drafting Directives* (a Dutch Drafting manual used by all civil servants working for the central Government) for instance, only recommend repealing the old act and replacing it with a totally new one in case of substantial amendments. See Directive 6.2.

[27] See, for instance, one of the first initiatives of the Tasmanian (Australia) Government. Timothy Arnold-Moore, 'Automatic Generation of Amendment Legislation' in *Proceedings of the 6th international conference on Artificial intelligence and law* (ICAIL, 1997) 56–72.

[28] *Kamerstukken* II 2009/10, 31731, nr 6 (Dutch Parliamentary Papers of the House of Representatives of the Netherlands).

advisory bodies, Members of Parliament and so on. The traditional legislative process is by its set up and very nature shut to 'outsiders'. For those in doubt, take a brief look at all the numbers and abbreviations, list of technical meetings, committee calendars, etc, to realise how inward-looking and 'in-crowded' the entrails of the legislative process really are. For the general public to really engage and participate in the Government's and Parliament's decision-making as part of the legislative process, transparency and openness are key.[29] Pin-pointed information and access to more are necessary in order to be able to truly participate in the decision-making process.[30]

Grimmelikhuijsen defines the concept of 'transparency' as follows: 'Transparency is the availability of information about an organisation or actor allowing external actors to monitor the internal workings or performance of that organisation.'[31] The meaning of transparency for public authorities is twofold: first, the Government must make information public and, second, citizens must be consulted. In Finland, Slovenia and the United Kingdom, the transparency of the legislative process is realised and promoted in several ways. The Finnish public sector is generally considered to be quite transparent. A code of consultation was adopted in order to support greater transparency, for instance, by specifying a minimum time limit for the consultation period. Finland – as we have seen – also employs ICT to improve the transparency of the legislative process. The Finnish administration operates three important websites which have the specific purpose of making the legislative process more transparent and improving consultation for citizens.[32] One of these is the abovementioned Kansanvalta.fi website, known as 'the democracy data bank'. This website is used by the Government and departments to inform the public on a number of topics, varying from specific legislation projects to general information about the functioning of the public sector. On Kansanvalta.fi, citizens can find, for instance, information on fundamental rights, democracy, political parties, etc. The website, inter alia, explains the ways a citizen can participate in the decision-making process. Furthermore, information about current issues can be found on the web portal. Notably, on a number of occasions the information is published with the contact details of the civil servants concerned.

There is, however, no total transparency in Finland, nor in any other jurisdiction we have studied. Sometimes negotiations or debates need to be – to varying degrees – kept from the public eye in order to be effective. Several interviewees (predominantly government-officials and Members of Parliament) emphasised that in Finland, for instance, there is a debate on the degree of transparency in the work of the Parliamentary Committees, and especially the perceived 'lack of

[29] SG Grimmelikhuijsen, *Transparency and Trust: An Experimental Study of Online Disclosure and Trust in Government* (Utrecht University, PhD thesis, 2012) 56.

[30] Raad voor het openbaar bestuur (n 15).

[31] Grimmelikhuijsen (n 29) 55.

[32] See www.kansanvalta.fi and www.otakantaa.fi.

transparency' in the hearings of the Committees. At present, these hearings are not open to the public: only invited persons can attend. This makes it difficult for journalists to find out what exactly happens in the Committees. Other interviewees seemed to be more in favour of closed Committee meetings: the fact that the hearings are conducted behind closed doors was believed to be essential for the quality of the process and for the 'consensus component' of it. This support mainly stems from the fact that the minutes are published afterwards and that the other phases are very much open once again. Most consulted actors had felt that the level of the political debate benefitted enormously from the reclusiveness of the meetings.

The British Government is currently trying to improve the overall transparency of the legislative process. All bills and Acts are published on the Internet, as are other relevant documents.[33] A lot of effort is put into the structure of the information on the websites of both the Government and Parliament. The Government publishes legislative calendars which allow the public to keep track of bills and their stage of passage through the Houses of Parliament.

Information on legislative initiatives and pending Bills is published as proactively as possible and in plain language. The House of Commons likewise uses social media in order to communicate to the general public. Some of the interviewees had mentioned that they regretted the absence of a lobby register, since it is not always clear how interest groups and stakeholders lobby MPs. Pilots to further improve drafting in plain English are underway as well, in an attempt to increase the public's understanding of bills. Some of the interviewees were critical of the way amendments are drafted. This is because bills are very difficult for laymen to read and understand. According to the interviewees, it would be a good idea to have consolidated versions that show what the effect of an amendment is to the amended text.[34]

E. Public Engagement and Participation

Finland, Slovenia and the UK are, in view of openness and inclusivity, introducing opportunities for citizens to participate directly in the legislative process. Although direct or participatory democracy, as opposed to representative democracy, is not very common as a part of the legislative process, elements of direct participation are being experimented with. In our day and age, direct democracy seems to be craved more and more as an integrated part of legislative decision-making. This cannot easily be achieved, with Cook and Morgan warning against unrealistic expectations of (direct forms of) participatory democracy:

The concept of participatory democracy seems to be in part a critique of the zealous democrat's utopia of only yesterday: universal suffrage for selection and

[33] www.legislation.gov.uk/new (last accessed 22 September 2019).

[34] See for an interesting concept in this respect R Cormacain, 'Keeling Schedules and Clarity in Amending Legislation' (2013) *European Journal of Law Reform*, Vol 15(1) 96–108.

control of elective officials claiming to be representative of the electors and meritocratic selection and control of other officials. Now the advocates of participatory democracy, seeking egalitarianism through a new and broader concept of citizenship as well as democracy, wish to enhance the citizen's role by reversing the trend toward concentration of political authority in the hands of elected representatives and appointed experts.[35]

The concept and success of participatory democracy require committed citizenship and strong motivation for involvement on the part of citizens. While both the organisation of citizens' participation and participation itself are easier nowadays in light of the Internet and ICT, this does not automatically entail commitment and active engagement in the voicing of opinion and a willingness for dialogue to influence government decision-making.[36] In the Internet and media age, ICT-facilitated participation easily risks simply becoming a form of so-called 'trolling'.[37]

Slovenia has implemented many means to enable citizens to directly partake in the legislative process, both by way of citizen consultations and by opening up the legislative process for citizens' initiatives. Consultation takes place in the form of government ministries' obligation to consult the public through the aforementioned e-democracy portal. Citizens' comments are sent to the Ministry responsible for the proposal, which is obliged to respond to all comments that are posted on the e-democracy portal. The documents accompanying a bill in the parliamentary phase include a summary of citizens' comments on the proposal.

Slovenian citizens are involved more actively in the legislative process by means of citizens' initiatives. For example, the website '*I suggest the government*'[38] offers the possibility to put forward ideas about new legislation and website users can vote on the ideas posted. If the idea can count on a certain level of support (50 per cent of the votes of which a minimum of five per cent are registered users), the Ministry is obliged to provide that idea with a reasoned answer and feedback. According to interviewees, this website is frequently used by interested members of the public. In the UK and the Netherlands, citizen's initiatives are made possible as well and suggestions for legislation are also allowed. However, a successful initiative does not automatically lead to a proposal – it only means that the subject matter of the initiative is put on the agenda of Parliament as a topic to be debated. There are no further accompanying obligations beyond a parliamentary debate.

[35] TE Cook and PM Morgan, *Participatory Democracy* (Canfield Press, 1971) 3–4.

[36] F Poletta, 'Participatory Democracy in the New Millennium' (2013) *Contemporary Sociology: A Journal of Reviews* Vol 42(1), 40–50, 42.

[37] Trolling is a skill, defined by a series of well-placed and thought-out arguments regarding a subject, that may or may not be refuted by persons who are of lesser intelligence than the troll himself.

[38] www.predlagam.vladi.si (last accessed 22 September 2019).

Participatory democracy is not only becoming increasingly popular at a national level, but also as a part of the legislative process (and decision-making process) of the European Union. Although the Treaty on the European Union (TEU) provides that the functioning of the Union shall be founded upon *representative* democracy (Article 10, paragraph one), several provisions concerning participatory democracy have been added, most notably, the citizen's initiative procedure.

V. Speeding Up, or Rushing to Results? By Way of Conclusion

Modern demands for a more efficient legislative process which is also transparent and encourages citizen participation are not easily satiated. Efficient lawmaking is not only focused on creating laws as quickly as possible. Efficiency implies that the available means, including the methods to plan the process and the use of ICT, are used optimally. The modern strive towards increased horizontalisation, citizen participation and government transparency, can contravene the need for more efficiency and accelerated legislative processes; the simultaneous realisation of both concepts is difficult. The traditional paper-based setup of the legislative processes is similarly not useful. The 2012 report that this contribution focuses on clearly shows how modern legislatures are grappling with different demands and trying to accommodate them.

The study confirms that an efficient, expedient legislative process does not automatically rule out high levels of transparency and participation, nor that it contravenes careful preparation, consultation and consideration. Indeed, informing and involving citizens in the legislative process is very much a part of efficiency and expediency in modern legislative processes. By investigating the performance of different legislative processes (in Finland, Slovenia, UK and other European countries) framed in terms of efficiency and transparency and inclusiveness, the report has demonstrated that the increased transparency and participatory nature of government organisations does not necessarily have a negative influence on the efficiency and speed of the legislative process. In these countries, the procedure is organised in such a way, particularly by utilising intelligent ICT applications, that there is enough room for citizen participation, while the overall efficiency and the speed of lawmaking are not negatively affected. It also proved that attitudes as regards public engagement in the legislative processes have changed and that legislative processes themselves are affected and changing. Although the clock speed of process does not seem to have changed much, the operation of the process has. Increased transparency and levels of 'outsider' (ie citizen's) participation within the legislative process may well have added value in the form of a new and modern check on government. Increased and direct involvement of citizens in the legislative process might boost the overall legitimacy of it, improve the fit to the problem at hand, increase the overall quality, help implementation and compliance, and, potentially, ensure a longer legislative life span.

On the other hand, the positive effects of transparency and inclusiveness should not be overestimated. Transparency is a complex and multifaceted concept.[39] Moreover, there is, to our knowledge, no necessary causal link between increased transparency and openness, on the one hand, and citizens' trust in public authorities, on the other.[40] As Juan Carlos de Martin, recently and rightly noted:

> In criticising – often justifiably – the Italian political system it should not be forgotten in particular that political activity is an essential democratic art, as Bernard Crick wrote in 1963 in a now classic book, *In Defence of Politics*, an art based on virtues such as prudence, conciliation, compromise and adaptability.[41]

His conclusion reads that

> [i]t might be better, then, to look at how we can encourage representative democracy to evolve towards more participatory forms, towards what we might call, in the words of Stefano Rodotà, the 'continuous democracy'. (…) In other words, it is neither direct democracy nor the defence of the status quo that will allow us to emerge from the current crisis, but an evolution – led by parties that have been thoroughly overhauled (or completely new parties) – of representative democracy towards forms that are more participatory: perhaps there is someone (…) up to the challenge?[42]

This, indeed, chimes well with our findings from 2012.

[39] E Scholtes, *Transparantie, icoon van een dolende overheid* [Transparency, Icon of a Wandering Government] (Boom/Lemma, 2012).

[40] Grimmelikhuijsen (n 29) 55.

[41] Juan Carlos de Martin, 'Cyber-Democracy is just the First Step' (*VoxEurope*, 16 April 2013) https://voxeurop.eu/en/content/article/3683241-cyber-democracy-just-first-step.

[42] *Ibid.*

12

Legal Schizophrenia: Rethinking the Dichotomy in Distinguishing between Retroactive Criminal and Civil Legislation

YANIV ROZNAI*

The wave that has gone by cannot be called back,
the hour that has gone by cannot return.[1]

I. Introduction

Much has been written about the distinction and boundaries of civil and criminal law.[2] Generally, while civil law deals with granting and enforcing rights and obligations, criminal law deals with the punishment of crimes. The distinction is usually based on their subsequent consequences: in civil law, the penalties granted by the court are monetary or a mandatory injunction, while criminal law includes financial or imprisonment punishments. A sharper distinction (and the least researched), discusses the dichotomy in legal systems' approach towards civil legislation compared to criminal legislation on matters of retroactivity. Generally, retroactive criminal legislation is forbidden, while retroactive civil legislation

* I would like to thank Suzie Navot, Guy Seidman, Nadiv Mordechay, Nir Yamin and Tamar Waldman for useful remarks of earlier drafts. I would also like to thank Sara Hava, Adam Taharani and Aya Dvir for editing and research assistance. This is part of a much larger project on retroactive legislation. The first part that focused on the distinction between retroactive and retrospective legislation and why the rule against retroactivity should be considered a basic constitutional one, was published in Hebrew as: Yaniv Roznai, 'Retroactivity – Not Only a Matter of Time! Thoughts on Analyzing Retroactive Legislation Following Genis' (2008) 9 *IDC Law Review* 395.

[1] Ovid, *The Art of Love* (Loeb eds, translated by JH Mozley) (Loeb Classical Library, 1929) 122.

[2] See, eg Franklin E Zimring, 'The Multiple Middlegrounds Between Civil and Criminal Law' (1992) *Yale Law Journal* 1901.

is permitted. Before delving into this distinction, it is worth explaining the key concepts of retroactivity and retrospectivity.

Legislation is considered *retroactive* if it changes the previous legal status, characteristics or consequences of an act committed before the legislation's entry into force. This is different than *retrospective* legislation that prospectively changes the legal status, characteristics or consequences of situations that ended, or actions committed or occurred prior to the law's entry into force.[3] In other words, *retroactive* law operates ex post facto and retroactively changes the law existent prior to the legislation, while a *retrospective* law acts prospectively but looks backwards and provides different present and future consequences for actions that occurred in the past. For convenience, unless explicitly stated otherwise, this chapter regards both of these types of legislation as retroactive. Retroactive criminal legislation includes both a later offence-creating norm that applies to acts committed before its entry into force, and a later norm requiring a more severe punishment than expected upon committing the act itself. Of course, a later norm may also retroactively abolish an offence or alleviate the offender's penalty compared to the relevant legal consequences under the norm at the time of the offence, yet these are considered alleviating retroactive applications. When non-retroactivity of a criminal norm is discussed, it refers to the more stringent norm.

So what is the dichotomy between retroactive criminal legislation and retroactive civil legislation? All sources of law agree that the principle of non-retroactive criminal norms is an 'inalienable asset of a cultured society'.[4] The prohibition on criminal punishment for an offence not illegal when committed: *reformation in peius* or *nullum crimen sine lege* is the most common prohibition throughout various constitutions worldwide.[5] The roots of this principle can be seen in Article 39 of the Magna Carta Libertatum, granted by the British Crown in 1215, stipulating that no free man shall be seized or imprisoned except by the law of the land.[6] However, retroactive civil legislation is more common and generally permitted.[7] For example, American courts' approach to retroactivity has been described as

[3] For various definitions, see Daniel E Troy, *Retroactive Legislation* (The AEI Press, 1998) 6–8; Paul Roubier, *Le Droit Transitoire: Conflicts Des Lois Dans le Temps*, 2nd edn (Année, 1960); Charles Sampford, *Retrospectivity and the Rule of Law* (Oxford University Press, 2006); CS Bobett, 'Retroactive or Retrospective? A Note on Terminology' (2006) *British Tax Review* 15. These definitions release us from the concept of vested rights, which was associated with retroactivity. See, eg, Bryant Smith, 'Retroactive Laws and Vested Rights' (1926–1927) 5 *Texas Law Review* 231; Bryant Smith, 'Retroactive Laws and Vested Rights II' (1927–1928) 6 *Texas Law Review* 409.

[4] Elmer A Driedger, 'Statutes: Retroactive Retrospective Reflections' (1978) *Canadian Bar Review* 231.

[5] See, generally, Jerome Hall, 'Nulla Poena Sine Lege' (1937) 47(2) *Yale Law Journal* 165; James Popple, 'The Right to Protection from Retroactive Criminal Law' [1989] 13 *Criminal Law Journal* 251; Csongor Herke and Csenge D Tóth, 'Prohibition of Reformation in Peius and the Principle of Constitutionality' (2012) 5(2) *Journal of Politics and Law* 49.

[6] '*Nullus liber homo capiatur vel imprisonatur nisi per legale judicium parium suorum vel per legem terrae*'; see Marc D Falkoff, 'Back to Basics: Habeas Corpus Procedures and Long-Term Executive Detention' (2009) *Denver University Law Review* 961.

[7] Norman S Marsh, *The Rule of Law in a Free Society – A Report on the International Congress of Jurists* (International Commission of Jurists, 1959).

almost schizophrenic: while courts firmly reject retroactive legislation in criminal contexts, in similar civil contexts they are nearly dormant, allowing a great degree of retroactivity. In other words, when civil and economic interests are at stake, courts prefer the democratic-lenient considerations. However, in the criminal context, considerations of the rule of law prevail.[8]

The purpose of this chapter is to challenge this distinction between criminal and civil law regarding retroactive legislation and examine whether it is justifiable.[9] Section II examines the sharp worldwide distinction between retroactive criminal and civil legislation. Section III points to the underlying considerations at the core of the opposition to retroactive legislation in order to determine whether they justify the alleged distinction, while section IV examines the common features distinguishing between criminal and civil law. These sections demonstrate that the distinction between the two legal systems is unjustified in the context of retroactive legislation. Section V concludes.

II. The Worldwide Distinction between Retroactive Criminal and Civil Legislation

In Europe, European law treats retroactive civil legislation differently from retroactive criminal legislation.[10] Article 7 of the 1950 European Convention of Human Rights explicitly states that no one shall be held guilty of any criminal offence on account of any act or omission which did not constitute a criminal offence under national or international law at the time it was committed. Article 7 applies only to criminal law and does not discuss civil retroactivity. Nonetheless, regulatory offences can be considered criminal regarding the prohibition on retroactivity.[11]

For non-criminal provisions, the European Commission on Human Rights and the European Court of Human Rights, typically begin their judgments by stating that retroactive civil legislation is not expressly forbidden by the Convention and that Member States have broad discretion concerning economic issues.

[8] See Harold J Krent, 'The Puzzling Boundary Between Criminal and Civil Retroactive Lawmaking' (1995–1996) *Georgia Law Journal* 2143, 2149; Matthew P Harrington, 'Retroactivity of Law: Forward: The Dual *Dichotomy of Retroactive Lawmaking*' (1997) *Roger Williams University Law Review* 19, 28.

[9] While there may be variations between common law and civil law systems, as in civil law countries the principle of legal certainty plays an important role (for example, it is a general principle of EU law, see Juha Raitio, *The Principle of Legal Certainty in EC Law* (Springer, 2003) and Patricia Popelier, 'Legal Certainty and the European Courts: Accessibility and Legitimate Expectations as Standards of Reasonableness' in M Fenwick et al (eds), *The Shifting Meaning of Legal Certainty in Comparative and Transnational Law* (Hart Publishing, 2017) 29–54), still I aim to demonstrate that the different treatment to criminal and civil legislation is quite generally accepted.

[10] On the jurisprudence of the European Court of Justice on retroactivity, see Patricia Popelier's contribution in chapter 5 of this book.

[11] DJ Harris, M O'Boyle, EP Bates and CM Buckley, *Law of the European Convention on Human Rights*, 2nd edn (Oxford University Press, 2009).

Nonetheless, there are a number of tools to challenge retroactive civil legislation under the Convention, such as Article 6 (fair trial) and Article 14 (non-discrimination), and especially Article 1 of the Additional Protocol, which assures protection of the right to the peaceful enjoyment of possessions and use of property. This is especially relevant for retroactive taxation. Although this Article protects the right of states to enact legislation that ensure taxes, even tax laws can be judicially reviewed for insuring that the law pursues a legitimate public interest and corresponds with the Convention and with the principles of legal certainty and proportionality.[12] But again, the point of the departure is that 'retroactive tax legislation is not as such prohibited'.[13] Thus, although retroactive civil legislation may enjoy a limited protection, it is incomparable to the absolute prohibition against retroactive criminal legislation.[14]

The prohibition on retroactive legislation holds importance not only on the regional but also at the international level.[15] The vast majority of International Human Rights Law instruments ignore retroactive civil legislation as explicit stand-alone prohibition, treating it under the umbrella of protection of certainty and legitimate expectation and as a component within the analysis of proportionality when assessing rights violation. In contrast, it explicitly refers to the criminal principle stating no one shall be held guilty of a penal offence on account of any act or omission which did not constitute a penal offence, under national or international law, at the time when it was committed.[16] A similar approach can be found in the Law of Armed Conflict. In the realm of belligerent occupation, Article 65 of the Fourth Geneva Convention states that new penal provisions, enacted by the Occupying Power, shall not come into force before they have been published and brought to the knowledge of the inhabitants in their own language, and that they shall not be retroactive.[17] The prohibition on retroactive criminal legislation also exists in International Criminal Law. For instance, the Rome Statute of the International Criminal Court states that no person shall be criminally responsible under the Statute for conduct prior to the Statute's entry into force.[18] Finally, the

[12] See Philip Baker, 'Taxation and the European Convention on Human Rights' (2000) *European Taxation* 298; Philip Baker, 'Retroactive Tax Legislation and the European Convention On Human Rights' (2005) *British Tax Review* 1; Melvin RT Pauwels, 'Retroactive Tax Legislation in View of Article 1 First Protocol ECHR' (2013) 22(6) *EC Tax Review* 268. And see, generally, Werner Haslehner, Georg Kofler, Alexander Rust (eds), *Time and Tax: Issues in International, EU, and Constitutional Law: Issues in International, EU, and Constitutional Law* (Kluwer Law International BV, 2018).

[13] Pauwels, *ibid*, at 276.

[14] See Raitio (n 9).

[15] See generally, JT Woodhouse, 'The Principle of Retroactivity in International Law' (1955) 41 *Problems of Public and Private International Law* 69.

[16] See, eg, Universal Declaration of Human Rights 1948, Art 11; International Covenant on Civil and Political Rights 1966, Art 15; African Charter on Human and Peoples' Rights 1981, Art 7.

[17] Protocol Additional to the Geneva Conventions of 12 August 1949 and Relating to the Protection of Victims of International Armed Conflicts (Protocol I) 1977, Art 11, Art 75.

[18] Rome Statute of the International Criminal Court 2002, Art 24; Robert Cryer, Hakan Friman, Darryl Robinson and Elizabeth Wilmshurst, *An Introduction to International Criminal Law and Procedure* (Cambridge University Press, 2008) 66.

Vienna Convention on the Law of Treaties of 1969 provides some sort of reference to retroactive civil law. Article 4 states that the Convention itself does not apply retroactivity. Article 28 states that in absence of any implied intentions, a Convention's provisions are not binding on all act or fact which took place, or a situation which ceased to exist before the date of entry into force. This provision applies to all other provisions, whether criminal or civil. However, similar to the presumption of interpretation, this provision against retroactivity is not an unsurmountable obstacle.

The same approach exists in domestic national constitutions. The most common constitutional prohibition on retroactivity refers to criminal legislation.[19] In contrast, constitutional provisions prohibiting retroactive civil legislation are only found in a few constitutions, generally worded in a laconic manner, providing the court with broad interpretation.[20] Specific prohibitions are found in Article 18 of the Portuguese Constitution of 1976 stating that laws which restrict rights, freedoms and guarantees shall not possess a retroactive effect. Article 13(2) of the South Korean Constitution of 1948 states that no restrictions may be imposed upon the political rights of any citizen, nor may any person be deprived of property rights by means of retroactive legislation. Sometimes, special provisions regarding the retroactive tax laws are enacted, for example, in Article 78.2 of the Greek Constitution stating that a tax or any other financial charge may not be imposed by a retroactive statute effective prior to the fiscal year preceding the imposition of the tax. Article 103.3 of the Portuguese Constitution mentioned above also states that no one shall be obliged to pay retroactive taxes.

In Canada, Article 11(g) of the Canadian Charter of Rights and Freedoms holds an explicit prohibition on retroactive legislation. Canadian case law interpreted the constitutional prohibition as only applicable to retroactive criminal legislation,[21] and additionally, it was decided that legislation relating to life, liberty or security, may contradict Article 7 if it is retroactive. This Article was used by taxpayers

[19] See, eg, Constitution of India 1950, Art 20(1); Paraguay Constitution 1992, Art 14; Constitution of Romania 1991, Art 15; Constitution of the Federative Republic of Brazil 1988; Constitution of the Republic of Estonia 1992, Art 23; The Basic law for the Federal Republic of Germany 1949, Art 5XXXIX and XL, 103(2); Constitution of Croatia 1990, Art 31; Constitution of the Republic of Cyprus 1960, Art 12; Constitution of the Czech Republic 1993, Art 40; Constitution of the Italian Republic 1946, Art 25; Constitution of Japan 1946, Art 39; Constitution du Grand-Duche de Luxembourg 1868, Art 14; Constitution of the Republic of South Africa 1996, Art 35; Constitution of the Kingdom of Netherlands 1983, Art 16; Constitution of New Zealand 1990, Art 26; Constitution of the Russian Federation 1993, Art 54; Constitution of the Slovak Republic 1993, Art 50; Spanish Constitution 1978, Art 25; Constitution of Sweden 1975, Art 10; The Grand National Assembly of Turkey 1982, Art 38; Constitution of the Republic of Poland 1997, Art 42.

[20] Art 4 of the 1996 Constitution of the Kingdom of Morocco briefly states that legislation will not have retroactive application and Art 83 of the Constitution of Spain states that regular Spanish legislation will not be of retroactive character. Art 97 of the 1814 Constitution of Norway states that legislation will not have retroactive application.

[21] *Air Canada v British Columbia* [1989] 1 SCR; *British Columbia v Imperial Oil* [2005] 49 SCC.

to challenge a retroactive fiscal legislation, claiming that economic security is part of one's right to security; however, this claim was rejected by the Canadian courts.[22]

In the United States, Article 9(3) of the Constitution states that: 'No … ex-post-facto law shall be passed'. At first glance it seems that the vague 'ex post facto' clause expresses the technical aspect that the law shall be prospective, as the Latin term 'ex post facto' includes verbatim every law passed 'after the fact' and therefore is broad enough to include all types of retroactive legislation.[23] However, according to American law, 'ex post facto' legislation is necessarily retroactive, but not all retroactive legislation is necessarily 'ex post facto'.[24] In the *Calder v Bull* case, the Supreme Court ruled that only 'ex post facto' criminal legislation is prohibited and that this prohibition could not have intended to apply to civil legislation.[25] The majority opinion is far from convincing and has attracted some criticism.[26] Indeed, it is unclear this was indeed the intention of the Constitution's drafters; as opposed to the prohibition on retroactive criminal legislation found in some US state constitutions, there is no such explicit prohibition in the Federal Constitution, showing that the legislator overtly chose not to limit the prohibition only to criminal cases, although they knew how to create such an explicit prohibition. In the absence of a solution against retroactive civil legislation within the ex post facto clause, courts were forced to turn to other provisions, such as the Contracts clause, Takings clause, Due Process clause, and other principles such as Separation of Powers and Invested Rights doctrine.[27] In *Eastern Enterprises v Apfel*, Supreme Court Justices unanimously focused on the question of retroactivity, but were undecided as to its legal analysis.[28] The verdict reflects an interesting constitutional status – none of the Justices doubted the retroactive application of the law. However, the five Justices of the majority opinion, who all found the retroactive law was unconstitutional, were also not unanimous as to their legal analyses, and

[22] Geoffrey Loomer, 'Taxing Out of Time: Parliamentary Supremacy and Retroactive Tax Legislation' [2006] *British Tax Review* 64, 80–81.

[23] CWK, Jr, 'Changes in Procedural Law as Ex Post Facto Legislation' (1926) *University of Pennsylvania Law Review* 400.

[24] H Campbell Black, 'The Constitutional Objections to Retrospective and Ex Post Facto Laws' (1886) *American Law Register* 681. See also Jane Harris Aiken, 'Ex Post Facto in the Civil Context: Unbridled Punishment' (1992–1993) 81 *Kentucky Law Journal* 323.

[25] *Calder v Bull*, 3 US (3 Dall) 386 (1798). See Breck P McAllister, 'Ex Post Facto Laws in the Supreme Court of the United States' (1926–1927) *California Law Review* 269.

[26] Julian N Eule, 'Temporal Limits on the Legislative Mandate: Entrenchment and Retroactivity' (1987) *American Bar Association Foundation Research Journal* 379; Laura Ricciardi and Michael BW Sinclair, 'Retroactive Civil Legislation' (1995–1996) *Toledo Law Review* 301; Edward S Stimson, 'Retroactive Application of Law – A Problem in Constitutional Law' (1939) *Michigan Law Review* 38.

[27] See, eg, *Plaut v Spendthrift Farms* (1995) US 514.

[28] *Eastern Enterprises v Apfel* (1998) US 524. See James L Huffman, 'Retroactivity, The Rule of Law, and The Constitution' (1999–2000) *Alabama Law Review* 1095; Matthew A Schwartz, 'A Critical Analysis Of Retroactive Economic Legislation: A Proposal For Due Process Revitalization In The Economic Arena' (1999) *Seton Hall Constitutional Law Journal* 935.

the four Justices of the minority opinion, who all agreed that the law was constitutional, also agreed that retroactive civil legislation in certain circumstances creates a constitutional problem. The existing constitutional situation is inconsistent and suffers from analytical difficulties. Even though these constitutional provisions are somehow related to legislation's temporal element, they are not a part of the 'inter-temporal conflict of legislation'. As Jeffrey Usman concludes the US situation, 'The Ex Post Facto Clauses are limited in application to criminal matters and while the other provisions address civil legislation, they are nonetheless not significant barriers if a legislature decides to engage in retroactive civil lawmaking.'[29]

Within the US states, many states have included in their Constitution an explicit prohibition on retroactive civil legislation.[30] However, while the prohibition on ex post facto legislation is explicitly enacted in more than 30 state constitutions, retroactive civil legislation is only prohibited by a few of the individual states.[31]

This review shows that international law instruments and the constitutions of modern states only regulate the prohibition of retroactive criminal legislation, giving short shrift to civil retroactive legislation.

III. Theoretical Analysis of the Fundamental Considerations against Retroactive Legislation

In order to determine whether the distinction between retroactive criminal and civil legislation is justifiable, one must analyse the fundamental considerations against retroactive legislation, and their applicability in each area of law. As I argue, the fundamental considerations against retroactive criminal legislation apply – often with greater force – to civil legislation.

A. Law as Guiding Future Human Behaviour

The core of theory against retroactivity argues that the main function of the law is to guide future human behaviour. A legal obligation must relate to the future

[29] Jeffrey Omar Usman, 'Constitutional Constraints on Retroactive Civil Legislation: The Hollow Promises of the Federal Constitution and Unrealized Potential of State Constitutions' (2013) 14:1 *Nevada Law Journal* 63, 66.

[30] See Marshall J Tinkle, 'Forward into the Past: State Constitutions and Retroactive Laws' (1992) *Temple Law Review* 1253.

[31] Constitution of New Hampshire 1784, Art I, para 23; Constitution of Ohio 1851, Art II, para 28; Texas Constitution 1886, Art I, para 16; Constitution of Georgia 1982, Art I, para 1, subpara 10; Constitution of Missouri 1945, Art I, para 13; Constitution of the State of Colorado 1876, Art XI, para 12; Constitution of the State of Idaho 1890, Art XI, para 12; Constitution of the State of Montana 1889, Art I, para 13; Constitution of the State of Tennessee 1796, Art I, para 20.

and cannot guide future human conduct through retroactive legislation.[32] Is there a difference between the criminal and civil law on this matter? First, criminal law knowingly tries to influence the public's consciousness prior to committing offences. This intention can only be implemented if the prohibition applies prospectively. What about the civil law? Retroactive civil legislation similarly cannot guide one's behaviour. For example, one can wonder why a similar prohibition on retroactivity does not exist in tax law. Arguably, tax law is similar to criminal law because it constitutes a significant phase in planning one's behaviour and may therefore shape one's behaviour. When criminal retroactivity assigns a penalty to an action that previously did not bear a penalty but is not unacceptable by society, there is an equal amount of injustice done as when the Government imposes tax on a person for an activity that was previously tax free.[33] Additionally, would we not, as a society, want civil legislation to guide our behaviour just as criminal legislation does? Thus, this behavioural consideration does not provide a holistic solution for the dichotomy.

B. Public Trust in Governmental Institutions

Retroactive legislation averts one from relying on the law, and challenges one's ability to understand the consequences of one's actions. Therefore, as opposed to prospective legislation, unconstrained retroactivity causes a confidence crisis in the citizen-state relationship, an unwillingness to rely on the law and a perception of the law as an arbitrary use of governmental authority.[34] Similarly, this consideration does not provide a substantial foundation to distinguish between the criminal and civil spheres. Public confidence in the Government may be undercut not only by retroactive criminal law but also, sometimes even more so, by retroactive civil law, because citizens encounter civil legislation daily, while encountering criminal law is only experienced by a small number of citizens. For example, a tax law that was retroactively imposed would lead to a decline in the public's confidence in the Government. Even if retroactive tax law is economically efficient due to the influx of income for the state, it is impossible to quantify the cost resulting from the loss of the taxpayer's confidence in the taxing system due to the injustice and unfairness resulting from the retroactivity,[35] as well as the uncertainty and inability to plan an 'economic lifestyle'.[36]

[32] Geoffrey de Q Walker, *The Rule of Law – Foundation of Constitutional Democracy* (Melbourne University Press, 1988).

[33] Lon L Fuller, *The Morality of Law* (Yale University Press, 1963).

[34] Schwartz (n 28) 973; Harrington (n8) 29.

[35] Note, 'Setting Effective Dates for Tax Legislation: A Rule of Prospectivity' (1970) 84 *Harvard Law Review* 436, 441.

[36] On the challenging question of retroactive tax laws, see, eg, Julius H Amberg, 'Retroactive Excise Taxation' (1923–1924) 37 *Harvard Law Review* 691; Frederick A Ballard, 'Retroactive Federal Taxation' (1934–1935) 48 *Harvard Law Review* 592; Comment, 'Limits on Retroactive Decision Making by

C. 'Targeted Legislation'

Prospective legislation minimises the possibility that legislators can use the law to target particular individuals or groups because when it applies the law retroactively, it may know which groups will be harmed. Thus, the law ceases to see citizens as equal under the law, exposing the lawgiver to greater temptation to partiality and corruption. Therefore, giving the majority the ability to retroactively change rights and obligations is concerning and leads to the subjugation of the minority group, making it easier to identify the 'winners' and the 'losers'.[37]

Is there room for differentiation between the two branches of law on this matter? Dripps believes so, claiming that history teaches us that political temptation to disabling political opposition necessarily entails penal sanctions, because at that time, public trials for criminal acts were used for political elimination;[38] this situation can constitute a good distinction for those concerned about the use of retroactive civil legislation. However, this is not enough. Retroactive civil matters can be just as punitive as in criminal ones; the legislator may single out particular individuals or political opponents equally effectively by making disfavoured individuals bear the cost of regulatory or revenue measures. Moreover, there is no definitive rule stating that retroactivity within the criminal field provides a larger opportunity for vindictiveness than in the civil field. In both situations, retroactivity may expose groups or individuals to ill treatment. Indeed, even Dripps recognises that penal sanctions are not the evil regimes' only weapon, or even its most dangerous weapon.[39]

D. Stability, Certainty and Security in Social and Personal Relations

The theory opposing retroactivity gained significance at a time when confidence in the legal system held great importance because retroactive laws were seen as potentially harmful to the legitimate expectations of citizens.[40] Due to the fact

the Internal Revenue Service' (1976) 23 *UCLA Law Review* 529; Michael J Graetz, 'Legal Transitions: The Case of Retroactivity in Income Tax Revision' (1977) 126 *University of Pennsylvania Law Review* 47; Avishai Shachar, 'From Income to Consumption Tax: Criteria for Rules of Transition' (1984) 97 *Harvard Law Review* 1581; Saul Levmore, 'The Case for Retroactive Taxation' (1993) 22 *Journal of Legal Studies* 265; Daniel Shaviro, *When Rules Change: An Economic and Political Analysis of Transition Relief and Retroactivity* (University of Chicago Press, 2000).

[37] For an opposing position, see Bernard W Bell, 'In Defense of Retroactive Laws' (1999) *Texas Law Review* 264.

[38] Donald Dripps, 'The Exclusivity of the Criminal Law: Toward a "Regulatory Model" of, or "Pathological Perspective" on, the Civil-Criminal Distinction' (1996) *Journal of Contemporary Legal Issues* 202, 204–205.

[39] *Ibid*, 208–209, 214.

[40] René Demogue, 'Analysis of Fundamental Notions' in A Fouillée (ed), *Modern French Legal Philosophy* (BiblioBazaar, LLC, 2009) 456–58.

that retroactive legislation cannot be used as a tool for behavioural guidance, civilians lose the ability to predict the consequences of their actions with reasonable certainty, therefore, the legitimate expectations that society wants to protect are greatly harmed.[41] Harming these expectations causes citizens to contest their confidence and certainty in the law. Can this consideration constitute the basis for differentiating between criminal and civil law?

Certainty in the law is especially important in the criminal field because retroactively deeming an act as criminal may cause grave insecurity within social relations. However, certainty is important to all legal fields, including civil law, as the existing legal system constitutes a basis for relationships between different parties in society. For social relations, certainty constitutes a foundation of a properly functioning legal system. As Chaturvedi writes, 'perhaps the most fundamental reason why retroactive legislation is suspect stems from the principle that a person should be able to plan his conduct with reasonable certainty of legal consequences'.[42] Can the differentiation between civil and criminal retroactive legislation be based on consideration of reliance and certainty? In my opinion, once again the answer is no.

First, economic actors rely on legislative enactments in the civil field to a far greater extent than criminal actors 'rely' on the law. Retroactive civil law can dramatically harm financial expectations and personal investments.[43] Therefore, retroactive legislation, specifically with civil law, can greatly harm the reliance interest on existing law. Tax law is a perfect example in showing the importance of the reliance interest – the tax system is based on 'self-assessment', a concept which greatly relies upon the security of a stable system.[44]

Second, is it always appropriate to protect a defendant's reliability on the law against criminal retroactive legislation? For argument's sake, I use Waldron's differentiation between legislation creating new crimes on an act done in the past as opposed to a law that adds punishments or imposes burdens on an already recognised crime. Waldron considers the former as a retroactive law, while the latter is retrospective because attaching a new consequence to a criminal action will not change society's forbiddance on the act, while retroactively changing an act previously regarded as innocent to criminal has relevant legal implications.[45] Regarding retroactive law; when the legislator imposes a penalty on an act previously regarded as innocent, reliability is incredibly pertinent. However, also in

[41] D Woozley, 'What is Wrong with Retrospective Law?' (1968) *Philosophical Quarterly* 40.

[42] Kameshwar Nath Chaturvedi, 'Legislative Retrospectivity and Rule of Law' (2013) 34(3) *Statute Law Review* 207.

[43] See, eg, William L Fishman, 'Property Rights, Reliance, and Retroactivity under the Communication Act of 1934' (1997) 50 *Federal Communications Law Journal* 2; Popelier (n9).

[44] Williams, 'Retroactivity in the Federal Tax Field' (1960) *USC Gould School of Law Tax Institute* 128.

[45] Jeremy Waldron, 'Retroactive Law: How Dodgy was Duynhoven?' (2004) *Otago Law Review* 631, 635. For an opposing opinion, see Ben Juratowitch, *Retroactivity and the Common Law* (Hart Publishing, 2008).

these circumstances, absolute prohibition on an act is not necessarily enough. For example, an exception to the prohibition on retroactive punishment is the principle according to which society can punish a criminal who is tried for a crime that was previously an innocent act in domestic law, but was recognised as a crime in international law, such as crimes against humanity or war crimes.[46] In this case, the law recognises the necessity to 'waive' the defendant's reliance interests and does not accept the argument that one cannot be sentenced for committing genocide or crimes against humanity because the crime was not considered a crime domestically where and when the crime was committed.[47] In other words, in certain circumstances, we recognise that the interest of reliability is withdrawn even when a law retroactively criminalises an action.

Regarding retrospective law, it seems that the interest of reliability is weak, because criminals knew, or were supposed to know that their behaviour in these types of scenarios is illegal. The only concern that arises regarding reliability or deterrence is the issue of expected punishment. The obligations of the fairness principle provide that upon deciding to commit an act or not, a person shall be entirely aware of the expected punishment. For example, the Nobel Prize in Economics winner, Becker, stated that efficient deterrence is equivalent to the penalty multiplied by the probability of being caught.[48] However, it is doubtful that this type of knowledge would constitute a precondition for punishment because very few criminals understand the potential penalties for their actions (very little consider the expected penalties of their actions prior to their involvement in illegal activity, and regardless, this calculation is inefficient because criminals generally plan a certain crime and find themselves involved in additional unplanned crimes). Therefore, Waldron claims, that if general deterrence is the goal of criminal law, there should be no reason to prevent changing the legal status of a previously recognised criminal act, because as long as the punishment is still deterring others, the change is justified. According to Waldron, penalising an activity cannot be seen as an agreed upon rate between the legislator and a potential criminal that contemplates when the right time is to commit a crime or not. Can one imagine a man willing to murder for a ten-year incarceration but will not do so if the punishment is increased to a 13-year sentence? However, another question arises – why should the legislator not impose the harsher punishment?[49]

[46] Cherif Bassionni, *Crimes against Humanity in International Criminal Law*, 2nd edn (Kluwer Law International, 1999).

[47] For another example, see a similar discussion regarding the Nuremburg Trials in Glanville Llewelyn Williams, *Criminal Law: The General Part*, 2nd edn (Stevens & Sons Ltd, 1961); Hans Kelsen, 'Will the Judgement in the Nuremberg Trial Constitute a Precedent in International Law?' (1947) *The International Law Quarterly* 153.

[48] Gary S Becker, 'Crime and Punishment: an Economic Approach' (1968) *Journal of Political Economy* 169.

[49] Waldron (n 45) 635–36.

In any event, individuals need to be able to rely on the law when planning their affairs, also in a civil context. For example, the economic analysis of law shows that most people are risk-averters as opposed to financial players that are usually risk-neutral (because they can disperse it, inter alia).[50] Due to the fact that people are naturally risk-averse, there is even more importance to being able to rely on the law, especially in the financial-civil field.

Clear legal obligations divert the individual's freedom of action and it seems that the treat to reliance interests is far stronger in most civil contexts. To summarise, retroactive legislation generates severe danger to the reliance interest in civil contexts, excluding the criminalisation of behaviour, which also in this case, as we have seen, sometimes accepts harming reliance at the expense of punishing those who have committed an act that society prohibited. Therefore, preferring an absolute prohibition on criminal retroactivity rather than civil retroactivity cannot be justified based on reliance considerations.

E. Considerations of Justice

Taking into account that 'justice' is a problematic and subjective term, retroactive legislation is generally considered unjust in our society because people believe they deserve to know the rules of the game that they are subjected to before they start to play, and it is worrisome that the legislator can change the rules throughout the game itself. Retroactive legislation can have significant influence on contract law, family law, tax law, property law and more, in each of these fields, retroactively changing status or rights may be considered as 'unjust' by some of the parties. Injustice can be caused by retroactivity in both criminal and civil contexts and therefore, the principle of justice is not satisfactory in providing a basis for distinction between these two legal branches.

To summarise, it can be said that the considerations at the basis of the opposition to retroactivity cannot clearly support a distinction between the criminal and civil legal fields. Retroactive civil legislation can also be unjust, it cannot guide future behaviour, harms the stability, certainty and secure feelings of social and personal relations, damages the public's confidence in governmental institutions, greatly harms the civilian's reliance interest, and might even constitute a tool for 'targeted legislation' by enabling a law to be directed against a certain group or individual. Since the considerations behind the theory opposing retroactivity do not provide a solution, I now turn to the distinctions between criminal and civil law.

[50] Harold D Skipper and W Jean Kwon, *Risk Management and Insurance: Perspectives in a Global Economy* (Wiley-Blakewell, 2007); Les Coleman, *Why Managers and Companies Take Risks* (Springer, 2006).

IV. Theoretical Analysis of the Fundamental Considerations at the Core of the Distinction between Criminal and Civil Law

This section focuses on the general distinctions between civil and criminal law, for examining whether they might justify the distinction between retroactive criminal and civil legislation.

A. Characterising Behaviour

At first glance, it seems that individual deterrence of the law is more important in criminal rather than civil law, because the former requires intention to create fault. Criminal sanctions and their associated condemnations are based on the recognition of one's free will. Therefore, Slawson argues that free will is the essence of the distinction between criminal and civil retroactive law, because in civil law, one is not condemned for choosing to perform an act society deems wrongful.[51]

Is knowledge or free will unimportant in civil law? Prior to signing a contract, are we not aware of the legal status and consequences of our actions? In order to lose a Torts case, one needs to only prove negligence or unreasonable acts were committed, not intent. What happens if the legislator retroactively changes the conditions for negligence? Can it be argued that this change is reasonable because those under the new law would not have changed their behaviour if they had known the conditions for negligence would be changed and their acts do not require intent? Liability in nearly all acts in civil law do not require intent, however, the law does refer to the knowledge of that person. Slawson's explanation may be suitable for the distinction between civil and criminal prospective legislation, but not regarding retroactive legislation. Here, we are not discussing those who choose wrongful acts, rather those whom relied on the law and their acts were 'lawful' at that time, while the legislator converted their act to 'wrongful' ex post facto. Therefore, the characterisation of behaviour cannot explain the difference between retroactive criminal and civil law.

B. Type of Sanction

The sanctions imposed within criminal law include incarceration, various limitations on freedom and even death. Conversely, even if the civil system allows incarceration through a writ of execution, the system usually implements financial

[51] David W Slawson, 'Retroactive Lawmaking' (1960) *California Law Review* 216.

sanctions according to the accepted practice within civil law, of financial 'damages' instead of financial 'punishments'.[52] Some claim that there is no significant difference between civil or criminal financial sanctions, because a $100 fine has no financial difference than paying $100 of damages, and that the criminal penalties of incarceration are more severe than fines, but are not different in kind. It is all a matter of prices.[53] This economic approach, whereby fines, taxes and damages are functionally identical because they express the price that must be paid for non-compliance – undermines the distinction between civil and criminal liability.[54] Despite this economic approach, a physical type of sanction, that is not monetary, is certainly harsher in the criminal sense; retroactive criminal legislation threatens the freedom interest much more than civil legislation risks economic freedoms.[55]

Indeed, a vital consideration on the prohibition of retroactive criminal legislation entails individual fundamental rights. Individuals have the right to know *ex ante* which activities are prohibited, and the potential penalty if the prohibited act is committed. The prohibition on retroactivity broadens the domain in which individuals can determine their fate by choice.[56] Retroactively changing a previously innocent act to a criminal offence, unilaterally negatively changes an individual's legal bond with the state. As opposed to the narrow interpretation of 'freedom' that is jeopardised by retroactive criminal legislation, retroactive civil legislation jeopardises the broad interpretation of the right to freedom: retroactive legislation violates the legitimate expectations of civilians that constitute their ability to rely on the law, one of the fundamental elements of individual choice.[57] Disrupting expectations, violating acts committed in the past and hindering people's ability to form and execute plans, constitute a violation of personal will, economic freedom, freedom of employment and other rights.[58] Does the distinction between incarceration and economic harm justify the nearly absolute absence of the prohibition on retroactive civil legislation? Indeed, most people would agree that the risk of losing ten years of freedom is worse than potentially losing ten thousand dollars. This distinction helps explain the existence of various procedural defences in both regards, such as the difference in burdens of proof. Although this argument is generally correct, it is not absolute. Some people see the loss of a personal fortune

[52] Danielle S Allen, 'Democratic Dis-ease: Of Anger and the Troubling Nature of Punishment' in Michel Crozier, Susan A Bandes, Samuel P Huntington and Joji Watanuki (eds), *The Passions of Law* (New York University Press, 1999).

[53] Lawrence M Friedman, 'Two Faces of Law' (1984) *Wisconsin Law Review* 13.

[54] Brent Fisse and John Braithwaite, *Corporations, Crime and Accountability* (Cambridge University Press, 1994).

[55] Stephen R Munzer, 'A Theory of Retroactive Legislation' (1982) *Texas Law Review* 425.

[56] Dan M Kahan, 'Some Realism about Retroactive Criminal-Lawmaking' (1997) *Rodger Williams Law Review* 95.

[57] John Prebble, Rebecca Prebble and Catherine Vidler Smith, 'Retrospective Legislation: Reliance, the Public Interest, Principles of Interpretation and the Special Case of Anti-Avoidance Legislation' (2006) *New Zealand Universities Law Review* 271.

[58] *Kaiser Aluminum & Chem Corp v Bonjorno* 494 US (1990).

as more devastating than an additional six months to a ten-year prison sentence. In other words, financial and consequently reputational damages can often be of stronger punitive effect than minor criminal sanctions.

An additional distinction regarding the type of sanction worth discussing is the moral condemnation of the community only experienced towards acts deemed 'criminal' by the legislature,[59] adding an element of blame to the original physical sanction.[60] The language of criminal law reflects this view; we speak of 'crimes' not 'violations' and 'penalties' instead of 'remedies' or 'damages'. The terms 'crime' and 'punishment' carry the implication of moral condemnation that the civil law terms do not. Something 'criminal' is something 'disgraceful'. Indeed, one of the aims of the law is to tag human acts with moral standards: the assertion that prostitution is criminalised constitutes a clear declaration on the moral status of prostitution. 'Symbolic importance' is added to criminal punishment but lacks in other types of punishment.[61] Moral stigmas constitute one of society's most powerful tools to influence behaviour and define its moral standards, therefore providing a negative incentive to commit a criminal act. Due to the unique influence of stigmas associated with criminal sanctions, the constitutional requirements for procedural defences are necessary in criminal proceedings.[62]

The most deterring punishment is the public identification of an individual as a criminal, cutting off social and economic ties and sometimes even equating to the consequences of a similar civil ruling.[63] For example, a criminal record can harm the freedom of movement or the freedom of employment, by preventing access to certain professions. Many scholars have claimed that compliance can be explained through moral respect for the law (the desire to be perceived as a moral person and not be morally condemned) more than the fear of a certain sanction.[64] Accordingly, one can claim that from a moral viewpoint there is a substantial distinction whether an act is criminal or not.

To summarise this point, the type of sanction should provide a justification for the significant distinction at the core of the dichotomy between retroactive criminal and civil legislation. However, this justification is not clear-cut. First, this distinction may be true regarding imprisonment, but the argument

[59] Mary Cheh, 'Unified Field?' (1985) *Cambrian Law Review* 42. For an opposing opinion that associates certain social stigmas to acts not deemed 'criminal', see Susan R Klein, 'Redrawing the Criminal-Civil Boundary' (1998–1999) *Buffalo Criminal Law Review* 679.

[60] Henry M Hart, 'The Aims of the Criminal Law' (1958) *Law & Contemporary Problems* 401; Dripps (n 38) 199.

[61] Joel Feinberg, *Doing & Deserving: Essays in the Theory of Responsibility* (Princeton University Press, 1970).

[62] Earl C Dudley, Jr, 'Getting beyond the Civil-Criminal Distinction: A New Approach to the Regulation of Indirect Contempts' (1993) *Virginia Law Review* 1025.

[63] Aaron Xavier Fellmeth, 'Civil and Criminal Sanctions in the Constitution and Courts' (2005–2006) *Georgia Law Journal* 1.

[64] Paul H Robinson, 'The Criminal-Civil Distinction and the Utility of Desert' (1996) *Boston University Law Review* 201, 513.

significantly weakens when discussing monetary compensations. Second, regarding imprisonment sentences, the distinction is not absolute. As we have seen, a financial burden of $1 million can be harsher than a week-long imprisonment. Yet, the former sanction is generally allowed while the latter is strictly forbidden. Third, a major distinction can be based on criminal condemnation and the influence of holding a criminal record, consequences that are non-existent in the civil sector. But also, here, perhaps there are people who would prefer to absorb slight criminal condemnation rather than lose large amounts of money. And arguably, condemnation exists to some extent in the civil sector, for example stigmas concerning not accepting the violation of contracts. Moreover, the condemnation argument is only relevant to retroactive legislation making new crimes, not retrospective change of punishments. An accused given a seven-year prison sentence will probably not worry about additional condemnation if their punishment has retroactively grown to seven-and-a-half years or even an eight-year imprisonment. Therefore, the type of sanction, which seems prima facie as the strongest rationale for distinguishing civil from criminal retroactive legislation is not as clear as it may seem; the question of what sanction type exists seems to be a matter of degree (stronger or weaker) and not one of type.

C. The Sanction's Aim

The aim of criminal law is to punish those who committed a prohibited act, in order to settle the score with a criminal and deter others in society from committing similar acts.[65] Retroactive criminal legislation does not punish an individual of a crime committed because at the time of the event, the act was not illegal. Therefore, deterring an individual through retroactive legislation is clearly impossible. However, punishment is also one of the aims of punitive civil sanctions.[66] Regarding deterrence; the law uses various levels of deterrence, also within the civil field, especially during the new age of punitive civil sanctions and compensations. Regarding retrospectivity, Waldron argued that if general deterrence is the purpose of criminal law, it seems unreasonable to prevent the severity of a penalty on an act previously recognised as a crime, as long as the penalty still deters others,[67] and therefore, in effect, he makes a claim for applying retrospectivity in the criminal sector. The fact that criminal retroactivity does not fulfill the aim of criminal sanctions does not explain why retroactivity in the civil field must remain.

[65] John C Coffee, Jr, 'The Blurring of the Criminal and Civil Law Models – And What Can Be Done About It' (1992) *Yale Law Journal* 1875.

[66] Kenneth Mann, 'Punitive Civil Sanctions: The Middleground between Criminal and Civil Law' (1992) *Yale Law Journal* 1795; Leonard Williams Levy, *A License to Steal: The Forfeiture of Property* (The University of North Carolina Press, 1996).

[67] Waldron (n 45) 140.

D. The Body Initiating the Sanction

In criminal law, the Government is the initiator. From the moment the Government is involved, the temptation to use its criminal authority increases. When the Government passes a law, those who experience loss and benefit are private citizens and corporations; the state constitutes a neutral party (excluding situations as an employee). However retroactive criminal legislation is different. In this situation, the individual loses and the state benefits by receiving more power to enforce crimes that it previously could not. Therefore, in the criminal field, state legislating retroactive laws can be seen as negatively taking advantage of its legislative authority for its own benefit.

This argument does not lead to an unequivocal conclusion of a similar defence in the civil sector, because retroactive tax legislation provides the Government with greater economic advantages than prospective effects do for example. Therefore, regarding tax law, the Government is not a neutral party, rather a direct benefactor. Additionally, organs of the state are greatly involved throughout the civil field and therefore the state cannot consider itself a neutral party in the civil field.

E. The Interest Group Theory and Legislative Process

Professor Krent analyses the difference between retroactive criminal and civil legislation and uses interest group theory to depict a fine line between the two fields. He explains that retroactive legislation protects the political legislative process in the criminal context more than in the civil context, because the civil retroactive process holds a few characteristics that its criminal counterpart does not: first, those who believe they will be harmed by potential civil legislation can use lobbyists in order to prevent the passing of retroactive legislation negating their interests, unlike the individuals actually subject to retroactive criminal legislation, whom belong to an inferior echelon of society and have far less economic and social tools to organise and influence the legislative process. Second, retroactivity in the civil context may end up increasing the cost of lobbyist activities by decreasing the profits from every successful lobbying effort and therefore decrease lobbying altogether. Third, retroactive civil legislation can constitute a tool to battle political corruption, by using retroactive legislation to remind the legislator that every favour given to friends can always be revoked in the future. Therefore, according to Krent, there is good reason to be suspicious of retroactive criminal legislation and be more open to accepting limited retroactivity in the civil context.[68]

Krent's arguments are not entirely convincing. First, political lobbying procedures allow many large bodies to access and influence the legislator and financial

[68] Krent (n 8) 2145, 2167–73, 2174–83.

stability, allowing massive lobbyist acts to be affordable. However, many non-governmental organisations and human rights organisations exist in order to defend the weaker layers of society through legislative procedures. Moreover, Krent's argument is weakened by the fact that criminal legislation also refers to the rich groups of society, such as white-collar crimes. And of course, during the legislative process, members of the threatened group are not yet considered criminals. So why should it be even harder for them to influence the legislative process? Second, crime has become politicised, as many politicians compete with slogans based on 'law and order'. However, even if retroactive criminal law is dangerous due to its potential for vindictiveness, retroactive civil legislation raises the difficulty of 'self-dealing'; the problem where an official is simultaneously in charge of his own interest and the public interest. Third, the decreasing lobbyist activity argument is unrealistic. Retroactive legislation actually increases the various existing legislative options and therefore allows the broadening of lobbyist activity. Likewise, if it is easier to identify the group harmed through retroactivity, then interest groups will likely broaden their lobbying efforts in order to curb proposed legislation that are not of interest to them or alternatively, will promote friendly legislation rather than lobbyist efforts regarding prospective legislation whose consequences are uncertain.[69]

V. Conclusion

The legal world inherently distinguishes between the criminal and civil sectors. There are different hearing procedures, burdens of proof, penalties and more. Academia teaches criminal law separately from civil law. However, the distinction between criminal and civil law is not entirely clear, and in recent years, their fine distinction has blurred due to the growing tendency to punish anti-social behaviour through civil penalties. One clear distinction has remained over the years – the distinction between retroactive criminal legislation and retroactive civil legislation. While the latter is usually permitted, the first has historically been perceived as a prohibited damaging norm.

Nevertheless, certain retroactive legislation, even if not criminal, can be just as contemptable. As Bridge correctly notes, 'people wish to plan their lives around all laws, not only the ones that could lead to criminal penalties'.[70] The various considerations underlying the objection to retroactive legislation do not lead to

[69] Levmore (n 36).

[70] JW Bridge, 'Retrospective Legislation and the Rule of Law in Britain' (1967) *University of Missouri at Kansas City Law Review* 132, adding that 'Further, in modern times the distinction between criminal and civil matters has become eroded. Breaches of civil laws may trigger consequences that, while not strictly speaking "penalties", are more severe than many penalties of criminal law … Furthermore, differences between legal systems can mean that what one jurisdiction classifies as a civil matter may be a criminal matter in another jurisdiction.'

the unequivocal conclusion why the ban on retroactive criminal legislation is preferable to its civil counterpart. Although criminal cases pose a greater threat to liberty interests, the reliance interest is more prominent in civil cases. Considerations of hindering harm as the purpose of the law to guide human behaviour in future actions, stability, certainty and security in social and personal relations, public confidence and the fear of 'targeted legislation', apply to the same extent in both legal fields. The criteria to distinguish between civil and criminal legislation, such as the sanction's type, purpose and body initiating such a sanction, do not question this ruling, except perhaps the condemnation attached to a criminal sanction. Even the interest groups theory and the legislative process do not justify the aforementioned distinction.

As I have tried to illustrate, the current position leads to an absurd situation in which it is strictly forbidden to increase, even in one day, the punishment of a crime which existed on the day the crime was committed, but in contrast, negatively impacting one's financial situation by applying retroactive civil law, even if the loss is one million Euros, is permitted. Moreover, for many people, losing a large amount of money can have a much larger impact than a short period of incarceration.

Therefore, the fear of retroactive civil legislation is just as prominent in the criminal context and rule of law considerations should apply to the same extent in the civil and criminal sector. To clarify, this chapter does ignore the possible advantages or benefits of retroactive civil legislation.[71] It also does not intend to call upon the legislator to remove the prohibition on retroactive criminal legislation, since this is a pivotal foundation of society aspiring to fulfill the rule of law and justice. Just the opposite – if we strive to strengthen the rule of law, we must rethink the protection from retroactive civil legislation, apart from certain exceptions. Such protection will promote human rights, maintain the reliance interest and the legitimate expectations of citizens, and will enable economic prosperity and be respectful of the rule of law.

[71] See, eg. Donald T Hornstein, 'Resiliency, Adaptation, and the Upsides of Ex Post Lawmaking Adaptation and Resiliency in Legal Systems' (2010–2011) 89 *North Carolina Law Review* 1549.

PART IV

Technology and Time

13

Disruptive Innovation and Sunset Clauses: The Case of Uber and other On-Demand Transportation Networks

ANTONIOS KOUROUTAKIS

I. Introduction

This chapter concerns the changes and challenges faced by regulators due to the disruptive innovation caused by new technology and peer-to-peer (p2p) online platforms. In particular, this chapter will focus on the disruption caused in the taxi industry by 'on-demand transportation network companies', also known as 'ride-hailing apps' or as ride-sharing platforms. Indeed, Uber and other regional ride-hailing apps[1] have changed our perspective on transportation within cities.

Calling a taxi was never so simple and effortless, while the payment transaction was never so fast and convenient. Nowadays, such ride-hailing apps operate in the vast majority of big cities, and in some cities like New York and London, cars affiliated with such on-demand ride-sharing platforms, which are in reality 'private hire cars', outnumber taxis.[2] Most importantly, this took away millions of rides from taxi drivers. For instance, in New York, as of August 2018, ride-sharing companies

[1] For instance Lyft which operates mainly in the US; Cabify which operates mainly in Spain; Ola which operates in India; Careem which operates in the Middle East; Didi which operates in China; and Grab which operates mainly in Southeast Asian markets.

[2] In particular, in New York City, in August 2018, around 99,000 cars were recorded as affiliates of ride-sharing companies or private hire cars (in particular 63,000 vehicles of Uber, and 34,000 vehicles of Lyft) while there were only 16,000 yellow cabs, see TW Schneider, 'Taxi, Uber, and Lyft Usage in New York City: Open TLC Data Reveals the Taxi Industry's Contraction, Uber's Growth, and the Scramble for Market Share' (New York, 5 April 2016) www.toddwschneider.com/posts/taxi-uber-lyft-usage-new-york-city. On the other hand, in London, during the year 2016/2017, private hire cars were around 118,000, while the number of taxis was 21,000. See J Titcomb, 'Who Really are Uber's 40,000 Drivers and 3.5m Users in London?' *The Telegraph*, London, 26 September 2017, www.telegraph.co.uk/technology/2017/09/26/really-ubers-40000-drivers-35m-users-london.

conducted around 558,000 rides per day (Uber conducted around 436,000 and Lyft around 122,000 rides), while taxis conducted around 275,000 rides.[3]

Due to the noticeable drop in passengers, and the de facto circumvention of the licencing system, the value of taxi licences has drastically reduced.[4] As a result, such disruption in the taxi business has caused protests from taxi drivers around the world, and has created challenges to policy-makers on how to resolve the conflicting interests and put an end to the tension. Such challenges are expected to rise with the spread of autonomous cars (also known as self-driving cars) in the future.[5]

In theory, a number of articles have examined the legal framework, the challenges of sharing the economy, and disruptive innovation.[6] In practice there are two public policy options: first, to allow the ride-hailing apps to operate, and thus de facto to liberalise the on-demand transportation industry as is the case for most of the cities in the US, and second to ban the ride-hailing apps directly as is the case for Greece or indirectly by imposing onerous regulations, as is the case for Barcelona.

However, both public policy options have some immediate drawbacks. The former inhibits the legal and financial expectations of the taxi licence owners, while the second protects the taxi licence owners at the expense of the innovation and the consumer benefits. That said, this chapter aims to show that a third way is possible that creates a middle ground and aims to demonstate how time-limited laws, which are subject to sunset clauses,[7] might be useful in order to mitigate different interests in society.

Specifically, this chapter seeks to examine the utility of sunset clauses in resolving this issue. It will use the progressive prohibition of roaming charges in the European Union as a pilot case. The prohibition of roaming charges took place progressively with a number of laws subject to sunset clauses, as such clauses allowed the imposition of ex ante obligations by capping retail and wholesale voice roaming charges in combination with periodic market review evaluation;[8] this

[3] M Iqbal, 'Uber Revenue and Usage Statistics' (2018) www.businessofapps.com.

[4] For more details see below part II.

[5] In 2017–2018, more than 25 companies tested autonomous vehicles in California. For the most recent report on the companies that operate under a licence testing autonomous vehicles in California see https://thelastdriverlicenseholder.com/2019/02/13/update-disengagement-reports-2018-final-results/.

[6] See, eg, S Ranchordas, 'Does Sharing Mean Caring: Regulating Innovation in the Sharing Economy' (2015) 16 *Minnesota Journal of Law Science and Technology* 413; D Rauch and D Schleicher, 'Like Uber, but for Local Government Law: The Future of Local Regulation of the Sharing Economy' (2015) 76 *Ohio State Law Journal* 901; B Edelman and D Geradin, 'Efficiencies and Regulatory Shortcuts: How Should We Regulate Companies Like Airbnb and Uber' (2015) 19 *Stanford Technology Law Review* 293; NM Davidson and JJ Infranca, 'The Sharing Economy as an Urban Phenomenon' (2015–2016) 34 *Yale Law and Policy Review* 215; M Finck and S Ranchordas, 'Sharing and the City' (2016) 49 *Vanderbilt Journal of Transnational Law* 1299; O Lobel, 'The Law of the Platform' (2016–2017) 101 *Minnesota Law Review* 88.

[7] A sunset clause is a provision that requires a legal norm to automatically terminate at the end of a fixed period.

[8] Such state intervention in reality aimed to make the market more competitive on the roaming charges. For more details about the different forms of state interventions and their impact, see S Ranchordas, 'Innovation Experimentalism in the Age of the Sharing Economy' (2015) 19 *Lewis & Clark Law Review* 871, 902 ff.

paradigm shows how such clauses may deregulate progressively the taxi industry through temporary laws which create a predictable business environment for every stakeholder in the transportation industry.

Laws subject to sunset clauses may regulate the cohabitation of licensed taxis with private hire cars connected to ride-hailing apps by progressively allowing more private hire cars to operate. The ultimate aim is to deregulate the market of transportation industry by abolishing the 'numerus clausus' (closed number) of the licensing system.[9] Thus, this chapter aims to show how lawmakers may use time-limited provisions in order to adapt the existing legal framework to the technological change and accommodate innovation. The chapter is structured as follows. First, section II will examine the emergence of ride-hailing apps and will focus on their disruptive impact on the taxi industry. Then, section III will demonstrate how sunset clauses and progressive regulation and deregulation may lead to a predictable and stable business environment for every stakeholder – taxi drivers, drivers of private cars affiliated with the ride-hailing apps and consumers.

II. Disruptive Innovation and Ride-Hailing Apps

Taxis, regardless of their colour – yellow or black, blue or red –, are present in every big city around the world. From the appearance of the first on-demand horse-drawn coaches in 1605,[10] the so-called Hackney carriage, the regulation of such business did not take too long. While in 1625 there were 20 Hackney carriages,[11] in 1635, Charles I adopted a proclamation to limit the use of Hackney coaches in London and Westminster 'unless the passenger is making a journey of at least three miles'.[12] According to the text of the law, such carriages

> were not only a great disturbance to his majesty, his dearest consort the queen, the nobility, and others of place and degree, in their passage through the streets; but the streets themselves were so pestered, and the pavements so broken up, that the common passage is thereby hindered and made dangerous.[13]

Possibly the first major 'disruption'[14] of on-demand transportation within cities took place with the appearance of automotive cars that made horsepower

[9] The use of sunset clauses to regulate the sharing economy, and in particular the ride-hailing industry for experimental purposes, is proposed by Ranchordas. See *ibid.* However, this chapter highlights another reason behind the use of the sunset clauses that is their utility to facilitate transitions with the aim to deregulate the market.

[10] W Gilbey, *Early Carriages and Roads* (Vinton, 1903) 26.

[11] J Noorthouck, 'Book 1, Ch. 11: Charles I' in *A New History of London Including Westminster and Southwark* (R Baldwin, 1773) 154–74.

[12] Gilbey (n 10).

[13] Noorthouck (n 11). The first parliamentary act to regulate hackney coaches took place in 1654. See CH Firth and RS Rait (eds), 'June 1654: An Ordinance for the Regulation of Hackney-Coachmen in London and the Places Adjacent', Acts and Ordinances of the Interregnum, 1642–1660 (British History Online, 1911) 922–24.

[14] The term 'disruptive innovation' was coined by Christensen in 1997. See C Christensen, 'The Innovator's Dilemma' (Harvard Business School Press, 1997).

obsolete.[15] In fact, in 1907, an MP of the House of Commons, Mr Gladstone, confessed during the debates on the Bill on 'Cabs and Stage Carriages (London)' that

> Now the privileged men, by virtue of the privilege, were no doubt in one of the best and most lucrative parts of the business of cab driving in London, and, as the House knew, horse cab-drivers generally were now suffering severely from the competition with motor cabs.[16]

Furthermore, in 1912, Viscount Allendale in the House of Lords provided evidence about the dramatic decrease of the horse-cab licences, stating that 'in 1902, for instance, there were 11,382 cab licences; they were practically all horse cabs. [...] In 1911 there were 10,973 licences – 3,347 horse cabs and 7,626 motor vehicles'.[17]

Nowadays, many dimensions of on-demand transportation are regulated. States or cities regulate the entry into the market with a licencing system, which is in most of the cases 'numerus clausus' (closed number),[18] fix prices per km, and also set the requirements and the standards for drivers and vehicles.[19]

However, the emergence of ride-hailing apps, which are based on p2p online platforms, has disrupted the business of on-demand transportation once again. In particular, such companies, which act as an intermediary between drivers and passengers, outmanoeuvred the state and city regulations on the licencing system and the standards for drivers and vehicles.[20] Anyone could use their personal car and join the online platform, while clients could request a taxi, locate it on the map, calculate how quickly it will arrive, pay for it online with their credit cards without spending time looking for cash, and finally have the option to rate the driver's service.[21] Most importantly, that service was cheaper than the ordinary taxi ride.

Suddenly, the roads of cities around the world were filled with ride-hailing apps' cars or private hiring cars.[22] For instance, Uber expanded in one country

[15] Leontief discussed the impact of machines in agriculture and its impact on horses. See W Leontief, 'National Perspective: The Definition of Problems and Opportunities, The Long-Term Impact of Technology on Employment and Unemployment, a National Academy of Engineering Symposium' (National Academy Press, 1983) 6.

[16] HC Deb 07 August 1907, vol 180, c 206.

[17] HL Deb 22 July 1912, vol 12 c 607.

[18] According to an Organisation for Economic Co-operation and Development (OECD) report, the entry restrictions to the market are generally placed in many jurisdictions because during times of high unemployment, more unskilled workers would be willing to enter this industry. See OECD/International Transport Forum, 'Regulation of For-Hire Passenger Transport: Portugal in International Comparison' (2016) 14–15.

[19] Regarding the evolution of the regulatory framework in UK and USA, see LC Shaw, G Gilbert, C Bishop and E Pruitt, 'Taxicab Regulation in U.S. Cities' (US Department of Transportation, 1983).

[20] Regarding the disruption in London, see G Dudley, D Banister and T Schwanen, 'The Rise of Uber and Regulating the Disruptive Innovator' (2017) 88 *The Political Quarterly* 492.

[21] For more details for instance about how Uber works, see generally www.uber.com.

[22] It is estimated that in 2018 Uber for instance operated in 65 countries and over 600 cities worldwide, while 15 million daily Uber trips were completed. See generally: 'Uber Revenue and Usage Statistics' (2018) www.businessofapps.com.

after the other, and more ride-hailing apps copied the business model. For instance, Lyft appeared in the US and, in Spain, Cabify.

Consequently, taxi drivers' share in the market was under threat, and their income started to shrink. As a result, demonstrations and protests occurred, for instance, in a number of European cities.[23] Taxi drivers were seeking a complete ban or at least regulation of such companies.

Furthermore, the licence to drive a taxi, which was considered an investment more profitable than an investment in the stock market, has been proven inferior and, in some cases, came with great loss. In particular, according to a number of reports, the licence to drive a taxi lost its value in a number of cities although the number of licences remained stable. For instance, in New York, in 2013, the value of the licence was $1.3 million, whereas nowadays it has dropped to $200,000.[24] Likewise, in Paris, between 2009 and 2013, the taxi licence had reached a value of €250,000, but, in 2017, it fell to €140,000.[25]

On the other hand, in Madrid, the value of the taxi licence in 2014 – the year that Uber was launched – was €133,100 and, in 2018, it has remained relatively stable at €135,000.[26] In addition, in Barcelona, in 2014, the value of a taxi licence was €11,9300 and, nowadays, it is valued at approximately €13,0000.[27] Interestingly, it seems that in cities where authorities have implemented a strict regulatory framework for on-demand companies, like Madrid and Barcelona, the taxi licence has not lost its value.[28] Likewise, in cities where ride-hailing apps do not operate, like San Sebastian and Segovia,[29] the taxi licence was not affected.

[23] A Topping and A Kassam, 'Angry Cab Drivers Gridlock Europe in Protest at "Unregulated" Taxi App' *The Guardian*, Madrid, June 2014, www.theguardian.com/uk-news/2014/jun/11/cab-drivers-europe-protest-taxi-app-uber-london-madrid.

[24] G Fleishman, 'Uber, Lyft and Via Offered to Make a Fund for Taxi Drivers. The New York City Council Declined' (*Fortune*, 2 August 2018) www.fortune.com/2018/08/01/taxi-driver-fund-uber-lyft-via-new-york-city-council.

[25] N Louvet and A Verloes, 'Taxi Licence Prices are Stabilizing in Paris: Is the Market Mature?' (*6-t*, 31 October 2017) www.6-t.co/en/taxi-license-prices-stabilizing-paris-market-mature.

[26] See El Mundo, 'The Irruption of Uber and Cabify has not Devalued the Price of Taxi Licences' (2018), available at https://www.elmundo.es/economia/macroeconomia/2018/08/04/5b64b96ee5fdead0028b4689.html.

[27] *Ibid.*

[28] See Directorate for Financial and Enterprise Affairs Competition Committee, 'Taxi, Ride-Sourcing and Ride-Sharing Services – Note by Spain' (4 June 2018) [17]. 'However, many Public Administrations in Spain, instead of having adopted a more flexible and cohesive regulatory approach, have rolled old regulatory regimes onto new technologies and sectors, thereby shielding the traditional taxi and PHV sectors'.

[29] As the National Commission of Markets and Competition (CNMC) drew in June 2016, the average value of a taxi licence in the secondary market is even more exorbitant in other cities: €220,271 in San Sebastián, €215,000 in Santander or €205,957 in Palma de Mallorca, €190,000 in Segovia. See CNMC, 'Economic Report on the Restriction on Competition included in the Royal Decree 1057/2015 and in the Order FOM/2799/2015 on Private-Hire Driver Licences' (Madrid, 8 June 2016) UM/085/15, www.cnmc.es/file/107176.

A. Regulatory Framework and the 'Right to Possess a Licence to Drive'

While such ride-sharing companies tried to differentiate themselves from the existing legal framework,[30] taxi drivers around the world sought protection from national courts in order to block the function of such companies. For instance, in the UK, the Licenced Taxi Drivers Association and the Licenced Private Hire Car Association contended that Uber cars are equipped with taximeters, which is an offence under the Private Hire Vehicles (London) Act of 1998.[31] Likewise, in Italy, taxi drivers brought claims before the courts arguing that the ride-sharing platforms should be banned as they violate competition provisions.[32]

At the same time, some authorities adopted laws to respond to the emergence of ride-sharing platforms. While some authorities deregulated the taxi industry removing the barriers to entry, for instance the State of Georgia,[33] other authorities adopted strict regulations on the ride-sharing companies. For instance, in France, the so-called law 'Thévenoud' was adopted, which prohibited, among others, the travelling in a chauffeured car, or obliging them to stay off-road, and it regulated the pricing system by prohibiting charges based on time and distance.[34] In Spain, a Royal Decree was adopted in 2015 introducing a 'numerus clausus', as the number of licences for the chauffeured car rental service is limited and dependent on a ratio of just one permit for every 30 taxi permits.[35]

While the property rights of taxi drivers due to the licence are not protected, it seems that rent seeking allows them to protect their interests. In particular,

[30] For instance Uber sought to differentiate itself from the taxi companies and it claimed that in the EU it should be classified as 'an information society service' falling under the regulation of Directive 98/34 of the European Parliament and of the Council of 22 June 1998 laying down a procedure for the provision of information in the field of technical standards and regulations [1998] OJ L204/37 and not as a 'service in the field of transport' falling under Directive 2006/123 of the European Parliament and of the Council of 12 December 2006 on services in the internal market [2006] OJ L376/36. See C-434/15 *Asociación Profesional Elite Taxi v Uber Systems Spain SL*, Judgment of the Court (Grand Chamber) of 20 December 2017 [40].

[31] *Transport for London v Uber London Ltd, Licensed Taxi Drivers Association, Licensed Private Hire Car Association* [2015] EWHC 2918 (Admin).

[32] Reuters, 'Italian Court Overturns Uber Ban' (Rome, 2017), https://in.reuters.com/news/picture/italian-court-overturns-uber-ban-idINKBN18M22Q.

[33] *Abramyan et al v State of Georgia et al*, May 15 2017 301 Ga 308.

[34] LOI n° 2014-1104 du 1er octobre 2014 relative aux taxis et aux voitures de transport avec chauffeur, JORF n°0228 du 2 octobre 2014. However, parts of this law and in particular the prohibition of the pricing system based on time and distance was held unconstitutional by the *Conseil Constitutionnel*. See French Constitutional Court, Decision 2015-468/469/472 QPC, 22 May 2015, Société UBER France SAS et autres.

[35] Royal Decree 1057/2015, of 20 November 2015. However according to Reuters, this ratio is not implemented, as in 2018 there were 9,000 chauffeured car rental licences and 70,000 taxi licences which is far from the ratio of 1:30. See Reuters, 'Spain Taxi Drivers End Uber Strike after Licence Limits Agreed' (Spain, 2018), www.reuters.com/article/us-uber-spain-strike/spain-taxi-drivers-end-uber-strike-after-license-limits-agreed-idUSKBN1KN0JH. If the ratio were to be upheld it would mean at least 6,000 less chauffeured car rental licences.

the idea that property rights deriving from the taxi licence are not absolutely protected was emphatically held by Irish courts in *Hempenstall v The Minister for the Environment* in 1993.[36] Ireland deregulated the market of taxis, and the holders of such licences brought a complaint before the courts claiming that their property rights based on such licences were violated.[37] The Court on this issue held that

> property rights arising through licences created by law (enacted or delegated) are subject to the conditions created by law and an implied condition that the law may change those conditions. Changes brought about by law may enhance the value of those property rights (as the Regulations of 1978 enhanced the value of taxi plates by limiting the numbers to be issued and permitting their transfer), or they may diminish them.[38]

Likewise, in the State of Georgia, the Supreme Court of Georgia rejected a class action on behalf of the taxi drivers.[39]

On the other hand, public policy concerns can limit the right of non-professional drivers to participate in sharing the economy of on-demand transportation companies. In fact in a number of countries such as Bulgaria and Dennmark, ride-hailing apps are banned or, due to the onerous regulation, such apps were forced to withdraw.[40] Such public policy concerns may include the state interest to control pollution emissions caused to the environment, or to regulate the traffic and congestion in cities, as well as concerns about the quality of the transportation services and the safety of the passengers.

Having said that, and withstanding that the holders of taxi licences have an interest in safeguarding their profession and the value of their licence, the presence of such companies offers a more advanced and cheaper service to customers.

III. Deregulation via Sunset Clauses

A. The Rise of the Autonomous Car and the Inevitable Need for Deregulation

In fact, the matter of the deregulation of the taxi industry is not new. Interestingly, it is a topic that has puzzled academics and policy-makers alike for years, and it seems that divergent views – in favour of and against deregulation – have

[36] *Hempenstall v The Minister for the Environment* [1993] ILRM 318.

[37] For more details about the case law on the property rights of the holders of taxi licences in Ireland, see SD Barrett, 'Regulatory Capture, Property Rights and Taxi Deregulation: A Case Study' (2003) *Economic* Affairs 23(4), 34–40.

[38] *Hempenstall v The Minister for the Environment* (n 36) (Costello J).

[39] *Abramyan et al v State of Georgia* (n 33).

[40] Greg Dickinson, 'How the World is Going to War with Uber' *The Telegraph*, 26 June 2018. Available at www.telegraph.co.uk/travel/news/where-is-uber-banned/.

been expressed.[41] The case of the city of Fresno exemplifies this. In 1980, the city authorities deregulated the taxi industry, allowing anyone with a driver's licence to operate a taxi service, but within two years, they regulated both entry into the market and fares again.[42]

Ride-hailing apps de facto deregulated the market, as any holder of a driving licence owning a car could offer on-demand transportation. In addition, the presence of such companies led to a de jure deregulation of the market in some cities.[43]

As a response to such de facto deregulation of the taxi industry, authorities around the world adopted laws to regulate ride-hailing apps.[44] In Spain, for instance, private hire cars associated with ride-hailing apps are subject to a licence like the taxi, while, recently, Barcelona has adopted a stricter law imposing restrictions on the use of ride-hailing apps, 'requiring vehicles to be booked with at least 15 minutes' notice'.[45]

While the regulatory response to the rise of sharing the economy de facto protects the taxi industry, the rise of autonomous cars which is expected in the future will inevitably, and once again, distort the taxi industry. Already, a number of car industries have invested in autonomous cars, including Uber,[46] while Waymo, a subsidiary of Alphabet, has already launched its self-driving taxi service (robotaxis) in four Phoenix suburbs in Arizona.[47]

[41] A number of economists have conducted research on the impact of deregulation. See A T Moore and T Balaker, 'Do Economists Reach a Conclusion on Taxi Deregulation?' (2006) 3 *Econ Journal Watch* 109; SD Barrett, 'Regulatory Capture, Property Rights and Taxi Deregulation: A Case Study' (2003) *Economic Affairs* 23(4), 34–40; C Gaunt, 'The Impact of Taxi Deregulation on Small Urban Areas: Some New Zealand Evidence' (1996) *Transport Policy* 2(4), 257–62; T Garling, T Laitila, A Marell and K Westin, 'A Note on the Short-term Effects of Deregulation of the Swedish Taxi-cab Industry' (1995) *Journal of Transport Economics and Policy* 29(2), 209–14.

[42] L Shaw, G Gilbert, et al, 'Taxicab Regulation in U.S. Cities' (National Technical Information Service, October 1983) 69.

[43] For instance, authorities in Long Beach, see LJ Nelson, 'Long Beach Allows Taxis to Lower Fares as they Compete with Uber, Lyft' *LA Times*, California, 14 May 2015, www.latimes.com/local/california/la-me-long-beach-uber-20150514-story.html.

[44] Such reaction from authorities was a consequence of a number of demonstrations from taxi licence holders. For more details about how organised communities can advance their interests and rent seeking, see A Krueger, 'The Political Economy of the Rent-Seeking Society' (1974) 64 *American Economic Review* 291. Competition authorities raise concerns about the strict regulation of ride-sharing companies. For instance the 'CNMC is of the opinion that limiting or prohibiting the development of these new models and the entrance of new players in the market via restrictive regulation is not only problematic, but also counterproductive, as it has a negative impact on the market itself, by limiting its potential growth, innovation and its consumer welfare.' Directorate for Financial and Enterprise Affairs Competition Committee, 'Taxi, Ride-Sourcing and Ride-Sharing Services – Note by Spain' (4 June 2018) [8].

[45] S Jones, 'Uber and Cabify to suspend operations in Barcelona' *The Guardian*, Madrid, 31 January 2019, available at www.theguardian.com/world/2019/jan/31/uber-cabify-suspended-operations-barcelona.

[46] D Wakabayashi, 'Uber's Self-Driving Cars Were Struggling Before Arizona Crash' *The New York Times*, San Francisco, 23 March 2018, available at www.nytimes.com/2018/03/23/technology/uber-self-driving-cars-arizona.html.

[47] A Sage, 'Waymo Unveils Self-Driving Taxi Service in Arizona for Paying Customers' (Reuters, 5 December 2019), available at www.reuters.com/article/us-waymo-selfdriving-focus/waymo-unveils-self-driving-taxi-service-in-arizona-for-paying-customers-idUSKBN1O41M2.

B. Sunset Clauses and Predictable Business Environment

Policy-makers are quite often called to adopt policies that dissatisfy parts of society. However, what is more challenging is to adopt policies when different societal groups have opposing interests. While this is occasionally inevitable, the major challenge is to adopt such policies at the minimum cost.

In our case, there is the professional interest of the taxi drivers for a secure business on the one hand, with regulated minimum fares and a limited number of licences in order to control the supply of the service. On the other hand, there is the interest of the participants in sharing the economy and of the owners of private cars to use their recourses in order to maximise their income. Furthermore, there is the interest of the society for the most efficient and safe on-demand transport service in combination with the lowest possible fare.

Based on the assumption that the development of autonomous cars will eventually disrupt the taxi business, it would be prudent for policy-makers to intervene in the taxi industry market sooner rather than later, in order to enhance predictability in the transportation industry.

The intervention of EU policy-makers in the telecommunication industry by progressively eliminating roaming charges might be a paradigm for deregulation via temporary regulation.

In particular, at the European level, until 2007, telecommunication companies and mobile phone operators were still charging roaming charges (higher charges) when their customers were using their mobile phone to make or receive calls or to use data from abroad, even from other EU countries.[48] While on the one hand, this was a significant resource for profit for such companies, it also burdened consumers significantly.[49]

Hence, in 2007, the European Parliament and the Council adopted a Regulation on roaming charges and mobile networks in Europe to introduce price caps for mobile phone voice roaming charges across the Member States of the EU.[50]

[48] In particular, the Commission considered the charges excessive. See European Union Committee, 'Mobile Phone Charges in the EU: Follow-up Report' (HL 2008–2009, 42) [2].

[49] European Parliament and Council Regulation (EC) 717/2007 of 27 June 2007 on roaming on public mobile telephone networks within the Community and amending Directive 2002/21/EC [2007] OJ L171/32 (14). 'Regulatory obligations should be imposed at both retail and wholesale level to protect the interests of roaming customers, since experience has shown that reductions in wholesale prices for Community-wide roaming services may not be reflected in lower retail prices for roaming owing to the absence of incentives for this to happen. On the other hand, action to reduce the level of retail prices without addressing the level of the wholesale costs associated with the provision of these services could risk disrupting the orderly functioning of the Community-wide roaming market'.

[50] *Ibid*, Art 3(1): 'The average wholesale charge that the operator of a visited network may levy from the operator of a roaming customer's home network for the provision of a regulated roaming call originating on that visited network, inclusive inter alia of origination, transit and termination costs, shall not exceed EUR 0,30 per minute'.

This Regulation included a sunset clause, stating that the Regulation would expire on 30 June 2010.[51]

The primary aim of the Regulation was to intervene in the market and eliminate the difference between roaming and national prices.[52] However, as the market conditions were not met, in 2009, the European Commission decided to amend the Regulation of 2007 and further decrease the roaming changes.[53] In addition, the sunset clause was extended for two more years until June 2012.[54]

Moreover, in 2012, a new regulation was adopted to further reduce the roaming charges with a sunset clause until 2017.[55] Finally, in 2015, the European Parliament, the Council and the Commission reached an agreement in order to completely abolish roaming charges for intra-EU travels.[56]

Thus, the ending of roaming changes in the EU was a roadmap that started in 2007, with policy-makers gradually phasing out roaming charges for intra-EU travel, and such charges being completely abolished in 2017.

[51] European Parliament and Council Regulation (EC) 717/2007 of 27 June 2007 on roaming on public mobile telephone networks within the Community and amending Directive 2002/21/EC 27 [2007] OJ L171/32, Art 13.

[52] European Commission, 'Digital Agenda for Europe: What would it do for me?' (19 May 2010, MEMO/10/199).

[53] European Parliament and Council Regulation (EC) No 544/2009 of 18 June 2009 amending Regulation (EC) No 717/2007 on roaming on public mobile telephone networks within the Community and Directive 2002/21/EC on a common regulatory framework for electronic communications networks and services [2009] OJ L167/12, Art 1(4). 'The average wholesale charge referred to in paragraph 1 shall apply between any pair of operators and shall be calculated over a twelve-month period or any such shorter period as may remain before the end of the period of application of a maximum average whole sale charge as provided for in this paragraph or the expiry of this Regulation. The maximum average wholesale charges shall decrease to EUR 0,28 and EUR 0,26, on 30 August 2008 and on 1 July 2009 respectively and shall further decrease to EUR 0,22 and EUR 0,18, on 1 July 2010 and on 1 July 2011 respectively'.

[54] *Ibid*, Art 1(14). See also European Union Committee (n 48) [14].

[55] European Parliament and Council Regulation (EU) No 531/2012 of 13 June 2012 on roaming on public mobile communications networks within the Union [2012] OJ L172/10, Art 8(2). 'With effect from 1 July 2012, the retail charge (excluding VAT) for a euro-voice tariff which a roaming provider may levy on its roaming customer for the provision of a regulated roaming call may vary for any roaming call but shall not exceed EUR 0,29 per minute for any call made or EUR 0,08 per minute for any call received. The maximum retail charge for calls made shall decrease to EUR 0,24 on 1 July 2013 and to EUR 0,19 on 1 July 2014 and the maximum retail charge for calls received shall decrease to EUR 0,07 on 1 July 2013 and to EUR 0,05 on 1 July 2014. Without prejudice to Article 19 those maximum retail charges for the euro-voice tariff shall remain valid until 30 June 2017'.

[56] European Parliament and Council Regulation (EU) 2015/2120 of 25 November 2015 laying down measures concerning open internet access and amending Directive 2002/22/EC on universal service and users' rights relating to electronic communications networks and services and Regulation (EU) No 531/2012 on roaming on public mobile communications networks within the Union (Text with EEA relevance) [2015] OJ L310, Art 6(a): 'With effect from 15 June 2017, provided that the legislative act to be adopted following the proposal referred to in Article 19(2) is applicable on that date, roaming providers shall not levy any surcharge in addition to the domestic retail price on roaming customers in any Member State for any regulated roaming calls made or received, for any regulated roaming SMS messages sent and for any regulated data roaming services used, including MMS messages, nor any general charge to enable the terminal equipment or service to be used abroad, subject to Articles 6b and 6c'.

Likewise, regulators may use temporary laws in order to create a transitional period for the progressive abolition of the closed number of licences regarding on-demand transportation. In particular, regulators may progressively deregulate the taxi industry by granting licences to private hire cars in order to affiliate with ride-hailing apps. In order to protect the existing taxi licences, more and more private hire car licences shall be granted gradually and with the aim to remove the 'numerus clausus'[57] from the licencing system within a long-term period, for instance a ten-year period.

This would be particularly useful in cities where such ride-hailing apps do not operate. That way a transitional period will allow the taxi business to adapt to the new environment smoothly, without unexpectedly harming the investments of the taxi licence owners.

The merit of sunset clauses to temporarily regulate an issue with the aim to deregulate was already stressed by the Chairman of the Delegated Powers and Regulatory Reform Select Committee, Lord Dahrendorf, who stated that:

> [I]n many, many cases, we should give much more serious consideration to the possibility, feasibility and usefulness of sunset clauses or, at any rate, of procedures which move in the direction of sunset clauses; that is, reviews in set periods and at particular times. That is by far the most effective way of deregulating and reducing the continuing burden of regulation.[58]

That said, authorities around the world can adopt regulations to ease the transition from a regulated taxi licence system to a deregulated one. By gradually deregulating the taxi industry, they create a stable business environment, and they increase the legal certainty of the interested parties. Moreover, they can consider the changing business climate and adapt the regulations to the needs of the society.

IV. Conclusions

This chapter has argued that the on-demand transportation industry has faced two disruptive phases – first, when the Hackney coaches were replaced by the motor engines, and second when ride-hailing apps based on p2p online platforms made their appearance.

[57] The existing regulation of the taxi industry with the 'numerus clausus' has created a path dependence which is always a challenge for regulators to surpass. Ginsburg, Masur and McAdams have argued that '[i]f the problem is path dependence, a temporary law will often be both necessary and sufficient to move behavior to the more efficient outcome [...] [because] temporary law offers a new "starting" point, resetting the system to allow the emergence of the equilibrium with the lowest smoking levels': see T Ginsburg, JS Masur and RH McAdams, 'Libertarian Paternalism, Path Dependence, and Temporary Law' (2014) 81 *Chicago Law Review* 291, 296.

[58] HL Deb 6 June 2005, vol 672, col 754. For more details, see A Kouroutakis, *The Constitutional Value of Sunset Clauses* (Routledge, 2017) 123ff. Ginsburg, Masur and McAdams (n 57) have also stressed how sunset clauses may be useful in liberating the market from path dependence.

In reality, such ride-hailing apps bypass the existing state regulation on the taxi business, which sets paramount challenges for regulators around the world. The drop-in passengers and the circumvention of the licensing system has reduced the value of taxi licences, which have caused protests on behalf of the taxi licence owners. On the other hand, consumers have clearly shown a preference for the service provided by such ride-hailing apps.

This has created challenges for policy-makers on how to resolve the conflicting interests and tension. Regulators in some regions, for instance in Catalonia, intervened in the market by adopting protective measures and rules that hinder the use of the ride-hailing apps, such as the requirement that customers must book their ride 15 minutes in advance. Such policies in reality strip the ride-hailing apps of their innovative elements as their use will no longer be as effortless. However, the challenges on the existing taxi licensing system are expected to rise with the spread of autonomous cars (robotaxis) in the future, which will lead to the third disruptive phase in the on-demand transportation industry. And the adoption of protective measures in order to preserve the existing taxi licensing system in reality brushes the necessity for rules to accommodate innovation in the on-demand transportation sector under the carpet.

In line with the above, this chapter has demonstrated that time-limited laws due to a sunset clause might be useful in order to mitigate different societal interests. On the one hand, the owners of taxi licences rely on the existing licensing system and they see that the value of their licences is threatened by the circumvention of the system by the private hire cars affiliated with ride-hailing apps. On the other hand, ride-hailing apps offer an innovative business model which benefits consumers more. Hence, in order to solve this conundrum, lawmakers might find it useful to employ time-limited regulations. By progressively deregulating the taxi industry licensing system, lawmakers' intervention may create a more predictable business environment for every stakeholder in the transportation industry.

14

Law and Technology in the Dimension of Time

LYRIA BENNETT MOSES AND MONIKA ZALNIERIUTE

I. Introduction

The relationship between law and technology in the dimension of time is a popular theme in legal and policy circles, usually recurring as a critique of outdated laws.[1] In these debates, law is most often portrayed as falling behind technology as both travel together along the dimension of time. While such images simplify the relationship between law and technological change, they reflect some insight into the challenge faced in ensuring law remains relevant, appropriate, comprehensive, well-adapted and clear in the face of an ever-evolving socio-technical landscape.[2] Law's struggle in the face of socio-technical change has been referred to as the pacing problem[3] or the challenge of regulatory connection.[4]

A phenomenon that receives much less attention in these popular debates is the temporal impact of attempts to embed the law and social values into technological design. There are a variety of terms that capture ideas around design-based regulation, each with different foci and associated literature.[5] For example, 'value sensitive design' focuses on the design process, while 'compliance by design' focuses on extracting, modelling and implementing legal requirements but both are about using architecture and processes to achieve a particular effect (respecting values or ensuring compliance with law). The idea of embedding law, values or preferences

[1] This generally takes the form of mocking outdated or short-sighted legal requirements such as the Locomotive Act 1865 (UK) which required that self-propelled vehicles be accompanied by a man with a red flag walking ahead of it.

[2] L Bennett Moses, 'Recurring Dilemmas: The Law's Race to Keep Up with Technological Change' (2007) *University of Illinois Journal of Law, Technology and Policy* 239.

[3] GE Marchant, BR Allenby and JR Herkert (eds), *The Growing Gap Between Emerging Technologies and Legal-Ethical Oversight: The Pacing Problem*, The International Library of Ethics, Law and Technology 7 (Springer, 2011).

[4] R Brownsword and M Goodwin, *Law and the Technologies of the Twenty-First Century: Text and Materials* (Cambridge University Press, 2012) Pt V.

[5] See section III.B of this chapter and the literature cited therein.

into technical design choices and business processes is rarely subjected to similar time-inspired critiques despite the fact that technology and procedures can be designed around outdated understandings of legal requirements and policy goals. Whether technology design decisions are based on technical, commercial, legal or regulatory objectives and requirements (or combinations thereof), they may come to be seen as obsolete as those objectives and requirements evolve. Therefore, the challenge of staying up-to-date or continuing to fit in an evolving world is not only a *legal* challenge.[6] Technology can also fail to meet evolving legal requirements or fail to adapt seamlessly to other technical elements within systems as the broader socio-legal environment evolves. In particular, law can impose particular demands on technology, so that it is called on to catch up.

There is, however, no perfect symmetry here. Both law and technology are human processes; while technological change is sometimes framed as inevitable or exogenous, it is driven by humans (most obviously in laboratories and corporations, but also acting as consumers and regulators).[7] This means that lawmaking and technology design are constrained by human knowledge, inevitably based on data from the past and present. The pace of legal change makes it more predictable so that, while it is rare that lawmakers will foresee not-yet-existing technologies,[8] technological design may pro-actively pre-empt legal change long before its implementation (which may be years after it is first proposed). Because humans can only consciously influence the future, based on knowledge and action in the present,[9] technology can more easily be designed so as to pre-empt law than the converse. This explains why law lagging technology is a more common complaint than the converse.

Just as the past and the present can influence today's technical and legal agendas, so do our beliefs about the socio-techno-legal landscape of the future. For example, ideas about tomorrow's technologies will affect how laws are drafted today. What we can do, what we ought to do and what we in fact do is influenced by the socio-techno-legal landscape of the present. But we also design that landscape

[6] Generally on the pacing problem, see GE Marchant, 'The Growing Gap between Emerging Technologies and the Law'. *The Growing Gap between Emerging Technologies and Legal-Ethical Oversight* (Springer, 2011) 19–33; L Bennett Moses. 'How to Think about Law, Regulation and Technology: Problems with "Technology" as a Regulatory Target' (2013) *Law, Innovation and Technology* 5.1, 1–20.

[7] On the importance of lawyers taking account of both instrumentalist and substantivist understandings of technology, see generally A Cockfield and J Pridmore, 'A Synthetic Theory of Law and Technology' (2007) 8(2) *Minnesota Journal of Law, Science and Technology* 475; MAC Dizon, 'From Regulating Technologies to Governing Society: Towards a Plural, Social and Interactive Conception of Law' in H Morgan and R Morris (eds), *Moving Forward: Tradition and Transformation* (Cambridge Scholars Publishing, 2012) 115.

[8] An exemption is the case of feared technologies on the near-horizon, as in the case of human reproductive cloning which was banned before it was ever done.

[9] D Kutach, 'The Asymmetry of Influence' in C Callender (ed), *The Oxford Handbook of Philosophy of Time* (Oxford University Press, 2011).

with an eye to the future, concerned about, for example, the future-proofing of law through the attractive, if unachievable, goal of technological neutrality.[10] Here, a sense of the eternal failure of law to keep up with technology encourages the design of flexible laws. But what we need are adaptable and flexible socio-techno-legal *systems* that recognise all the ways in which its elements can fall out of sync with the broader landscape.

This chapter seeks to explore how an enhanced understanding of the evolution of socio-techno-legal landscapes over time can help to improve decision-making in the present. Section II summarises how legal scholars imagine technology and the future and how they characterise law's failures to adapt to socio-technical change over time. Section III discusses the various approaches that describe the use of technology as the primary regulator (so that design choices replace or supplement law's regulatory effect) and considers how such design-based regulation fares along the dimension of time. Section IV supplements these theoretical musings with concrete examples drawn from the context of automated driving systems and data privacy laws in Australia and Europe. Section V concludes.

II. Socio-Technical and Legal Change and the Dimension of Time

The idea of time plays an important role in policy and scholarship on the theme of law and technology. This importance manifests via the perceived chronologies of technology (fast) and law (slow), as well as considerations of the timing of legal interventions designed to regulate technology. The relationship between law and technology over time is often conceptualised as a race in which law is the perpetual loser, although the reality is far more complex than this metaphor would suggest. For instance, the fast pace of technology is often associated with the oversimplification of the innovation process and the idea that innovation is mostly disruptive, which is not always true. Actual chronologies of technology and law will vary. In times of perceived crisis or where it is politically opportune, legal change can be surprisingly quick. Conversely, the innovation process can be frustrating and experience extended periods of slowness in particular fields. As this section will demonstrate, law and technology do not take isolated journeys over the course of time and history, and developments in one area will often affect the other. On the one hand, law may prohibit certain paths of technological development, such as human reproductive cloning.[11] On the other hand, technology may

[10] Bennett Moses (n 2); A Chander, 'Future-Proofing Law' (2017) 51(1) *UC Davis Law Review* 1 (summarising mostly US scholarship in a special issue).

[11] Prohibition of Human Cloning for Reproduction Act 2002 (Cth).

change or render obsolete the social impact of law, as was the case, for instance, when barbed wire fences effectively overturned the law of the open range in the American West.[12] This section briefly sketches the role that time plays in understanding the relationship between legal and socio-technical change.

Both the content of law and features of socio-technical systems can be understood and mapped along the dimension of time. Legislation is passed, perhaps amended and possibly repealed, all by reference to explicit commencement dates and, more rarely, sunset clauses. While the law of the present comprises the cumulative effect of this alongside case law, one can also search legislation as at particular points in time.[13] Moreover, judges look to the past to construct the future, interpreting pre-existing statutes and case law to articulate the principles on which cases in the present will be decided. There is a sense of legal progress along the dimensions of time, as concepts are explained and refined, bad ideas are discarded, and statutes are adapted to new possibilities. Similarly, technological artefacts and possibilities generally expand over time due to the engines of invention and innovation. For example, computers began with ideas and designs in the nineteenth century, with the first physical electronic digital computer, the Atanasoff-Berry Computer, built between 1937 and 1942.[14] Before this, writing computer software was not an observable activity, whereas today, mobile phones have become more powerful computers than machines that once filled rooms.

With regards to technology, there is also a general sense that it is not only moving forwards over time but that the rate of change is increasing.[15] In articulating the challenge of this pace for law, references to Moore's law (that the performance of integrated circuits doubles over 18–24 months) are common, as are references to the rate at which new products and processes are introduced and the rapid growth of technologically driven enterprises.[16] In the world of technology, there is a sense of exponential acceleration; a sense that time is not only moving quickly but that the pace is simultaneously exhilarating and exhausting. Law, on the other hand, is rarely described in such fast-paced terms. The point here is not that there is such a thing as 'speed of change' that can be measured and compared across disciplines in a meaningful way, only that the literature suggests a perceived difference in relative velocity.

[12] G Calabresi, *A Common Law for the Age of Statutes* (Harvard University Press, 1982) 244.

[13] For example, the Australian Legal Information Institute, Austlii, allows for point-in-time searches of New South Wales legislation: http://portsea.austlii.edu.au/pit/xml/nsw/act/ (last accessed 24 September 2019).

[14] AR Burks and AW Burks, *The First Electronic Computer: The Atanasoff Story* (Ann Arbor, University of Michigan Press, 1988).

[15] RW Rycroft, 'Time and Technological Innovation: Implications for Public Policy' (2006) 28 *Technology in Society* 281.

[16] R Brownsword, E Scotford and K Yeung, 'Law, Regulation and Technology: The Field, Frame and Focal Questions' in R Brownsword, E Scotford and K Yeung (eds), *The Oxford Handbook of Law, Regulation and Technology* (Oxford University Press, 2017) 3, 3–4.

It is therefore unsurprising that the dominant discourse about law and technology envisages a perpetual 'catching up'. For example, 'Law, marching with medicine but in the rear and limping a little'[17] and 'The hare of science and technology lurches ahead. The tortoise of law ambles slowly behind'.[18] These metaphors place law clearly behind technology in their chronological orientations – law commands in the present but is directed to the socio-technical landscape of the past. Judges are often conservative, declining to develop or interpret law to extend its protective reach when older formulations fail to deal adequately with new situations.[19] Thus formulations and categories designed around an historic socio-technical landscape continue to operate into the present, to the detriment of those who may be harmed or have their values impinged on by newer technologies. It is not just formal law, but also policy and ethics that are said to struggle on a rapidly moving and risky technological frontier.[20] There are thus calls for disruptive and innovative thinking in government, legal and policy circles that mirror the mantras of technological innovation.[21]

The reification of law and technology as differently paced travellers moving along the dimension of time is often unhelpful. Aside from an arbitrary metric of 'progress', against which it would be difficult to compare such different entities as legal rules and technical systems, engineers and lawyers work on very different kinds of problems. Engineers' ability to create the future by sheer will is unmatched in the legal profession, outside of roles such as legislators and, some might argue, judges. Most of the time, lawyers are called on to predict the present based on texts written in the past. It is also not clear whether we want legal rules that are projected too far into the future – road rules that prescribe how to overtake vertically would be useless and unhelpful. The inability to predict the directions of technological change or the nature of its legal and social implications makes the creation of law directed at the future an extremely challenging task.[22]

Despite the prevalence of hare and tortoise metaphors in comparing technology and law,[23] it is not true that lawyers focus on the old and ignore the new. To the contrary, there is an obsession with new problems generated by new technologies, even as old systems receive less attention. For example, by 1858 when the railroad system was in its infancy, two treatises on railroad law had been published in the

[17] *Mount Isa Mines v Pusey* (1970) 125 CLR 383, 395.

[18] M Kirby, 'Medical Technology and New Frontiers of Family Law' (1987) 1 *Australian Journal of Family Law* 196, 212.

[19] See distinction in AJ Cockfield, 'Towards a Law and Technology Theory' (2004) 30 *Manitoba Law Journal* 383.

[20] Rycroft (n 15) 297; D Rejeski, 'Public Policy on the Technological Frontier' in Marchant, Allenby and Herkert (eds) (n 3); MH Shapiro, 'Is Bioethics "Broke"? On the Idea of Ethics and Law "Catching Up" with Technology' (1999) 33 *Indiana Law Review* 17.

[21] Rejeski, *ibid.*

[22] GN Mandel, 'Legal Evolution in Response to Technological Change' in Brownsword, Scotford and Yeung (eds) (n 16) 226, 238.

[23] L Bennett Moses, 'Agents of Change: How the Law "Copes" with Technological Change' (2011) 20 *Griffith Law Review* 263.

United States.[24] These dealt with a variety of then-current issues including property rights over track and eminent domain as well as liability for damages to employees, passengers, stock and land. No doubt, railroads continue to be guided by law and have legal officers or external firms advising on corporate responsibilities, liability for accidents, duties to employees and so forth. But there are no longer railroad lawyers, a term used to describe the pre-presidential career of Abraham Lincoln,[25] and railroad law courses are rare. In contrast, there are increasing numbers of courses and articles focusing on issues related to digital platforms, artificial intelligence, telecommunications, biotechnology and so forth. At least in selecting professional labels and areas of specialty, lawyers prefer to situate themselves on the technological frontier. Indeed, legal imagination of new technological possibilities is often wildly adventurous, as captured in Beebe's description of golden age space law scholarship.[26] Perceptions that law is falling behind are thus not based on outmoded fashions in legal scholarship.[27]

Rather, law is seen to fall behind because there are real challenges for those seeking to evaluate, interpret, enforce or obey laws applied to new technologies. There are always new legal problems that arise at the frontier of socio-technical change. Legal rules are designed for things, activities and relationships that are envisaged by their creators. Linked to the limits of legal interpretation, statutes and judgments are worded within the limits of their author's conceptions of what might be created, what might happen and how interactions might occur. While the human imagination can project some way forward on the dimension of time, forecasts are imperfect, and law sometimes stumbles when applied to situations that were unexpected. Even if words are interpreted generously in light of changed circumstances,[28] interpretation of statutory language in particular has limits – courts cannot change what a statute says to optimise outcomes for unforeseen circumstances.[29] Socio-technical change can thus result in calls for new law (to prohibit, regulate or encourage these new possibilities), clarification of rules (where interpretation in the light of new possibilities is unclear, re-targeting of rules (where new conduct falls inside or outside existing language in undesirable ways) and repeal of rules (where the justification for those rules no longer holds, or the rules are no longer necessary or cost-effective).[30]

[24] EL Pierce, *A Treatise on American Railroad Law* (Baker Voorhis, 1857); IF Redfield, *A Practical Treatise upon the Law of Railways*, 2nd edn (Little, Brown and Company, 1858).

[25] beyondthehistorytextbooks.com/2015/05/12/abraham-lincoln-railroad-lawyer-extraordinaire/ (last accessed 24 September 2019).

[26] B Beebe, 'Law's Empire and the Final Frontier: Legalising the Future in the Earlyorpus Juris Spatialis' (1999) 108 *Yale Law Journal* 1737.

[27] For a broader discussion of trends in the legal academy, see SK Stadler, 'Essay: The Bulls and Bears of Law Teaching' (2006) 63 *Washington and Lee Law Review* 25.

[28] Cockfield (n 19).

[29] See, eg, *National Rugby League Investments Pty Ltd v Singtel Optus Pty Ltd* (2012) 201 FCR 147, 170 [96] '... no principle of technological neutrality can overcome what is the clear and limited legislative purpose of s 111 [of the *Copyright Act 1968* (Cth)]. It is not for this Court to re-draft this provision to secure an assumed legislative desire for such neutrality'.

[30] Bennett Moses (n 2).

There is much discussion in the literature about the timing of legal interventions, mostly in the context of discussions about the best time to introduce new laws that prohibit, regulate or encourage new activities. Many of these refer to the 'Collingridge dilemma', an articulation of the challenge in regulating in a shifting space.[31] Move too early and there is a risk that new laws will be poorly designed, failing to match the final form that technology takes. Move too late and socio-technical systems are less flexible – the changes needed to comply with new regulation make them very expensive or even impossible.[32] The regulation of the Internet is a good example to illustrate this tension.[33] For instance, laws regulating copying using digital tapes failed to deal with Internet file-sharing and thus became irrelevant, while, by failing to incorporate early cyber-security standards, it is difficult to retro-fit the Internet for security today. As Murray points out, new regulations are influenced by an existing regulatory framework rather than operating from scratch.[34] There is thus path dependence in socio-technical and regulatory states and some regulatory options prove difficult in practice once socio-technical systems have reached particular states of development. The choice of regulatory timing thus ought to be made with awareness of the diffusion pattern of the technology concerned so that there is, for example, greater urgency where technology diffuses exponentially after reaching a critical mass of users.[35]

It is not only law that is called on to adapt to socio-technical change, but also technologies that are called on to adapt to (that is, comply with) changing laws. Compliance with new regulations will often require adjustments, as where the law requires the installation of seatbelts or better fuel efficiencies in automobiles. Such laws may not only require changes in the physical artefact (the car) but also in the physical factories, human workers and research laboratories of automobile manufacturers. Just as law may be based on an old understanding of socio-technical systems, so too socio-technical systems may have been designed for obsolete legal requirements or be based on designs that did not account for new legal requirements. This is discussed further in the following section.

Institutionally, there are a variety of players whose mission includes navigating the socio-techno-legal frontier.[36] From the legal side, Law Reform Commissions (popular in the Anglo-Australian sphere) have, as part of their mission, the task of ensuring the adaptation of law to manage socio-technical change. As one former

[31] D Collingridge, *The Social Control of Technology* (St Martin's Press, 1982).

[32] *Ibid.* See also A Knie, 'Yesterday's Decisions Determine Tomorrow's Options: The Case of the Mechanical Typewriter' in M Dierkes and U Hoffmann (eds), *New Technology at the Outset: Social Forces in the Shaping of Technological Innovations* (Campus Verlag, 1992).

[33] M Mueller, *Networks and States: The Global Politics of Internet Governance* (MIT Press, 2010); J Goldsmith, 'Regulation of the Internet: Three Persistent Fallacies' (1997) *Chicago-Kent Law Review* 73, 1119.

[34] AD Murray, *The Regulation of Cyberspace: Control in the Online Environment* (Routledge-Cavendish, 2007) 240.

[35] G Bernstein, 'The Paradoxes of Technological Diffusion: Genetic Discrimination and Internet Privacy' (2006) 39 *Connecticut Law Review* 243.

[36] Bennett Moses (n 23).

Commissioner observed, '[a]lmost every task of the Commission evidenced the impact of science and technology on the law'.[37] From the technological side, technology assessment bodies, particularly in the European Union, seek to understand and develop policy around new technologies.[38] European institutions, such as the European Commission, have also played a strong role in developing policy in response to emerging possibilities. Other bodies with different expertise also emerge or take on challenges. In the context of the case study in section IV, the National Transport Commission (NTC) has taken the role of developing policy for advances in automated driving systems and related technologies.[39]

As time moves forward, law and technology are called on to adjust to the expectations of the other. While much of the literature focuses on law needing to keep up with technology, this is but one thread of a broader need for cross-adjustment. In particular, technological design is often shaped by law and may itself become a regulatory force that furthers a particular policy. The next section explores these issues, both at a general conceptual level and through the lens of time.

III. Design-Based Regulation in the Dimensions of Law and Time

A. Regulatory Effects of Technology

Law and technology do not simply exist at particular time-states, they are forces that exert influence and shape social affairs. One way of expressing this is that law and technology are forms of regulation, which can be defined as 'the sustained and focused attempt to alter the behaviour of others according to standards or goals with the intention of producing a broadly identified outcome or outcomes, which may involve mechanisms of standard-setting, information-gathering and behaviour-modification'.[40] The regulatory nature of law is widely acknowledged. There are also a variety of ways in which technology can be regulatory through all three mechanisms articulated within Black's definition above. In the context of automated driving systems, for example, data analytics can be used to optimise safety features for standard-setting, embedded cameras could be used for information-gathering, and eye movement trackers may sound an alarm when a driver seems drowsy for behaviour-modification. Similarly, technology can force

[37] M Kirby, 'Law, Technology and the Future' (1989) 21 *Australian Journal of Forensic Sciences* 112, 112.

[38] European Parliamentary Technology Assessment, eptanetwork.org/ (last accessed 24 September 2019).

[39] NTC, *Roads/Technology/Automated Vehicles in Australia*, www.ntc.gov.au/roads/technology/automated-vehicles-in-australia/ (last accessed 24 September 2019).

[40] J Black, 'Critical Reflections on Regulation' (2002) 27 *Australian Journal of Legal Philosophy* 1.

compliance through, for example, automatic braking at stop signs or vehicle software that prevents manual entry while in motion. These examples demonstrate that while law specifies what is permitted, encouraged, controlled or prohibited, technology can de facto shape what is easy, possible, difficult or impossible.

B. Conceptual Lenses on the Regulatory Effect of Technology

Scholarship discusses regulatory effects of technology through a variety of conceptual lenses and approaches, such as 'code is law', 'design-based regulation,' 'values sensitive design', 'compliance by design' and 'compliance through design'. In addition to such general approaches that we discuss below, context specific labels and terms such as 'RegTech'[41] and 'privacy by design',[42] are also abundant. The meanings of these terms are subtly different, and each is also multivocal. Nevertheless, they all describe a common idea, namely that technological choices, like just law, have the capacity to shape societal behaviour.

i. Code is Law

Lessig was the first to bring these ideas to the forefront of legal scholarship.[43] Lessig's core idea was that there are different modalities of regulation – legal rules, social norms, market and architecture or, in the world of cyberspace, code. Hence the famous statement that 'code is law', which expressed the position that computer code controls interactions in cyberspace in a potentially more powerful manner than traditional statutes. Benkler expanded on this idea by describing the ways in which networked information infrastructure enhances possibilities for broad, diverse communication and cooperative enterprises, which law may oppose but only at great cost.[44]

The role that technology can play in a regulatory system, however, extends well beyond computer code, and may include, for example, speed humps that encourage drivers to travel at slow speeds on residential roads, which can be more effective than a sign displaying a legal speed limit.[45] Approached through such lens, societal behaviour can thus be regulated not simply through legal prohibitions but also through technological prescriptions, and technology companies

[41] DW Arner, J Barberis and R Buckey, 'FinTech, RegTech, and the Reconceptualization of Financial Regulation' (2016) *Northwestern Journal of International Law and Business* 37, 371.

[42] A Cavoukian, 'Privacy by Design: The 7 Foundational Principles'. Information and Privacy Commissioner of Ontario, Canada (2009) 5; Marc Langheinrich, 'Privacy by Design – Principles of Privacy-Aware Ubiquitous Systems' International Conference on Ubiquitous Computing (Springer, Berlin, Heidelberg, 2001).

[43] L Lessig, *Code and Other Laws of Cyberspace* (Basic Books, 1999).

[44] Y Benkler, *The Wealth of Networks: How Social Production Transforms Markets and Freedom* (Yale University Press, 2006) 9, 130, 139, 149, 212, 271–72, 385.

[45] B Latour, 'On Technical Mediation – Philosophy, Sociology, Genealogy' (1994) *Common Knowledge* 29, 38–41.

as well as states can be regulators. In the literature, the ability of law and code to regulate is often described in terms of relative power, fluctuating over time. Johnson and Post believed that the power of national laws would wane in the context of the Internet,[46] while Wright and De Philippi believe that, despite the potential for blockchains to operate as autonomous systems, there are ways in which law's power can be preserved.[47] There is a time dimension to such arguments – new techniques of legislative intervention and new technological means to work around laws or avoid detection and enforcement are in constant competition, with different victors at different moments in time. The extent to which regulation by law or regulation by code dominates may thus be time-dependent.

ii. Value Sensitive Design

The idea of 'value sensitive design' is related to the 'code is law' approach, but the focus is less on attaining a particular regulatory outcome and more on 'building in' or 'embedding' values as an aspect of technological design. Value sensitive design has been defined as a 'theoretically grounded approach to the design of technology that accounts for human values in a principled and comprehensive manner through the design process'.[48] The focus thus seems to be on process (in particular the design process) rather than on power to influence behaviour. Nevertheless, value sensitive design will have regulatory effects and will also influence society's relationship to the embedded values.

iii. Compliance by Design

Another related conceptual lens is the notion of 'compliance by design'. This involves extracting and modelling legal requirements, then integrating them within business processes.[49] There is thus a literature seeking to identify and interpret legal requirements in order to convert them into machine-readable conditions.[50] Compliance by design assumes that law already exists at a particular point in time, and the goal is to ensure that systems support, or even guarantee, compliance with that law. In the financial services industry, this is often marketed as 'RegTech', although that term also encompasses the use of technology *by regulators*

[46] DR Johnson and DG Post, 'Law and Borders – the Rise of Law in Cyberspace' (1996) 48 *Stanford Law Review* 1367.

[47] P De Filippi and A Wright, *Blockchain and the Law: The Rule of Code* (Harvard University Press, 2018).

[48] B Friedman, PH Kahn and A Borning, *Value Sensitive Design: Theory and Methods* (University of Washington, Technical Report 02-12-01).

[49] P Casanovas, J González-Conejero and L de Koker, 'Legal Compliance by Design (LCbD) and through Design (LCtD): Preliminary Survey' *Proceedings of the 1st Workshop on Technologies for Regulatory Compliance* (2018).

[50] *Ibid.*

to enhance compliance and reconceptualise the nature of financial regulation itself.[51] It also encompasses the development of software or software specifications *by regulators*, where use of the software replaces rules written in ordinary English as the compliance requirement.[52]

iv. Compliance through Design

Beyond compliance by design, 'compliance through design' encompasses social and institutional aspects of legal compliance, recognising that there is an element of choice in how institutions respond to legal rules.[53] For example, some institutions may wish to go further than formal legal requirements for risk management reasons. Others may wish to do no more than is necessary due to competing policy or business objectives. Compliance through design builds these choices into the design process, also incorporating into the compliance process manual elements where human interpretation is required.

v. Design-Based Regulation

In 'design-based regulation', the influence of design of technology is not limited to situations where design elements are imposed to achieve legal compliance or even promote shared social values. Design can also 'nudge' users to make choices in directions preferred by the architect of technology.[54] The idea of using design architecture to nudge users was made famous by Thaler and Sunstein, who described how 'choice architects' could influence people's behaviour in ways that were (in the choice architect's view) beneficial to those people.[55] In their conception, nudges were beneficial but also cheap, easy to avoid and transparent. This is not always the case where design is used to influence behaviour. An example of behaviour-altering technology that has no alignment with a legal rule or beneficent objective is the Mosquito which emits sound at a particular pitch uncomfortable for young people, thus discouraging them from loitering nearby.[56] There is potential for such design choices to be non-transparent, manipulative and/or deceptive and generally

[51] Arner, Barberis and Buckey (n 41).

[52] E Micheler and A Whaley, 'Regulatory Technology: Replacing Law with Computer Code' (LSE Law, Society and Economy Working Papers 14/2018). Available at SSRN: https://ssrn.com/abstract=3210962 or http://dx.doi.org/10.2139/ssrn.3210962.

[53] Casanovas et al (n 49).

[54] K Yeung, '"Hypernudge": Big Data as a Mode of Regulation by Design' (2017) 20(1) *Information, Communication and Society* 118, 119. See generally R Thaler and C Sunstein, *Nudge* (Penguin Books, 2008).

[55] RH Thaler and CR Sunstein, *Nudge: Improving Decisions about Health, Wealth and Happiness*, updated edition (Penguin Books, 2009).

[56] B Bowling, A Marks and C Murphy, 'Crime Control Technologies: Towards an Analytical Framework and Research Agenda' in R Brownsword and K Yeung (eds), *Regulating Technologies: Legal Futures, Regulatory Frames and Technological Fixes* (Hart Publishing, 2008) 51.

advantage powerful actors (for example, corporations developing platforms) over users.[57] However, design-based regulation may also be deployed in line with goals based on shared social values, law or policy.

C. Overlap of Approaches: Example of Privacy by Design

There is an important overlap among the various approaches and associated literature. This overlap is well illustrated by the notion of 'privacy by design', where privacy becomes a 'default setting' internalised in the design of systems for preventative, end-to-end lifecycle protection.[58] On the one hand, this emerges as a particular case of value sensitive design, where the value being promoted is privacy. However, because data privacy laws exist in most jurisdictions, privacy by design straddles the 'value sensitive design', 'compliance by design' and 'regulation by design' scholarship.[59] Embedding the value of privacy in technologies bolsters the regulatory objectives of data privacy law and enhances compliance with that law. Our focus in this chapter is limited to what we refer to as 'data privacy' or the 'informational' subset of privacy, also known as 'data protection', rather than on the general question of privacy. A general right to privacy has a history spanning centuries,[60] and has been recognised as a constitutional right in many countries and in most prominent international human rights instruments.[61] *Data* privacy as a policy issue, however, has emerged only in the wake of the information technologies revolution in the late 1960s and early 1970s.[62] The term 'data protection' is mainly used in Europe, where it became a prominent epithet since the enactment of initial data protection laws in the 1970s. In other parts of the world, such

[57] Yeung (n 54) 123–31.

[58] A Cavoukian, *Privacy by Design: The Seven Foundational Principles*, revised edn (Information and Privacy Commissioner of Ontario, 2011), iapp.org/resources/article/privacy-by-design-the-7-foundational-principles/ (last accessed 24 September 2019).

[59] D Klitou, *Privacy-Invading Technologies and Privacy by Design: Safeguarding Privacy, Liberty and Security in the 21st Century*, Information Technology and Law Series 25 (TMC Asser Press, 2014) 262.

[60] See M Zalnieriute, 'An International Constitutional Moment for Data Privacy in the Times of Mass-Surveillance' (2015) *International Journal of Law and Information Technology* 23.2, 99–133, at 6–7, explaining that already in 14th century England, the Justices of The Peace Act Of 1361 (34 Edw 3 C.1) established some sense of individual privacy. Also by the end of the 19th century in the USA, Warren and Brandeis in their famous 'The Law of Privacy' (1890–91) 4 *Harvard Law Review* 195 formulated the right to privacy in terms of 'The Right to be Let Alone'. A constitutional right to privacy was formulated by the US Supreme Court in *Griswold v Connecticut*, 381 US 479 (1965). For more on the history of privacy, see JQ Whitman, 'The Two Western Cultures of Privacy: Dignity Versus Liberty' (2004) *Yale Law Journal* 1151–221.

[61] See universal instruments: Art 12 of the The Universal Declaration of Human Rights 1948; Art 17 of the International Covenant on Civil and Political Rights 1966; regional instruments: Art 8 of the European Convention on Human Rights 1950; Art 11 of the American Convention on Human Rights 1969. Note, however, that the African Charter on Human and People's Rights 1981 is exceptional in omitting an express right to privacy or private life.

[62] See M Zalnieriute (n 60) 99–133, at 6–7.

as the USA, Canada and Asia-Pacific, policy debate is about 'privacy' protection. Both definitions are argued to be problematic and, in part, misleading.[63] Thus, the newer term 'data privacy' has been gaining traction among scholars since the mid-2000s, and 'its use can be seen as an attempt to signal more accurately than the other two terms the focus, thrust, and rationale of the relevant norms'.[64] We therefore, use the term data privacy in this chapter.

With an optimistic lens, privacy by design suggests that 'Technology …, in theory, can apply privacy laws and principles constantly, consistently, objectively, mechanically and without errors, improving the rate, quality and effectiveness of privacy compliance.'[65] But the idea of privacy by design, if not its enactment as part of the EU data privacy law, also extends beyond legal requirements.[66]

Various types of privacy-enhancing technologies can be deployed as part of privacy by design. The first type is essentially target hardening, employing cyber security techniques to make it more difficult for criminal actors to obtain data to which they are not entitled. This is sometimes also called 'security by design'.[67] Another example is where systems require users to comply with privacy laws when collecting, accessing or using data in those systems.[68] This enhances compliance with the existing law (to the extent it can be built into design), possibly to the level where non-compliance becomes practically impossible or extremely difficult. This is the aspect of privacy by design that most closely resembles the idea of compliance by design. A third type consists of techniques that encourage data minimisation, for example forms that do not collect data that are not essential to a particular transaction, systems that enable anonymous transactions, and software that allows consumers to manage multiple pseudo-identities.

D. Limitations of Design-Based Regulation

Design-based regulation suffers from similar limitations as legal rules. For example, design-based regulation may be over- or under- inclusive with respect to a regulatory goal and this may be exacerbated by new technologies. Yeung gives the example of enforcing a quiet carriage law through a device that blocks mobile

[63] LA Bygrave, *Data Protection Law: Approaching its Rationale, Logic and Limits* (Kluwer Law International, 2002) 21–23.

[64] LA Bygrave, *Data Privacy Law: An International Perspective* (Oxford University Press, 2014) 26. See also G Greenleaf, *Asian Data Privacy Laws: Trade & Human Rights Perspectives* (Oxford University Press, 2014); C Kuner, *Transborder Data Flow Regulation and Data Privacy Law* (Oxford University Press, 2013).

[65] Klitou (n 59) 295.

[66] A Cavoukian, 'A Regulator's Perspective on Privacy by Design' (Future of Privacy Forum, 2012), www.futureofprivacy.org/privacy-papers-2012 (last accessed 24 September 2019).

[67] LA Bygrave, 'Hardwiring Privacy' in Brownsword, Scotford and Yeung (eds) (n 16) 754, 756–57.

[68] Klitou (n 59) 267.

phone reception.[69] Noise is not generated exclusively or necessarily by mobile phones, and changes in telecommunications networks may alter the operation of the device. Deployment of the device is thus both over- and under-inclusive with respect to its regulatory goal of maintaining silence in a carriage, and risks becoming obsolete as telecommunications technologies evolve.

Over-inclusiveness of design-based regulation can be harmful to the protection and exercise of fundamental rights. One example is in the context of Internet content filtering technologies, designed to work as effectively as legal regulation of off-line content, enhances censorship of legitimate speech.[70] Over-inclusiveness can also compound with legal rules as where technologies over-protect copyright works (for example, without allowing for copyright exceptions) and those technological measures are then protected by law.[71]

As well as being potentially over-inclusive, design-based solutions will often fail to resolve all compliance challenges. For example, in the context of privacy by design, any translation between words (found in statutes or case law) and design is likely to be imperfect. Privacy law is typically principles-based and human interpretation is thus an important aspect of its application, whereas built-in processes must of necessity be unambiguous.[72] The compliance through design scholarship recognises this dilemma but only by simultaneously recognising the need for human interpreters in the process so that technical design is not the sole mechanism for achieving compliance.

There are also additional concerns about design-based regulation, such as Brownsword's concerns about the loss of moral community that comes from eliminating the practical option of doing the wrong thing.[73] In the context of privacy by design, Brownsword raises additional concerns including the impossibility of moral balancing (where agents may perceive a greater moral good from sacrificing privacy to another value or interest) and removal of the option to reach different individual or collective conclusions about what privacy means and the relative value of privacy and other rights and interests.[74] In addition to general limitations of design-based regulation, there are more specific questions about how such systems fare along the dimension of time, and we discuss them in the following section.

[69] K Yeung, 'Towards and Understanding of Regulation by Design' in Brownsword and Yeung (eds) (n 56) 79, 92.

[70] L Bennett Moses, 'Creating Parallels in the Regulation of Content: Moving from Offline to Online' (2010) 33 *University Of New South Wales Law Journal* 581.

[71] L Lessig, *The Future of Ideas: The Fate of the Commons in a Connected World* (Random House, 2001) 200–202.

[72] Bygrave (n 67) 767–68; Klitou (n 59) 284.

[73] R Brownsword, *Rights, Regulation and the Technological Revolution* (Oxford University Press, 2008) 258–82.

[74] R Brownsword, 'Law, Liberty and Technology' in Brownsword, Scotford and Yeung (eds) (n 16) 41, 62–64.

E. Design-Based Regulation Along the Dimension of Time

Race-based metaphors of law and technology uniformly place technology in the lead. Nevertheless, there are circumstances where technology can fall behind law and policy. Just like the law, design has a temporal dimension. Science and Technology Studies – widely known as STS – are helpful in understanding the inflexibility of technological design. In particular, once design choices are made, they may be difficult to reverse due to 'technological momentum' where costs associated with changes in technological choices, particularly when embedded into broader socio-technical systems, create a path dependence.[75] In other words, a technological choice made today may be hard to reverse even if future socio-technical or legal change renders that choice redundant or counter-productive. For example, the Anglosphere still uses QWERTY keyboards despite the fact that keys no longer jam – a choice that reduced typist muscle strain in the past may now be increasing it.[76] Similarly, it is unlikely that every data collection, data storage and data processing system operating in Europe complied with the new rules in the General Data Protection Regulation ('GDPR') on the day that law was enacted; it inevitably took some companies a long time to redesign or replace older systems.[77] As time passes, technologies themselves can become obsolete and design choices made to achieve particular functional, legal or policy objectives may not meet new objectives.

There are technological features that help predict whether a design choice is more or less adaptable to future change. For example, features may be static (like a speed hump that prevents all cars travelling quickly) or dynamic (like software that automatically controls the speed of a car with reference to its location, weather conditions, time of day, vehicle type, traffic).[78] Software is also more 'plastic' or adaptable than either law or physical architecture; it can be updated relatively easily.[79] Software may be updatable centrally (through remote updates) whereas hardware updates will generally require action on the part of a user (although interchangeable physical parts may be shipped from a central location). Diffusion patterns may help predict the persistence of technological design in the face of regulatory and legal pressures.[80] The extent to which technology is inflexible or adaptable will thus depend on context.

[75] TP Hughes, 'Technological Momentum' in Merritt Roe Smith and Leo Marx (eds), *Does Technology Drive History? The Dilemma of Technological Determinism*' (MIT Press, 1994) 112.

[76] PA David, 'Clio and the Economics of QWERTY' (1985) 75 *The American Economic Review* 332.

[77] Regulation (EU) 2016/679 of the European Parliament and of the Council of 27 April 2016 on the protection of natural persons with regard to the processing of personal data and on the free movement of such data, and repealing Directive 95/46/EC (General Data Protection Regulation or GDPR) [2016] OJ L119/1.

[78] Yeung (n 54) 121–22.

[79] J Grimmelmann, 'Note: Regulation by Software' (2005) 114 *Yale Law Journal* 1719, 1723.

[80] Bernstein (n 35).

Where it exists, inflexibility in technological design means that technological choices, like legal ones, may be a poor fit for the future. Technologies built to be compliant need to adapt both to changes in the rules with which they seek to comply, changes in user behaviours that affects the situations in which non-compliance is likely, and changes in other technologies (as where 'data security by design' can only be achieved by continual updates as new methods of compromising security are discovered). Technologies built to have a regulatory effect need to build in feedback loops in order to monitor the regulatory effect over time and adjust course as required, not only in response to the matters raised above, but also to changes in regulatory objectives as broader social and political objectives evolve.[81] Such feedback loops take account of the fact that design choices made at one point in a socio-technical system may cease to be optimal as the broader system evolves.

As this section has sought to demonstrate, the evolution of socio-techno-legal landscapes over time involves complex interactions. At a given point in time, particular elements may seem to fall behind others (in that they fit best with activities, technologies or legal frameworks of the past rather than the present). There are, however, legal and technical choices that can be made in the present to maximise adaptive behaviour over time. For example, centrally updateable software can ensure that necessary adaptations are implemented swiftly, just as administrative regulations are generally easier to update than legislation. The ways in which law and technological design combine to achieve particular regulatory effects and interact over time can be explored in practical settings. In the following section, we draw on a more concrete example, self-driving cars, to illustrate some of the issues discussed above.

IV. Self-Driving Cars and the Complexity of Regulation over Time

Self-driving cars or automated vehicles ('AVs') are vehicles that include an automated driving system capable of performing the entire dynamic driving task (steering, acceleration, braking and monitoring the driving environment) on a sustained basis.[82] Research suggests that such vehicles will have some initial benefits, such as reducing the risk of accidents and improving fuel efficiency, although broader and longer-term benefits and risks remain unclear.[83] What is apparent is that policy-makers and regulators have recognised that existing laws are not able

[81] Murray (n 34).

[82] NTC, *Regulating Government Access to C-ITS and Automated Vehicle Data: Policy Paper* (August 2019) 10.

[83] D Milakis, B van Arem and B van Wee, 'Policy and Society Related Implications of Automated Driving: A Review of the Literature and Directions for Future Research' (2017) 21 *Journal of Intelligent Transportation Systems* 324.

to cope with these new technologies and that new regulation may be necessary to ensure safety and assuage potential user and societal concerns. The full range of legal issues is too extensive to be covered in this chapter; rather we draw out some examples of the phenomena discussed above.

In particular, our focus is on the complexity of building a regulatory system, which consists of both legal and design features, and which could withstand the waves of time. We are interested in how decisions made in one area (such as safety) can impact on a separate policy debate (such as data privacy), resulting in constant evolution of law over time to adapt to the changing needs of the present. These shifts in law also affect technological design, with design features introduced to comply with law at one point in time needing to keep up with subsequent legal change. Technology falls behind the law just as law is constantly struggling to keep up with new technology. The entire socio-techno-legal system is in a constant state of flux and adaptation to the evolving needs of the present.

A. Law 'Out of Time' and Behind Technology

The development of AVs raises a variety of challenges to the current legal regime, which are illustrative of how law and technology could be 'out of time' and behind one another. For example, the NTC notes that '[c]urrent Australian transport legislation assumes there is a human driver'.[84] Existing law, including driving laws, safety regulations and rules regulating government access to data, lag behind the development of new AV technologies. For example, driver licensing is the primary mechanism of ensuring that vehicles are driven safely under existing laws, but there is no equivalent regulatory mechanism in the absence of a human driver.[85] The NTC is thus engaged in an extended process of identifying legal challenges and making concrete recommendations for reform.[86]

On the safety side, various options have been proposed, most of which (including the ultimate recommendation) will require an entity bringing automated driving technology into the Australian market to self-certify against 11 principles-based safety criteria in a Statement of Compliance.[87] In addition, such entities will be required to outline the data that will be recorded and how it will be provided to relevant parties, including to ensure real time monitoring of driving performance and incidence, including event data records collected in the lead-up to a crash.[88]

[84] NTC, *Safety Assurance for Automated Driving Systems: Decision Regulation Impact Statement* (November, 2018), www.ntc.gov.au/Media/Reports/(A7B4A10F-22A5-2832-1A11-6E23E1BB7762).pdf (last accessed 24 September 2019) 8. See, in particular, Australian Road Rule 297, which provides that 'a driver must not drive a vehicle unless the driver has proper control'.

[85] *Ibid*, 22.

[86] The full scale of this can be seen in a figure representing 'Creating an end-to-end post-trial regulatory system' in *ibid*, 10.

[87] *Ibid*, 49.

[88] *Ibid*, 69.

Whether for compliance with safety criteria and related reporting requirements or for operational reasons (providing equivalent functionality to human eyes and ears), automated driving systems are likely to collect, process, store and share across the transportation system significantly more personal data than is the case for traditional cars. Such data is likely to include video and audio data from sensors within and outside a vehicle, data concerning vehicle location and movement, data concerning operation of vehicle features (such as airbags), and even biometric data of the individual in control of a vehicle.[89] Such enhanced data collection, processing, storage and transmission raise significant data privacy challenges.

Australia's data privacy regime is a combination of federal law (the Privacy Act 1988, applying to the Federal Government and corporations over a particular size) and state and territory laws.[90] Like other areas of law, there is a perception that data privacy law is 'increasingly becoming outdated' in the context of modern digital technologies.[91] In particular, existing laws may provide insufficient protections against corporate and government surveillance, defined as the 'focused, systematic, and routine attention to personal details for the purposes of influence, management, protection or direction'.[92] There is also concern that simplistic distinctions between personal and non-personal information fail to take account of developments in re-identification techniques that demonstrate that these categories lie along a scale rather than in two buckets. In Australia, the Privacy Act 1988 (Cth) defines personal information as 'information or an opinion about an identified individual, or an individual who is reasonably identifiable' and information is either regulated (because it is personal information) or completely unregulated (because it is not personal information). The challenge is that the identifiability of an individual in a data set is relative, so that some entities may be able to identify an individual in a data set whereas others will not.[93] Australian law seems to assume an understanding of personal information (as an objective category) that pre-dates developments in re-identification techniques that allow some entities to re-identify information that others might have thought was not personal.

The NTC has expressed a tentative view that reform is needed to limit government collection, use and disclosure of AV information due to the insufficiency of existing legal framework.[94] In particular, it has proposed new principles for

[89] NTC (n 82).

[90] The primary privacy legislation is: Privacy Act 1988 (Cth); Information Privacy Act 2014 (ACT); Privacy and Personal Information Protection Act 1998 (NSW); Information Act 2002 (NT); Information Privacy Act 2009 (Qld); Personal Information Protection Act 2004 (Tas); Privacy Data and Protection Act 2014 (Vic). In SA, see Information Privacy Principles Instruction (IPPI) published as Premier and Cabinet Circular No 12, June 2016. Some jurisdictions have separate privacy regimes for particular categories of data (for example, health data).

[91] Klitou (n 59) 3.

[92] D Lyon, *Surveillance Studies: An Overview* (Polity, 2007) 14.

[93] Australian Computer Society, *Data Sharing Frameworks: Technical White Paper* (September, 2017).

[94] NTC (n 82) Ch7.

regulating government access to AV information data and cooperative intelligent transport system data specifically, where those terms would need to be clearly defined.[95] These principles contemplate that 'Australian governments will need to develop laws and aligned standards that support lawful collection, use and disclosure' of such data as well as additional privacy protection 'to appropriate limit the collection, use and disclosure' of this data.[96] Because of the NTC's limited jurisdiction, the question of broader privacy reform remains open.

B. Technology 'Out of Time' and Behind Law

In the context of AVs, it is not only the law that needs to catch up with technology, but also technology that is 'out of time' and behind law in certain contexts or jurisdictions. For example, in the European Union, data privacy laws are often assumed to be 'ahead' of Australian laws in the sense that they provide stronger protections against corporate and government surveillance. The GDPR creates an obligation on data controllers to comply with principles of 'data minimisation' and 'data avoidance' which require that personal data be adequate, relevant and *limited to what is necessary* in relation to the purposes for which those data are processed.[97] This means that those manufacturing or importing to Europe automated driving systems must limit personal data collection only to what is needed for their legitimate business purposes and delete it when it is no longer 'necessary'. Complying with this provision may require manufacturers to change vehicle design to keep up with law, in particular by reducing the volume of data collected and processed.

Moreover, the GDPR articulates the so-called 'right to be forgotten', formally known as the 'right to erasure', which entitles individuals to require data controllers to delete their data when it is no longer necessary for the purposes for which it was collected, or when the individual withdraws their consent and there is no other legal ground for processing personal data.[98] Right to erasure, which does not have an equivalent under the Australian Privacy Act 1988, could cover image data internal to the vehicle, audio data, and data covering biometric and biological factors in particular, which could be argued to be not necessary after a short period of time after their collection. If not already built into system design, systems would need to be modified to ensure prompt deletion.

These examples suggest that technology *can* fall behind evolving legal requirements, and thus be 'out of time'. Whether or not this is the case for any particular automated driving system would need to be tested against technical specifications. Of course, it is usually possible to change the design of vehicles and infrastructure

[95] *Ibid*, Ch 8.
[96] *Ibid*.
[97] GDPR (n 77) Rec 39, Art 5(1)(c).
[98] GDPR (n 77) Art 17.

to adapt to new legal rules such as the GDPR (subject to economic and practical considerations). The point is merely that this requires something *of technology* – whether that takes the form of software updates, new hardware components or refitted factories. Before those changes are made, technology lags behind the law, just as the law lags behind technology in the period relevant statutes are amended. This is so even if design decisions were made at an earlier point in the process in order to comply with earlier versions of the law.

C. Surfing the Waves of Change: Legal and Technical Solutions in the Dimension of Time

As demonstrated in subsections A and B above, socio-techno-legal systems are constantly changing, with changes in one sphere, such as law, requiring changes in others, such as technology. In this section, we consider the relative flexibility and longevity of proposed solutions to existing challenges.

i. Context-Specific Legal Rules in the Dimension of Time

One potential response to the data privacy challenges raised by AV technologies, which appears to be favoured by the NTC in Australia, is to introduce a new, context specific data principles. However, this proposal encounters two main challenges.

The first challenge is one of fragmentation. If data protection law is adapted for all of the different domains for which this is proposed or developed, including regulation of AVs and consumer rights in the banking sector,[99] there will be an increasing complexity of the legal landscape. The consequences of this complexity are likely to increase over time, particularly as technologies often tend towards convergence. Just as it has become less helpful to think about content regulation separately for radio, television and the Internet, it will become less useful to regulate data differently in different domains. This is particularly so as data will flow across distinct domains, as where vehicle data accessed by banks who lent money to purchase vehicles and is thus incorporated into a customer's profile.

The second challenge relates to the discussion in section II of this chapter, namely that laws designed to operate optimally in a narrow domain defined by reference to particular technological practices are more likely to face obsolescence as those practices evolve over time. This has been experienced with many formerly new technologies, where regulations on digital tapes, nanotechnology,

[99] See the consumer data right proposed by the Australian Competition and Consumer Commission: *Competition and Consumer (Consumer Data) Rules 2019* (Exposure Draft, 29 March 2019), www.accc.gov.au/system/files/Exposure%20draft%20CDR%20rules%2029%20March%202019.pdf (last accessed 24 September 2019).

assisted reproductive technologies have become obsolete or artificially narrow as those technologies have themselves evolved.[100] A similar fate could await laws designed specifically for data privacy issues relating to specific types of driving systems although the NTC seeks to ameliorate this risk by adopting inclusive and technology-neutral definitions of the relevant data categories (albeit still limited to the vehicle technology context).[101]

ii. Broad Legal Rules in the Dimension of Time

Rather than reforming Australian privacy law in the context of a particular technology, such as AVs, one could update generally formulated data privacy laws with awareness of, but not an exclusive focus on, technological challenges that need to be addressed. For example, the GDPR regulates processing of personal data in a technologically neutral way and will govern the processing of data generated by automated driving systems. In the context of the challenge for which context-specific laws are proposed in Australia, both the GDPR and the Law Enforcement Directive (protecting personal data processed for law enforcement purposes) are relevant.[102] Essentially, the GDPR explicitly allows for exceptions to its provisions when they are necessary for 'the prevention, investigation, detection or prosecution of criminal offences …',[103] so that any further processing of such data by the law enforcement agencies for criminal purposes is addressed separately in the Law Enforcement Directive.[104]

The GDPR and Law Enforcement Directive only apply to 'personal data', which is defined as any information relating to an identified or identifiable natural

[100] See generally L Bennett Moses, 'Sui Generis Rules' in Marchant, Allenby and Herkert (eds) (n 3).

[101] NTC (n 82) 60.

[102] Directive (EU) 2016/680 of the European Parliament and of the Council of 27 April 2016 on the protection of natural persons with regard to the processing of personal data by competent authorities for the purposes of the prevention, investigation, detection or prosecution of criminal offences or the execution of criminal penalties, and on the free movement of such data, and repealing Council Framework Decision 2008/977/JHA, [2016] OJ L119/89. Member States are also offered a derogation allowing legacy processing systems to remain in place until 6th May 2023, with the option of a three-year extension to 2026, where there is a 'disproportionate effort' required to bring them into compliance. Additionally, a new ePrivacy Regulation proposed in 2017 (not yet adopted) would cover machine to machine communications, and thus may become relevant in the future, see European Commission, 'Proposal for a Regulation of The European Parliament and of The Council concerning the respect for private life and the protection of personal data in electronic communications and repealing Directive 2002/58/EC (known as the 'e-Privacy Directive')' COM/2017/010 final – 2017/03 (COD).

[103] GDPR (n 77) Art 23.

[104] The GDPR and Law Enforcement Directive do not apply to data processing by the institutions, bodies, offices and agencies of the European Union (as an organisation), whose data processing is governed by a separate EU legal framework under Regulation 45/2001 of the European Parliament and of the Council of 18 December 2000 on the protection of individuals with regard to the processing of personal data by the Community institutions and bodies and on the free movement of such data [2000] OJ L8/1. The processing of personal data by special EU law enforcement agencies such as Europol and Eurojust are governed by tailor-made data protection regimes and processing outside the scope of EU law such as in the context of national security are not included here.

person ('data subject').[105] Personal details such as the name of a vehicle occupant, address and contact details, as well as occupants' biometric and biological data will be uncontroversially classified as personal data. It also includes geo-location data,[106] and it is irrelevant whether such data is technical, generated by the technology or provided by the data subject. Contrary to European automotive industry association arguments,[107] data generated by vehicle sensors such as information about speed, acceleration and use of brakes, which could be either supporting the operation of automated functions, communication with other vehicles and infrastructure or collected by Event Data Recorders (EDR), *could* constitute personal data in the opinion of the EU Court of Justice,[108] the EU Commission,[109] EU data protection authorities and the Article 29 Working Party.[110]

Unlike in Australia, the definition of 'personal information' in the European Union is contextual and adapted to changes in re-identification techniques. In particular, classification of data as 'personal data' depends on whether the data subject is identifiable which in turn depends on the knowledge of the

[105] GDPR (n 77) Art 2.

[106] GDPR (n 77) Art 4(1) explicitly clarified the status of geo-location data by expressly stating that an individual can be identified directly or indirectly by reference to 'location data'. See Article 29 Data Protection Working Party – *Opinion 13/2011 on Geolocation services on smart mobile devices*, available at ec.europa.eu/justice/data-protection/article-29/documentation/opinionrecommendation/files/2011/wp185_en.pdf 22 (last accessed 25 September 2019), which states location data collected by smartphones is considered personal data because individuals can be directly or indirectly identified through their patterns of movement.

[107] German Association of the Automotive Industry – *Verband der Automobilindustrie (VDA)*, 'Data Protection Principles for Connected Vehicles' (3 November 2014), available at www.vda.de/en/- topics/innovation-and-technology/network/data-protection-principles-for-connected-vehicles.html (last accessed 25 September 2019); VDA, in 'Position – Access to Vehicle and Vehicle Generated Data' (19 September 2016) p 1, available at www.vda.de/en/topics/innovation-and-technology/network/access-to-the- vehicle.html (last accessed 25 September 2019). See also European Automobile Manufacturers Association – (ACEA), 'ACEA Principles of Data Protection in Relation to Connected Vehicles and Services' (September 2015) p 4, available at www.acea.be/publications/article/acea-principles-of-data-protection-in-relation-to-- connected-vehicles-and-se (last accessed 25 September 2019); ACEA, 'ACEA Strategy Paper on Connectivity' (April 2016) p 4, available at www.acea.be/uploads/publications/ACEA_Strategy_Paper_on_Connectivity.pdf (last accessed 25 September 2019); British Society of Motor Manufacturers and Traders (SMMT), 'Connected and Autonomous Vehicles – Position Paper' (February 2017) p 6 et seq, available at www.smmt.co.uk/wp-content/uploads/sites/2/SMMT-CAV-position-paper-final.pdf (last accessed 25 September 2018).

[108] C-582/14 *Patrick Breyer v Bundesrepublik Deutschland* [2016] ECLI:EU:C:2016:779.

[109] EU Commission, 'A European strategy on Cooperative Intelligent Transport Systems, a milestone towards cooperative, connected and automated mobility' COM(2016) 766 final, 30 November 2016, available at ec.europa.eu/energy/sites/ener/files/documents/1_en_act_part1_v5.pdf, 8 (last accessed 25 September 2019). See also EU C-ITS Platform Final Report (September 2017), available at https://ec.europa.eu/transport/sites/transport/files/2017-09-c-its-platform-final-report.pdf, 28 (last accessed 25 September 2019).

[110] Article 29 Working Party, Opinion 03/2017 on Processing personal data in the context of Cooperative Intelligent Transport Systems (C-ITS), ec.europa.eu/newsroom/article29/item-detail.cfm?item_id=610171, 6 (last accessed 25 September 2019). The Working Party was set up under Article 29 of Directive 95/46/EC (n 77). It is an independent European advisory body on data protection and privacy. Its tasks are described in Article 30 of Directive 95/46/EC and Article 15 of Directive 2002/58/EC (n 102).

data controller[111] and the reasonable means they are able to deploy to re-establish the identity of the data subject.[112] This recognises that classification of information as personal is contextual: as long as an entity may link a data item to a unique identifier that can be associated with a person, the data qualifies as personal data in relation to that specific manufacturer or government agency.

The EU framework under the GDPR thus articulates a technology-neutral approach for addressing data privacy challenges, including in the context of AVs. This achieves similar regulatory objectives to those identified by the NTC through a general rather than context-specific legal framework, offering greater likelihood that it will be relatively future-proof over time.[113] The GDPR also introduces a requirement of privacy by design, which we discuss below.

iii. Mandatory Compliance by Design in the Dimension of Time

The GDPR explicitly incorporates an obligation for data controllers to comply with principles of 'privacy by design' and 'privacy by default'.[114] Privacy is one of the most prominent fields in which a 'by design' solution operates alongside statutes. As it was discussed earlier in this chapter, it is seen as a way of moving the regulatory focus from data processors to those manufacturing or developing systems for data processing.[115] It aims at ensuring that privacy protections are built into the design and development of new technologies and services, as opposed to being implemented subsequently as part of a legal review process. In relation to automated driving systems, 'privacy by design' would mean, for example, an interactive dashboard, allowing the driver to customise and turn on/off the technology's ability to collect different types of personal data, thereby giving the data subject more control over their personal data. It may also include elements like automatic deletion of data no longer relevant to the operation of a vehicle or automatic anonymisation of data communicated to transportation infrastructure. Similarly, a 'privacy by default' approach would mean having sensors that collect personal data switched off by default (a so-called 'opt-in' approach). This would help to ensure that data subjects' personal data are not processed automatically

[111] Article 4(7) of the GDPR (n 77) defines 'controller' as 'the natural or legal person, public authority, agency or other body which, alone or jointly with others, determines the purposes and means of the processing of personal data; where the purposes and means of such processing are determined by Union or Member State law, the controller or the specific criteria for its nomination may be provided for by Union or Member State law'.

[112] GDPR (n 77) Art 2; *Breyer* (n 107). The Court in *Breyer* further states, that for a qualification of data as personal it is not required 'that all the information enabling the identification of the data subject must be in the hands of one person'.

[113] See chapter 16 of this book by Ranchordás and 't Schip on future-proofing.

[114] GDPR (n 77) Recital 78. The Recital requires that the producers of the products, services and applications be encouraged to take into account the two principles to make sure that the controllers and processors are able to fulfil their obligations.

[115] Klitou (n 59) 4.

without their consent. In other words, the GDPR is not technology-neutral but demands that technological practice conforms to legal standards.[116]

'Privacy by design' elements can be embedded in hardware or software and in components of each that are updated with different frequencies. This choice has an important impact on the ability of systems to remain adaptable and protective over time. For example, when there are changes to transportation infrastructure that impacts how data is collected and used, this could be communicated to users through a software update that requires users to make new decisions on the interactive dashboard. Where data collection systems are embedded in hardware, such as cameras, it may be more difficult to ensure that updates are incorporated over time.

If manufacturers want to avoid technological design choices falling behind evolving legal requirements, they need to be proactive. Just as lawmakers need to understand the risk of framing law in terms of particular technologies in order to ensure ongoing adaptability to technological change, designers need to consider the ease with which systems can be changed in order to ensure ongoing adaptability to legal change. Allowing for remote software updates is one means that manufacturers are using to achieve this.

V. Conclusion

Scholars working on the issues at the intersection of law and technology have recognised the need for an interdisciplinary socio-techno-legal approach to problems brought about by new technologies.[117] This chapter makes a similar plea, anchoring its argument on the need to consider the evolution of socio-techno-legal landscapes over time in making decisions about the appropriate form of regulation in the present. As both the theory and the example of AVs discussed in this chapter suggest, the design of regulatory frameworks (through law and technology) that takes place in the present should reflect an awareness of the need for adaptability in the future. Legal and technical solutions that solve the problems of the past or keep up with the present may themselves become obsolete in light of future legal, technological or social change. Both technical design and legislation need to adjust over time in response to the rapid evolution of the socio-techno-legal systems as a whole. In other words, while the extent and nature of issues is not symmetric, technology and law can both lag behind. Both choices in how law is formulated and how design choices are enacted will affect adaptability over time.

[116] See generally M Hildebrandt and L Tielemans, 'Data Protection by Design and Technology Neutral Law' (2013) 29(5) *Computer Law & Security Review* 509.
[117] Dizon (n 7).

15

Back to the Future: Waves of Legal Scholarship on Artificial Intelligence

CATALINA GOANTA, GIJS VAN DIJCK
AND GERASIMOS SPANAKIS

I. Time and Technology: The Rise of Artificial Intelligence as a Topic of Legal Inquiry

The fact that nowadays more and more attention is focused on digitalisation is well-known. Even national research policies include, at the moment, a preference for themes such as Digital Legal Studies.[1] Inspired by the diversification of societal phenomena resulting from technological innovation, the past decade has seen a newfound interest in the emerging field of law and technology. Legal research – both doctrinal-dogmatic and multidisciplinary – on topics such as reputation systems,[2] fake news and democracy,[3] content moderation,[4] platform governance,[5] algorithmic gatekeeping,[6] or distributed ledger technologies,[7] has been on the rise.

[1] For instance, in the Netherlands, 'Digital Legal Studies' is one of the themes universities can choose to focus on in order to benefit from an extra investment in social sciences and humanities totalling €10 million made by the Ministry of Education, Culture and Science. See Netherlands Organisation for Scientific Research, 'Samen sterker. Beeld van het SSH Domein' (*Sectorplan SSH*, November 2018) www.sectorplan-ssh.nl/artikelen/documenten.

[2] Sofia Ranchordás, 'Public Values, Private Regulators: Between Regulation and Reputation in the Sharing Economy' (2019) 13 *Law & Ethics of Human Rights* (forthcoming).

[3] Alberto Allemano, 'How to Counter Fake News? A Taxonomy of Anti-fake News Approaches' (2018) 9(1) *European Journal of Risk Regulation* 1. See also Siva Vaidhyanathan, *Antisocial Media: How Facebook Disconnects Us and Undermines Democracy* (Oxford University Press, 2018).

[4] James Grimmelmann, 'The Virtues of Moderation' (2015) 17 *Yale Journal of Law and Technology* 42.

[5] Jack M Balkin, 'Information Fiduciaries and the First Amendment' (2016) 49(4) *UCD Law Review* 1183.

[6] Mireille Hildebrandt, 'Algorithmic Regulation and the Rule of Law' (2018) 376 *Philosophical Transactions of the Royal Society A: Mathematical, Physical and Engineering Sciences* 1.

[7] Primavera de Filippi and Benjamin Loveluck, 'The Invisible Politics of Bitcoin: Governance Crisis of a Decentralised Infrastructure' (2016) 5(3) *Internet Policy Review*.

One of the topics enjoying increased popularity, partly due to its media hype,[8] is artificial intelligence. Both praised and feared, the technological progress allowing machines to mimic cognitive functions that are inherently human (eg learning) has a much richer history than popular analyses often acknowledge, and is developing at an astounding pace.[9] In the next section, we will first outline a brief history of artificial intelligence as a field of computer science research. Legal audiences – especially those interested in artificial intelligence – can benefit from such an exposé, which is rare in legal scholarship.[10] Subsequently, this information will be overlapped with the Corpus of legal literature, consisting of 3931 academic journal articles obtained from HeinOnline, which will be explored using descriptive statistics.

A. A Brief History of Artificial Intelligence

In the words of Bruce Buchanan, a computer scientist from the University of Pittsburgh, 'the history of AI is a history of fantasies, possibilities, demonstrations and promise'.[11] The fantasy part of this history has left its traces in literary fiction.[12] The scientific path – which led to the consolidation of artificial intelligence as a field of research – lies in the history of computer science.

In 1955, mathematicians John McCarthy, Marvin Minsky, Nathaniel Rochester and Claude Shannon initiated the Dartmouth summer research project on artificial intelligence for the following year, based on the assumption that '[…] every aspect of learning or any other feature of intelligence can in principle be so precisely described that a machine can be made to simulate it.'[13] While earlier notable work on the idea of machines behaving intelligently had been theoretically designed

[8] See for instance Martin Giles, 'Ian Goodfellow GANfather' (2018) (*MIT Technology Review*, 2018) www.technologyreview.com/s/610253/the-ganfather-the-man-whos-given-machines-the-gift-of-imagination/; Hugh Howey, 'How to Build a Self-Conscious Machine' (*Wired*, 4 October 2017) www.wired.com/story/how-to-build-a-self-conscious-ai-machine/; Gideon Lewis-Kraus, 'The Great A.I. Awakening' *The New York Times*, New York, 14 December 2016 /www.nytimes.com/2016/12/14/magazine/the-great-ai-awakening.html; Richard Socher, 'AI's Next Great Challenge: Understanding the Nuances of Language' (*Harvard Business Review*, 25 July 2018) https://hbr.org/2018/07/ais-next-great-challenge-understanding-the-nuances-of-language.

[9] Vishal Maini, 'Machine Learning for Humans' (*Medium*, 19 August 2017) https://bit.ly/2wrDJTO.

[10] In the very few papers where it is included, the history of artificial intelligence mostly refers to the specific subfield of legal informatics, as an expression of artificial intelligence applied to issues or questions mostly relevant for legal practice (eg e-discovery, etc). See Sergio David Becerra, 'The Rise of Artificial Intelligence in the Legal Field: Where We Are and Where We Are Going' (2018) 11 *Journal of Business, Entrepreneurship & the Law* 27; Richard E Susskind, 'Artificial Intelligence, Expert Systems and Law' (1990) 5 *Denning Law Journal* 105.

[11] Bruce G Buchanan, 'A (Very) Brief History of Artificial Intelligence' (2005) 26 *AI Magazine* 53.

[12] See for instance Frank Baum, *Tiktok of Oz* (Reilly & Britton, 1914).

[13] John McCarthy and others, 'A Proposal for the Dartmouth Summer Research Project on Artificial Intelligence' (31 August 1955) http://jmc.stanford.edu/articles/dartmouth/dartmouth.pdf.

and published,[14] the 1956 Dartmouth conference is generally considered to be the moment when artificial intelligence was born as a concept and was shaped as a mission.[15]

The following decade was so full of optimism due to breakthroughs in natural language processing and machine vision:[16] researchers started speculating that it would only be a matter of time before computers would be able to beat humans in chess, or discover and prove new mathematical theorems.[17] The optimism fuelled and was fuelled by an injection of funding into research on artificial intelligence. Between 1963 and approximately 1970, the Massachusetts Institute of Technology was heavily funded by the Advanced Research Projects Agency (later known as DARPA) with up to $3 million per year.[18] However, problems such as the lack of data and computational power, as well as design issues in the application of reasoning or logic systems, soon began to slow down the initial enthusiastic efforts of creating human-like intelligence in machines. This led to what is known as the first AI winter (see **Figure 1**), and it manifested itself through the cutting of financial support to artificial intelligence research projects and the general loss of interest in the topic.

Figure 1 A Brief History of Artificial Intelligence[19]

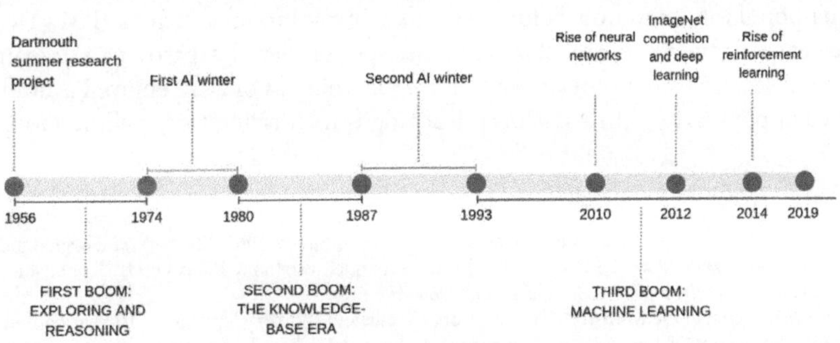

[14] Alan Turing, 'Computing Machinery and Intelligence' (1950) 59 *Mind* 433; Claude E Shannon, 'Programming a Digital Computer for Playing Chess' (1950) 41 *Philosophy Magazine* 356.

[15] Daniel Crevier, *AI: The Tumultuous Search for Artificial Intelligence* (Basic Books, 1993) 49; Pamela McCorduck, *Machines Who Think: A Personal Inquiry into the History and Prospect of Artificial Intelligence*, 25th anniversary edn (AK Peters Ltd, 2004) 114.

[16] Crevier, *ibid*, 84–102; 164–72; McCorduck, *ibid*, 300–305. See also Marvin L Minsky, *Semantic Information Processing* (MIT Press, 1968); Marvin L Minsky, 'Steps Toward Artificial Intelligence' (1961) 49 *Proceedings of the Institute of Radio Engineers* 8.

[17] Stuart J Russell and Peter Norvig, *Artificial Intelligence: A Modern Approach* (Prentice Hall, 2003); see also Arthur L Samuel, 'Some Studies in Machine Learning Using the Game of Checkers' (1959) 3 *IBM Journal of Research and Development* 210.

[18] Crevier (n 15) 64–65.

[19] Original scheme based on Daniel Crevier, *AI: The Tumultuous Search for Artificial Intelligence* (Basic Books, 1993) 49; Pamela McCorduck, *Machines Who Think: A Personal Inquiry into the History and Prospect of Artificial Intelligence*, 25th anniversary edn (AK Peters Ltd, 2004) 114; Tokio Matsuzaki, 'Ethical Issues of Artificial Intelligence in Medicine' (2018) 55(1) California Law Review 255; and Karen Hao, 'We analyzed 16,625 papers to figure out where AI is headed next' (*MIT Technology Review*, 25 January 2019) https://engt.co/30VFQfn.

In addition to the lack of funding, researchers from other disciplines (eg philosophy) started voicing concerns regarding the goals and vision of artificial intelligence projects. One was philosopher Hubert Dreyfus, who was already emphasising the threats of reliance on machines for reasoning in his 'Alchemy and AI' paper dating from 1965: 'if, using only digital programs, we try to process the ill-structured data in which real-life problems are posed, we face an infinite regress'.[20]

In spite of growing criticism, the early 80s saw renewed funding opportunities, giving researchers the possibility to approach artificial intelligence in a new manner.[21] There was a shift away from large, abstract concepts and towards knowledge-based approaches by zooming into specific domains – leading to the rise of expert systems.[22] Once again penning over-ambitious goals[23] and also finding itself under the pressure of external factors,[24] the research community could not consistently hold the attention of public and private investors since it seemingly failed to achieve commercial impact. The end of the 80s saw a new AI winter that required not only redesigning methods of research in artificial intelligence, but even the terminology used to describe the discipline as a whole (eg knowledge engineering, informatics, etc).

As of the early 1990s, the third boom of artificial intelligence research saw more media popularity than ever before, with new breakthroughs such as IBM's Deep Blue defeating the world's most famous chess player, Gary Kasparov, in a six-game match in 1997.[25] The promising rise of deep learning as of 2012 enjoyed a similar degree of popularity.[26] In 2019, deep learning is the driving force behind Google

[20] Hubert L Dreyfus, 'Alchemy and AI' (*RAND Corporation Memo*, 1965) 70 www.rand.org/content/dam/rand/pubs/papers/2006/P3244.pdf. See also the expanded version of the paper, later published as a book, Hubert Dreyfus, *What Computers Can't Do* (MIT Press, 1972).

[21] In 1981, the Japanese Ministry of International Trade and Industry launched a 10-year computer technology project focusing on natural language and knowledge-based information processing, totalling $850 million. See Mikel Olazaran, 'History of the Neural Network Controversy' (1993) 37 *Advances in Computers* 335, 387.

[22] McCorduck (n 15) 421. See also P Jackson, *Introduction to Expert Systems* (Addison-Wesley Longman Publishing, 1986) 2–10.

[23] Crevier (n 15) 212; McCorduck (n 15) 441.

[24] Progress in artificial intelligence research has always been dependent on the hardware used for development. The late 80s saw a revolution in personal computing that affected the development of computers specialised to support artificial intelligence. See McCorduck (n 15) 435.

[25] Steven Levy, 'What Deep Blue Tells Us about AI in 2017' (*Wired*, 23 May 2017) www.wired.com/2017/05/what-deep-blue-tells-us-about-ai-in-2017/.

[26] A turning point in the use of deep learning is marked by the 2012 edition of the ImageNet Competition, where three researchers from the University of Toronto submitted a deep convolutional neural network architecture called AlexNet, which was the first to get a less than 25% error rate for the visual recognition challenge. Using the same method in the subsequent five years, competing teams managed to further bring down the error rate to less than 5%. See Dave Gershgorn, 'The Data that Transformed AI Research – and Possibly the World' *Quartz* (*Quartz*, 26 July 2017) https://qz.com/1034972/the-data-that-changed-the-direction-of-ai-research-and-possibly-the-world/. See also the original paper, Alex Krizhevsky and Ilya Sutskever and Geoffry Hinton, 'ImageNet Classification with Deep Convolutional Neural Networks' (2012) *NIPS*.

searches, Facebook news feeds and Netflix recommender systems, although it is predicted that it may soon be replaced by other approaches.[27] In a study of all 16625 papers available in the 'artificial intelligence' section of the database arXiv,[28] it is clear that interest in artificial intelligence is growing (see **Figure 2** below), and three main trends describe research on this topic since the early 2000s:[29] a paradigm shift from knowledge-based systems towards machine learning in the early 2000s; the rise of neural networks in the early 2010s; and the popularity of reinforcement learning since 2014.[30]

Figure 2 arXiv Papers in the Section 'Artificial Intelligence'[31]

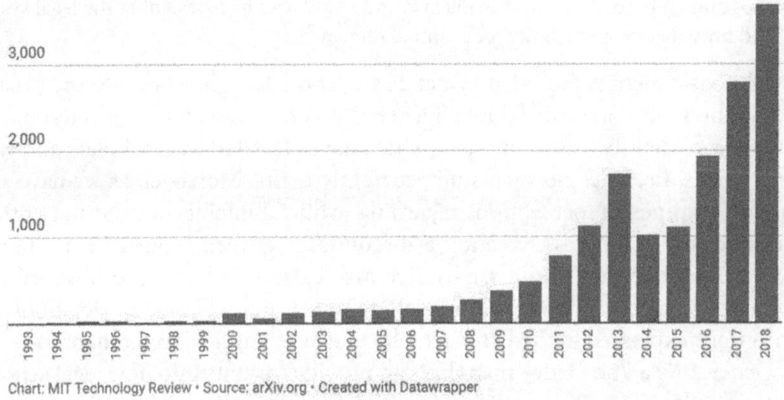

It is difficult to assess whether the current boom focused on machine learning will at some point result in disillusionment and consequently yet another AI winter. Well-known deep learning experts such as Andrew Ng claim artificial intelligence will benefit from stabilised financing in the years to come, given it has such a 'strong underlying driver of real value that it won't crash like it did in previous years'.[32]

[27] Karen Hao, 'We analyzed 16,625 papers to figure out where AI is headed next' (*MIT Technology Review*, 25 January 2019) https://engt.co/30VFQfn. See also Nils Nilsson, *The Quest for Artificial Intelligence: A History of Ideas and Achievements* (Cambridge University Press, 2009).

[28] arXiv is an open access database owned and operated by Cornell University, currently storing 1545774 electronic papers on physics, mathematics, computer science, quantitative biology, quantitative finance, statistics, electrical engineering and systems science, and economics, available at https://arxiv.org.

[29] Hao (n 27). The research takes into account papers indexed on arXiv on 18 November 2018.

[30] Hao (n 27).

[31] Figure from Karen Hao, 'We analyzed 16,625 papers to figure out where AI is headed next' (*MIT Technology Review*, 25 January 2019) https://engt.co/30VFQfn.

[32] Will Knight, 'AI Winter Isn't Coming' (*MIT Technology Review*, 7 December 2016) https://www.technologyreview.com/s/603062/ai-winter-isnt-coming/.

B. Legal Literature on Artificial Intelligence: The HeinOnline Corpus

Legal interest in artificial intelligence can be traced as far back as the late 1950s and early 1960s, when journals such as *MULL: Modern Uses of Logic in Law*[33] were publishing reports on conferences akin to the Law and Electronics Conference. Participants to the Conference were discussing how

> data processing could effect accurate and rapid handling of vast amounts of data, improve legal research through more ready accessibility to repositories of the law, reveal gaps and contradictions in existing law, and could lead to redesigning the legal system and provide the opportunity for eventual reform.[34]

Beyond basic metrics provided by databases, and a few glimpses into the existing body of literature on artificial intelligence,[35] it is not easy to answer questions on how we can identify trends in topics which have attracted legal scholars and what these trends are, since no such study currently exists. Moreover, as we have seen above,[36] commercial perceptions regarding artificial intelligence did not entirely match scientific progress, creating a disconnect between industry expectations and research output. A similar narrative may exist with respect to how artificial intelligence has been perceived in legal literature, and whether expectations and projections expressed in the latter actually match scientific developments.

Generating a knowledge map that can provide insights into how legal scholars relate to artificial intelligence brings multiple benefits:

- It helps legal scholars get a faster grasp on legal literature available on artificial intelligence;
- It helps legal scholars understand on a macro scale what research questions have been asked in the past, therefore offering a way to reduce the risk of repetitive research;
- It helps computer scientists, and researchers from other disciplines, to get a simplified, bird's eye view over legal concerns regarding artificial intelligence;
- It allows for comparisons between literature trends based in different disciplines;
- It has the potential of generating additional research questions regarding the path of legal scholarship on artificial intelligence.

[33] The *MULL: Modern Uses of Logic in Law* journal, dating from 1959, became known as the *Jurimetrics Journal* in 1966, and was later renamed *Jurimetrics* in 1978.

[34] Martha Gibbs and Eldridge Adams, 'A Report on the Second National Law and Electronics Conference' (1962) 3 *Modern Uses of Logic in Law* 215, 222.

[35] See n 10. See also Anne von der Lieth Gardner, *An Artificial Intelligence Approach to Legal Reasoning* (MIT Press 1987) 67–76.

[36] See n 21.

To create such a knowledge map, we have designed a two-part project, consisting in an automated analysis of a body of legal literature that ought to reveal how the latter developed over time. Part one focuses on the descriptive and normative underpinnings of this analysis and finds itself at the core of this chapter. Part two focuses on the actual data analysis, and will result in a separate computer science publication. In what follows, we will discuss the Corpus, the collection methodology, and the characteristics of the Corpus as revealed using descriptive statistics and data visualisation.

The Corpus includes a total of 3931 journal articles obtained from HeinOnline as a result of a Data and Text Mining Agreement. Absent a centralised, comprehensive, open access repository for international legal scholarship,[37] we focused on one of the commercially-available databases: we chose HeinOnline as one of the leading international databases on legal materials, containing over 170 million pages of literature and indexing over 2700 law journals.[38] In the section 'Law Journal Library', section type 'Articles',[39] the Corpus covers literature available in the database between 1960 and 2018. Unlike arXiv, HeinOnline does not have a section on 'artificial intelligence'. The total number of retrieved articles reflects the results of a Boolean search using the keywords 'artificial intelligence'. Resulting articles therefore discuss a wide array of aspects relating to artificial intelligence, and do not, as such, focus on specific technical or legal issues. We do, however, account for the number of pages with keywords in a given article, and are thus able to distinguish between different parts of the Corpus (see **Figure 3** below):

Figure 3 A Classification of the HeinOnline Corpus

	Number of pages with keywords	Number of journal articles
Journal articles focused on artificial intelligence	40-2	1463
Journal articles mentioning artificial intelligence	1	2468
TOTAL		**3931**

[37] The problems arising out of publication models for legal research are well-known. See Gary Wolf, 'Who Owns the Law' (*Wired*, 1 May 1994) www.wired.com/1994/05/the-law/. See also Graham Lawton, 'Paywall: The Business of Scholarship Review – Analysis of a Scandal' (*New Scientist*, 5 October 2018) www.newscientist.com/article/2181744-paywall-the-business-of-scholarship-review-analysis-of-a-scandal/.

[38] See https://home.heinonline.org.

[39] All additional sections have been omitted, namely: 'Comments', 'Notes', 'Reviews', 'Legislation', 'Cases', 'External', 'Miscellaneous', 'Index' and 'Editorial'.

As it can be seen in **Figure 3**, the articles indexed by HeinOnline under the keywords 'artificial intelligence' may refer to the keywords between one and 40 times. While this metric may tell us for sure that an article referring to the keywords 40 times will certainly be a legal article on artificial intelligence,[40] the opposite may not always hold true. For instance, a paper dealing with copyright and Google Books which only refers to the keywords once is not necessarily *not* a paper on artificial intelligence, seeing as the latter is the foundation of the technology discussed within such a paper.[41] Given the inherent method of legal analysis, namely the doctrinal method, it is difficult to draw strong lines and categorise to what degree a paper discusses artificial intelligence. However, for the purposes of this study, we assume that even one reference to the keywords is sufficient to include an article in the Corpus. This is because sometimes articles refer to the shortened version of the term (eg AI), or in other cases they refer to different sub-categories of artificial intelligence, such as machine learning, and one reference to the keyword is sufficient to define its use in the body of the article. Moreover, excluding single references to the keyword would not show the full extent of the interest shown by legal scholars in the topic. **Figure 3** shows this category to include most of the articles in the Corpus. Some of these articles are very much related to artificial intelligence and on rare occasions are even written by computer scientists,[42] or belong to the wider umbrella of law and technology scholarship, whether in its earlier iterations[43] or in a more contemporary format.[44] One possible classification on the basis of this metric is to divide the papers between papers that focus on artificial intelligence (between two and 40 references to the keywords) and papers that refer to artificial intelligence more casually (one reference).

The lack of indexed books is an additional drawback, since a lot of legal literature is published in edited books. More data sources can be added to the Corpus, whether from HeinOnline or externally, leaving room for the expansion of this initial framework and dataset in order to facilitate further comparisons (eg what are the differences between the body of literature on artificial intelligence published in journals, as opposed to that published in books, etc). The distribution of journal articles in the Corpus can be observed in **Figure 4** below.

[40] Henry H Jr Perritt, 'Artificial Intelligence Techniques for Evaluating Employee Terminations on a Personal Computer' (1987) 13 *Rutgers Computer & Technology Law Journal* 341.

[41] See for instance Maurizio Borghi and Stavroula Karapapa, 'Non-Display Uses of Copyright Works: Google Books and Beyond' (2011) 1 *Queen Mary Journal of Intellectual Property* 21.

[42] Daniela Rus, 'The Robots Are Coming' (2015) 94 *Foreign Affairs* 2.

[43] Anthony D'Amato, 'Can/Should Computers Replace Judges' (1977) 11 *Georgia Law Review* 1277; Alan Wolfe, 'Algorithmic Justice' (1990) 11 *Cardozo Law Review* 1409.

[44] Jonathan Zittrain, 'Engineering an Election' (2013–2014) 127 *Harvard Law Review* F 335.

Figure 4 The Time Distribution of the Corpus per Journal per Year[45]

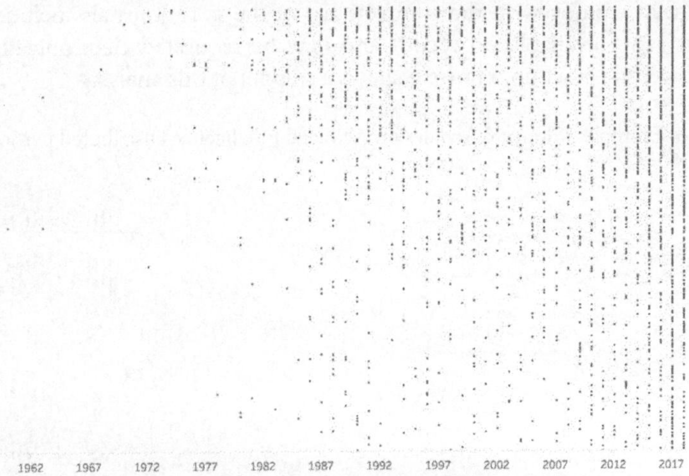

Figure 4 shows an increase in the popularity of artificial intelligence in the Corpus, especially between 2016 and 2018. A clearer overview can be seen in **Figure 5**, depicting the number of journal articles with keywords per year, and in **Figure 6**, showing the number of pages with keywords per year:

Figure 5 Number of Journal Articles with Keywords per Year and Number of Pages with Keywords per Year[46]

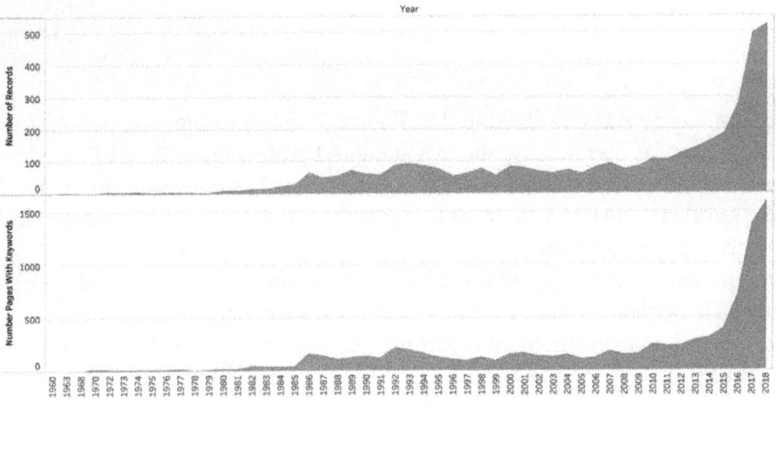

[45] Data visualisation done using Tableau.
[46] Data visualisation done using Tableau.

A detailed overview of the number of publications on artificial intelligence per journal – along with the complete list of the 1113 journals included in the analysis – are available on record and may be requested electronically until a project website is set up. **Figure 6** shows a snippet of this analysis.

Figure 6 Example of Journal Articles on Artificial Intelligence Distributed per Journal per Year[47]

The Corpus, especially as illustrated in **Figure 5**, shows similar patterns of growth in the volume of literature as the arXiv dataset from **Figure 2**. While computer science literature on artificial intelligence is obviously more voluminous than its legal equivalent,[48] between 2016 and 2018 we can identify a clear boom in scholarship pertaining to both disciplines.

In the next section, we will further the overlap between law and computer science. Through the framework of the history of artificial intelligence described above, we will map examples of topics tackled by legal scholars in various parts of the Corpus.

[47] Data visualisation done using Tableau.

[48] In 2018 alone arXiv indexed 3976 papers on artificial intelligence, compared to the 524 papers indexed by HeinOnline for the same year.

II. Topics of Legal Research on Artificial Intelligence

The mapping of topics in this Section is a qualitative exercise that proposes using the three AI waves identified in section I (see **Figure 1**) to highlight some parts of the Corpus, with a view to discussing selected topics found at its core.

A. The First AI Boom: Exploring and Reasoning (1956–1974)

Early interest in artificial intelligence and law was a shy undertaking. The Corpus features 20 articles published between 1956 and 1974, most of them published in the *Jurimetrics Journal*; topics included digital legal research,[49] the use of logic in the systematisation of law,[50] or predictive legal analytics.[51] During this period, one of the earliest comprehensive reflections on the impact of artificial intelligence on legal reasoning was authored by Bruce Buchanan and Thomas Headrick.[52] In this paper, interdisciplinarity was already regarded as a desirable but difficult to attain goal for two disciplines projecting misconceptions about one another.[53] In trying to achieve this goal, the two authors focus on the characteristics of the legal thought process, identifying four legal problem-solving processes: (i) finding conceptual linkages in pursuing goals; (ii) recognising facts; (iii) resolving rule conflicts; and (iv) finding analogies.[54]

Given its scarcity, this early work can be characterised even without further insights from data analysis. Its contributors show a high level of information literacy, due to interdisciplinary collaborations, like the one between Buchanan and Headrick, and also in the light of the other fields (eg logic or linguistics) necessary to conceptualise and systematise legal needs. This likely determined a concentration of thematic expertise, signalling the formation of a highly specialised research environment. Lastly, there seems to be a focus on processes relevant to the legal discipline (eg research and argumentation), and not so much on the content of theoretical legal concepts. From this perspective, this wave of law and artificial intelligence – while raising some normative questions on the role of technology in

[49] See for instance Philip Slayton, 'Electronic Legal Retrieval – A Report Prepared for the Department of Communications of the Government of Canada' (1974) 15 *Jurimetrics Journal* 108; Kenneth Katz, 'Computerized Research: An Editorial' (1973) 14 *Jurimetrics Journal* 25.

[50] Peter B Maggs and Cary G deBessonet, 'Automated Logical Analysis of Systems of Legal Rules' (1972) 12 *Jurimetrics Journal* 158.

[51] Gerald Fleischmann and Philip Scaletta, 'Use of Computer Technology in Legal Analysis and Prediction' (1974) 11 *American Business Law Journal* 251.

[52] Bruce G Buchanan and Thomas E Headrick, 'Some Speculation about Artificial Intelligence and Legal Reasoning' (1970) 23 *Stanford Law Review* 40.

[53] *Ibid.*

[54] *Ibid*, 53.

performing human functions associated with the legal profession –[55] focused on specific applications and not so much on the normative questions arising out of a potential paradigm shift.[56]

B. The Second AI Boom: Exploring and Reasoning (1980–1987)

This era already sees an increase in the interest attached to law and artificial intelligence. In this period, the Corpus lists 208 articles published in a much broader array of publications, from law and technology journals such as the *Rutgers Computer and Technology Law Journal*, the *Journal of Law and Information Science*, the *Yearbook of Law, Computers and Technology*, to regular law journals such as the *International Labour Review*, the *Southern California Law Review* or the *Modern Law Review*.

Interestingly, during this period, parallels are drawn between novel issues posed by technology and other similar situations arising beyond the use of technology.[57] Among the topics discussed by authors writing between 1980 and 1987 we can find: management concerns with microcomputing in police activity;[58] reflections on the law applicable to robots as machines performing human tasks;[59] legal information processing;[60] or text retrieval systems.[61]

In addition to furthering the work on legal reasoning and information retrieval by discussing knowledge-based expert systems,[62] we see some emerging work

[55] Buchanan and Headrick (n 52) at 61 speak about the vision of 'removing the "dog work" that underlies quality legal service, the computer system might make good service available to a larger segment of society'.

[56] For instance, at the time, Maggs and deBessonet (n 45) at 1 claimed the ultimate goal of research in law and artificial intelligence is '… to enable machines to perform [the tasks of receiving, retaining information, and rendering cognitive responses to natural language questions] at levels of sophistication sufficient to warrant the use of formal expressions instead of natural language expressions in recording information, thereby making possible certain automated inferential and communicative process involving legal textual materials'.

[57] See for instance the comparison between a machine endowed with artificial intelligence and an organisation, Michael McDonald, 'The Personless Paradigm' (1987) 37 *University of Toronto Law Journal* 212.

[58] Armand P Hernandez, 'Police Microcomputing: Strategies and Concerns' (1986) 5 *American Journal of Police* 1.

[59] Michael Gemignani, 'Laying Down the Law to Robots' (1984) *21 San Diego Law Review* 1045.

[60] Cary G Debessonet and George R Cross, 'An Artificial Intelligence Application in the Law: CCLIPS, a Computer Program That Processes Legal Information' (1986) 1 *High Technology Law Journal* 329. See also Richard Craswell, 'Interpreting Deceptive Advertising' (1985) 65 *Boston University Law Review* 657.

[61] Jon Bing, 'Third Generation Text Retrieval Systems' (1981) 1 *Journal of Law Information and Science* 183.

[62] Richard E Susskind, 'Some Preliminary Considerations Concerning Expert Systems in Law' (1987) 14 *Northern Kentucky Law Review* 211; Richard E Susskind, 'Expert Systems in Law: A Jurisprudential Approach to Artificial Intelligence and Legal Reasoning' (1986) 49 *Modern Law Review* 168.

on normative legal questions that do not focus on how artificial intelligence can be applied to law, but rather on whether artificial intelligence should be used at all in various societal spheres and, if so, what the legal repercussions ought to be.[63] At the same time, artificial intelligence is also mentioned in scholarship on jurisprudence.[64]

C. The Third AI Boom: Machine Learning (1993–Present)

By far the largest part of the Corpus, this period brings together a total 3353 articles on law and artificial intelligence. In this period, the number of journals on law and technology is on the rise (eg *Law, Innovation and Technology*; *Richmond Journal of Law and Technology*; *Journal of Business and Technology Law*; etc), and so is the interest of legal scholars who are not necessarily specialised in legal reasoning or information retrieval.

In the 90s, automated logic applied to legal information was still *en vogue*,[65] and had spread well beyond the United States,[66] yet it was soon replaced with applications of artificial intelligence – and digitalisation, more broadly speaking – pertaining to legal sub-fields such as intellectual property,[67] data protection,[68] fundamental rights[69] or Internet governance.[70] Philosophical considerations became more prevalent.[71] The tone of a lot of the writings in this period became more sceptical, perhaps due to the initial scepticism clouding the commercial success and proliferation of personal computing and communication infrastructures, such as the Internet. Writings were focused on the limitations of existing technologies, and the need to maintain human control in certain aspects of the

[63] Gemignani (n 59).

[64] See for instance the criticism of the goal of the artificial intelligence community to 'articulate a complete formalization of the mental processes involved in thought and action', George Wright, 'On a General Theory of Interpretation: The Betti-Gadamer Dispute in Legal Hermeneutics' (1987) 32 *American Journal of Jurisprudence* 191, 204.

[65] Antonio A Martino, 'Artificial Intelligence and Law' (1994) 2 *International Journal of Law and Information Technology* 154; Kevin W Saunders, 'What Logic Can and Cannot Tell Us about Law' (1998) 73 *Notre Dame Law Review* 667; Edward S Adams and Daniel A Farber, 'Beyond the Formalism Debate: Expert Reasoning, Fuzzy Logic, and Complex Statutes' (1999) 52 *Vanderbilt Law Review* 1241.

[66] Henry Prakken, 'Logica en Recht: Wat Heeft de Kunstmatige Intelligentie ons Geleerd' (1993) 22 *R & R* 198; Giovanni Sartor, 'Defeasibility in Legal Reasoning' (1993) 24 *Rechtstheorie* 281.

[67] Samuel Oddi, 'An Uneasier Case for Copyright than for Patent Protection of Computer Programs' (1993) 72 *Nebraska Law Review* 351.

[68] Eric Thiry, 'Personal Medical and Social Data: Their Processing and Legal Protection' (1993) 12 *Medicine & Law* 643.

[69] Jacques Fremont, 'Computerized Administrative Decision Making and Fundamental Rights' (1994) 32 *Osgoode Hall Law Journal* 817.

[70] Klaus W Grewlich, 'Access to Global Networks – European Telecommunications Law and Policy' (1998) 41 *German Yearbook of International Law* 9.

[71] Brett Lilly, 'Resistance is Futile: Nietzsche, Foucault, and the Aporia of Modern Legal Thought' (1995) 19 *Legal Studies Forum* 139. See also Louis E Raho, James A Belohlav and David R Drehmer, 'Expert Systems in Organizations: Legal and Ethical Considerations' (1994) 3 *Law Computers & Artificial Intelligence* 47.

application of justice (eg judging).[72] While these trends seem to be maintained throughout the 2000s, with the explosion of literature as of 2016, writings start to become more granular and topics become more diversified, with some skyrocketing in popularity.[73]

Precisely because of this popularity, methods of analysis often employed in legal research (eg qualitative content analysis) fall short of being able to generate an accurate overview of exactly what topics attract the interests of which legal scholars, and when. This is the main justification behind the need for part two of this study; in this section, we have merely given some illustrations of the type of literature included in the Corpus.

D. The Need for Further Research

Part two of the research project focuses on applying quantitative data analysis methods to the Corpus, particularly topic modelling. In the field of empirical legal research, text analysis has so far been done with the help of human coding, a time-consuming and sometimes inaccurate method, especially when it needs to be applied to a high volume and high variety of unstructured text. Topic models are a category of statistical algorithms that address such shortcomings, as they are used to 'summarize, explore and index large collections of text documents'.[74] Given the assumption that observable data in the body of text may be clustered according to topics using probability theory, they are called 'generative models'.[75] Topic modelling has been widely used in computer science,[76] but is increasingly used in the emerging field of computational social sciences, with the goal of augmenting social science analysis.[77]

Together with natural language processing (NLP), topic models, as statistical algorithms, represent the two main approaches to computer-based text analysis. In the case of NLP, grammar plays a very important role, whereas statistical models are based on the so-called 'bag-of-words' (BoW) assumption: text documents are

[72] See for instance Jeanne Lee, 'The Era of the Computer Judge' [1995] *UCL Jurisprudence Review* 249.

[73] See for instance Jessica S Brodsky, 'Autonomous Vehicle Regulation: How an Uncertain Legal Landscape May Hit the Brakes on Self-Driving Cars' (2016) 31 *Berkeley Technology Law Journal* 851; Madeleine de Cock Buning, 'Autonomous Intelligent Systems as Creative Agents under the EU Framework for Intellectual Property' (2016) 7 *European Journal of Risk Regulation* 310; Andrew M Brown, 'Blame It on the Machines: How Autonomous Vehicles Will Impact Allocation of Liability Insurance and the Resulting Impact on the Legal Community' (2016–2017) 95 *North Carolina Law Review Addendum* 29.

[74] Ryan Wesslen, 'Computer-Assisted Text Analysis for Social Science: Topic Models and Beyond' (University of North Carolina, 2018) https://arxiv.org/pdf/1803.11045.pdf, 1.

[75] *Ibid.*

[76] See also David M Blei, 'Probabilistic Topic Models' (2012) 55 *Communications of the ACM* 7784.

[77] Justin Grimmer and Brandon M Stewart, 'Text as Data: The Promise and Pitfalls of Automatic Content Analysis Methods for Political Texts' (2013) 21 *Political Analysis* 267.

introduced in a document-term matrix (DTM), which will count the occurrence of specific words in specific documents. While the performance of topic models in problems such as question-and-answer is not great, for collections of documents that are large enough, 'the BoW assumption provides the theoretical foundation for a richer set of statistical methods (mixture models) by the assumption of exchangeability'.[78] One of the most often used topic models is the latent Dirichlet allocation topic model.[79] A more concrete explanation of how this topic model is applicable to legal literature is included in part two of the project, which will be published separately as a computer science publication.

This section provided selected illustrations of the legal scholarship included in the Corpus, and highlighted some trends which remain to be proved by the analysis of the dataset using the latent Dirichlet allocation topic model. A lot of normative questions arise, however, from this brief descriptive assessment of the Corpus. In the following section, we will raise questions regarding the meaning and future of legal research on artificial intelligence.

III. Time and Research: Questions for the Legal Community

In 1970, Buchanan and Headrick, a computer scientist and a legal scholar from Stanford University, were writing: '[…] research proceeding in computer science could enhance our understanding of the processes by which lawyers work and think. So far lawyers have not attempted to explore its relevance. They should.'[80] And so they did. Yet what has the legal community achieved so far in doing so, and where is this research headed? We will endeavour to reflect upon this question by going through a handful of observations based on the analysis above.

First, it seems there is a trend for legal research on artificial intelligence to generally become less specialised. As we could see from the early body of legal work on artificial intelligence, only a handful of researchers were interested in the topic, driven by the same appreciation of formal reasoning and the goal of applying artificial intelligence to very practical questions of legal research. Even without the results of topic modelling, the decentralisation of legal research interests in artificial intelligence is notable when becoming more familiar – albeit qualitatively – with the body of legal literature on law and technology from the past decades. The crowd of researchers covering artificial intelligence is the largest ever, yet what is equally visible is that technology literacy has dropped rather than improved,

[78] Wesslen (n 74) 2.
[79] David M Blei, Andrew Y Ng and Michael I Jordan, 'Latent Dirichlet allocation' (2003) 3 *Journal of Machine Learning Research* 993.
[80] Buchanan and Headrick (n 52) 62.

in spite of the increase in the volume of literature and the granularity of technical and legal issues. A sign of this shift is the coverage of law and technology questions by not only specialised journals such as *Jurimetrics*, dedicated to a niche audience of legal informatics aficionados, but also by general legal journals serving as outlets for legal communities with much broader interests. This may have hindered, to some extent, the process of authors specialising in topics that involves understanding highly complex technologies, especially in light of the fast pace at which these technologies are developing. Of course, content experts – especially found in the emerging field of law and technology – who have been writing about artificial intelligence for more than the past couple of years, are the exception. However, the noticeable explosion of literature as of 2016 could very well have resulted in the dilution of the level of expertise necessary when dealing with the intricacies of artificial intelligence.

Second, the legal community must seriously reflect upon the cyclicity of questions in research on artificial intelligence and law. As we have seen above,[81] the idea of robots fulfilling human tasks considerably pre-dates any hype surrounding Sophia, the first robot to be offered citizenship of a country.[82] While the field of robotics may have seen considerable technological progress since the mid 80s, the matrix of legal questions – as well as answers – which may be given in response to the challenge of integrating robots in society is rather finite. Should a robot, or any type of autonomous system, be held liable for losses caused by it? The answer to this question is either a negative or a positive one. Surely there can be many different variations built on each of these answers, however the factors that ought to be taken into account when dealing with this matter (eg personhood, insurance, etc) do not raise an unlimited number of permutations. From this perspective, legal scholars must very carefully consider the added value of contributing to topics where thorough research unveils that most – if not all – of these permutations have already been addressed. This does not mean that writing one article on such a complex question as liability for autonomous agents will suffice, but we question whether too many articles on this topic actually advance palpable solutions or make considerable contributions to existing inquiries. From the perspective of time, we ought to consider what it means for the same legal questions to be asked for decades over and over again: does this mean the questions are so complex that a solution has simply not been found yet? It could be the case, for instance, when dealing with the early efforts of using computer-based logic to represent legal argumentation. The approach and the technology available to deploy it were simply not ripe enough in the 1960s for legal scholars to be able to apply automated content review software to contracts, but it is being applied with some

[81] Gemignani (n 59).

[82] Emily Reynolds, 'The agony of Sophia, the World's First Robot Citizen Condemned to a Lifeless Career in Marketing' (*Wired*, 1 June 2018) www.wired.co.uk/article/sophia-robot-citizen-womens-rights-detriot-become-human-hanson-robotics.

commercial success today.[83] In this case, subsequent iterations of the question may contribute to solving the problem. However, there is also the option of spending decades asking questions that have the same (limited amount of) answers: should we allow for automated decision-making in a courtroom? This topic may be trending right now, but authors like D'Amato and Lee have been asking this question since 1977 and1994 respectively,[84] and a lot of the points they make in their papers are currently circulated in literature on algorithmic discrimination. If the latter does not build on earlier work but rather reiterates it, we are left wondering why that would be the case. While there may be more explanations, at least two points are noteworthy: there can be a disconnect between earlier and current research (eg no citations between different time clusters), or perhaps the incentives behind asking these questions might be different. The latter point will be further explored under the fourth observation below.

Third, legal research ought to include more meta-studies and literature reviews. The legal issues arising out of artificial intelligence are becoming more and more technically and societally sophisticated (eg recommender systems; targeted advertising; etc), thus driving the quest to understand and manage human behaviour from a legal point of view. From this perspective, legal researchers ought to be able to embrace law and technology – and artificial intelligence in particular – without necessarily having had to do a PhD on the topic. However, this brings us back to the first observation made in this section: the extent to which contributions on this topic are specialised given that, as opposed to being experts on narrow yet interdisciplinary topics such as information retrieval or expert systems, authors increasingly originate from other fields of law (eg criminal law, contract law, administrative law) and look into various applications of artificial intelligence impacting the field in question. Specialisation, as a benefit arising out of expertise, is accrued when a researcher reaches an in-depth understanding of a given topic. For this reason, expertise takes time. However, the design of the legal research landscape itself hinders an efficient process of gaining sufficient information in order to develop meaningful expertise: legal literature is generally decentralised, paywalled and often not linked; the legal community has not managed to harmonise its research indexing (eg Web of Science); databases with legal information – wherever available – are rarely user friendly or grant access to meta data; many researchers do not use Boolean searches; legal journals generally do not publish literature reviews; and the list may continue. On the one hand, all these factors increase the time and complexity attached to legal research activities. On the other hand, this is a perfect environment to design and test new optimisation strategies. One example to this end is the generation of knowledge maps, as undertaken by the research project underlined in this chapter. Knowledge maps help researchers gain a complete picture of a specific field, topic or question; depending on the

[83] See for instance www.lawgeex.com/aboutus/.
[84] D'Amato (n 41); Lee (n 72).

number of variables used to build it, researchers may apply those to understand a defined field at a faster pace, or uncover new patterns or ideas in a given dataset. Undoubtedly, this means researchers need to start attaching more value to building, maintaining, linking and reusing datasets. Similarly, this also means the legal field needs to embrace a paradigm shift with respect to how information is perceived and managed, as researchers can no longer keep track of knowledge by reading every single article available on a given research question – this will simply become physically impossible in the next decade, due to the exponential propagation of academic literature. In this context, legal information in general has the characteristics of big data and may benefit from more automated, systematic data analysis and visualisation.[85]

Lastly, the research community could also benefit from reconsidering publication incentives. Doing legal research, just as any other kind of research, is ideally about furthering the boundaries of knowledge, and contributing to either the theory or practice represented by this knowledge. What hypes in legal research reveal – at least during the most considerable growth between 2016 and 2018 – is that sometimes, published legal research does not fulfil this goal. Instead, scholars may sometimes choose to focus on topics that are popular because they may increase their reputation or their funding chances. To the extent that each of these considerations has a direct bearing on the quality of legal research, current publication incentives may be detrimental to the type of consolidated and coordinated efforts that are necessary for a field to experience knowledge breakthroughs.

IV. Conclusion

In this chapter, we have embedded the theme of the book – time and law – in the context of mapping several waves of legal literature on artificial intelligence. To do so, we centred the chapter around a dataset obtained from HeinOnline. This dataset reflects all legal journal articles indexed by the database between 1960 and 2018 referencing 'artificial intelligence'. The dataset (called throughout the chapter 'Corpus') involves a twofold analysis: the current chapter in a legal publication, where we contextualise it, describe it and use examples to depict its content; and a subsequent computer science publication where the dataset is analysed using a statistical model called the latent Dirichlet allocation topic model, one of the two main approaches to computer-based text analysis, together with NLP.

We started this chapter by outlining a brief history of artificial intelligence, and then proceeded to describe the dataset using descriptive statistics and data visualisation made with the help of the Tableau software. Some parts of this statistical analysis are available on record and may be requested via email until a project

[85] Catalina Goanta, 'Big Law, Big Data' (2017) October (Special Issue) *Law and Method* www.bjutijdschriften.nl/tijdschrift/lawandmethod/2017/10/lawandmethod-D-17-00007.

website is set. We then overlapped the insights from the historical overview on artificial intelligence with literature available in the Corpus, and discussed examples of topics included therein. This analysis enabled a general reflection on the meaning and future of legal research on artificial intelligence.

Overall, in the chapter we proposed the generation of knowledge maps, such as the one to be made using topic modelling, because of a number of benefits. These include: helping legal scholars to get a faster grasp on legal literature available on artificial intelligence; helping legal scholars to understand on a macro scale what research questions have been asked in the past, therefore potentially reducing the risk of repetitive research; helping computer scientists, and researchers from any other discipline, to get a simplified, bird's eye view over legal concerns regarding artificial intelligence; allowing for comparisons between literature trends based on different disciplines; and generating additional research questions regarding the path of legal scholarship on artificial intelligence. Additional research may complement the current project by adding new datasets, in order to better explore the potential of knowledge maps in legal research on artificial intelligence.

16

Future-Proofing Legislation for the Digital Age

SOFIA RANCHORDÁS AND MATTIS VAN 't SCHIP

I. Introduction

When the Swedish teenager Greta Thunberg initiated her school strike in front of the Swedish Parliament in 2018, her demands for laws and policies accounting for the needs of future generations were not immediately taken seriously. Two years later, Greta Thunberg has amassed hundreds of thousands of followers on social media and has been invited to speak at the European Parliament, several national and international institutions, and the world's most prestigious events on climate change. Moreover, thousands of students throughout the world have joined her strike with the 'Fridays for Change' demonstrations. These young activists ask their governments to stop global warming, climate change and devote more attention to the interests of future generations.[1] They demand forward-looking laws and policies that will reverse climate change and promote sustainability.[2] Their claims require the design of sustainable and adaptable solutions with the ability to resist future environmental challenges and uncertainties. This approach to sustainability and the idea that it is possible to legislate for the future is one of the several dimensions of the so-called 'future-proof approach'.[3]

'Future-proofing' is an interdisciplinary perspective that offers a systematic framework to deal with future requirements and uncertainties while accommodating innovation.[4] A future-proof solution is resilient to novel events in years to

[1] Empirical research has demonstrated that different generations differ in their environmental awareness, see Eliana A Severo, Julio C Ferro de Guimaraes, Eric CH Dorion, 'Cleaner Production, Social Responsibility and Eco-Innovation: Generations' Perception for a Sustainable Future' (2018) 186 *Journal of Cleaner Production* 91.

[2] Certain fields of law and instruments have always required a forward-looking approach, this is the case of competition law where regulators predict the future impact of a merger on the market, see Deborah L Feinster, 'The Forward-looking Nature of Merger Analysis' (2014) *Business Law Today* 1.

[3] Michael J Ryan and Mahmoud Efatmaneshnik, 'Future Proofing Process' 27th Annual INCOSE International Symposium (2017), DOI: 10.1002/j.2334-5837.2017.00403.x.

[4] Obaid Ur Rehman and Michael J Ryan, 'On the Dynamics of Design of Future-Proof Systems' 25th Annual INCOSE International Symposium (2015), DOI: 10.1002/j.2334-5837.2015.00050.x.

come and it is durable.[5] It aims to ensure that a system or product will not become obsolete or useless because of changes that could have been anticipated ahead of time or because they were not designed with the required flexibility to adapt to new circumstances.[6] This approach has been discussed in numerous fields such as design, architecture, engineering and medicine.[7] Future-proofing products and processes have been associated in different sciences with the study of resilience to climate change, uncertainties and the need to develop resistant responses to novel challenges.[8] For example, future-proofing a building means that constructions will be tested against potential future risks of overheating due to climate change.[9] In urban planning, a future-proof city is one that is resilient and thus adaptable to complex urban transformations.[10] Future-proofing is thus an approach with two dimensions: anticipation and adaptability. It seeks to provide forward-looking solutions that defer the obsolescence process, and by doing so, this approach aims to extend the life of products, processes and policies.[11]

In the last few years, lawmakers have also become intrigued by the question whether laws and policies could be future proof.[12] While the vast majority of laws and regulations tend to last decades, there is a longstanding academic debate on the obsolescence of laws, the ineffectiveness of regulations, the need to improve the quality of legislation and regulation and ensure that new rules do not contradict existing rules.[13] A future-proof approach adds a novel and comprehensive layer

[5] Joe Kerr, *Future-Proof Design: Must all Good Things Come to an End?* (University of Calgary, 2011, PhD thesis), available at www.joekerr.ca/site/files/joe-kerr---fpdesign-2011.pdf. In the field of energy, see Lewis Evans, 'The Electricity Spot Market: Is it Future-Proof?' (2017) 30 *The Electricity Journal* 25.

[6] Kristine R Gøeg and others, 'A Future-Proof Architecture for Telemedicine using Loose-Coupled Modules and HL7 FHIR' (2018) 160 *Computer Methods and Programs in Biomedicine* 95.

[7] See, eg, Bethany Jill Williams, David Bottoms, Darren Treanor, 'Future-Proofing Pathology: The Case for Clinical Adoption of Digital Pathology' [2017] *Journal of Clinical Pathology* 10; Tariq Masood, Duncan McFarlane, Ajith Parlikad, John Dora, Andrew Ellis & Jennifer Schooling, 'Towards the Future-Proofing of UK Infrastructure' (2016) 3 *Infrastructure Asset Management* 1, https://doi.org/10.1680/jinam.15.00006.; David Coley, Tristan Kershaw, Matt Eames, 'A Comparison of Structural and Behavioural Adaptations to Future Proofing Buildings against Higher Temperatures' [2012] *Building and Environment* 55.

[8] See, eg, Christos Kossoudis and others, 'Future-Proof Crops: Challenges and Strategies for Climate Resilience Improvement' [2016] *Current Opinion in Plant Biology* 30.

[9] Mark F Jentsch and others, 'Climate Change Future Proofing of Buildings – Generation and Assessment of Building Simulation Weather Files' (2008) 40 *Energy and Buildings* 12, 2148.

[10] Ward Rauws, 'Embracing Uncertainty without Abandoning Planning' (2017) 53 disP – *The Planning Review* 32.

[11] Obaird Ur Rehman and Michael J Ryan, 'A Framework for Design for Sustainable Future-Proofing' (2018) 170 *Journal of Cleaner Production* 715.

[12] See, eg, European Economic and Social Committee, 'Future-Proof Legislation' SC/045.

[13] See, eg, Luzius Mader, 'Evaluating the Effects: A Contribution to the Quality of Legislation' (2001) 22 (2) *Statute Law Review* 119; Helen Xanthaki, 'Quality of Legislation: Focus on Smart EU and Post-Smart Transposition' (2014) 2 *Theory and Practice of Legislation* 329; Jyrki Tala, 'Better Regulation through Programs and Quality Standards – Are New Perspectives Needed?' (2010) 4 *Legisprudence* 193; Stephen Weatherill (eds), *Better Regulation* (Hart Publishing, 2007); Sacha Garben and Inge Govaere (eds), *The EU Better Regulation Agenda: A Critical Assessment* (Hart Publishing, 2018).

to this debate which encompasses several regulation and governance instruments which thus far have been discussed in a fragmented way (eg, anticipatory governance, policy flexibility, experimentation, temporary legislation). Yet, the legal literature has devoted little attention to the application of this forward-looking approach to law.[14]

This chapter delves into the origins of future-proofing and its broader application to law and policymaking. We inquire whether future-proofing can also be employed to prolong the longevity and effectiveness of legislation, take into account the interests of future generations, and promote legislative and regulatory flexibility and innovation. Thus far, similar perspectives have been posited in the literature through other instruments such as the adoption of goal regulation or principle-based approaches.[15] According to these perspectives, the adoption of indeterminate concepts could allow legislators to adapt to future circumstances.[16] Nevertheless, it is clear that not all fields of law can be regulated through principles and goals as legal uncertainty tends to generate high social costs and risks for markets, transactions and legal structures.[17]

This chapter discusses the complex balance between flexibility and legal certainty in future-proofing legislation.[18] We explain that future-proofing is closely connected to the debate on the promotion of innovation which favours the adoption of a flexible approach to rulemaking.[19] For legal scholarship and legislation, the novelty of future-proofing resides in the fact that this perspective emphasises

[14] Some of the few exceptions are the special issue of the *UC Davis Law Review* with an introduction by Anupam Chander, see Anupam Chander, 'Future-Proofing Law' (2017) 51 *UC Davis Law Review* 1; and Alexandra B Klass, 'Future-Proofing Energy Transport Law' (2017) 94 *Washington University Law Review* 4.

[15] Pauline Westerman, 'Goal Regulation, Democracy and Organised Distrust' in S Comtois and K De Graaf (eds), *On Lawmaking and Public Trust* (Eleven International Publishing, 2016); Westerman, *Outsourcing the Law* (Edward Elgar, 2018); Julia Black, 'The Rise (and Fall?) of Principles Based Regulation' in K Alexander and N Moloney (eds), *Law Reform and Financial Markets* (Edward Elgar, 2011); Ian Ayres and John Braithwaite. *Responsive Regulation: Transcending the Deregulation Debate* (Oxford University Press, 1992).

[16] On legislative indeterminacy, see Timothy AO Endicott, 'Linguistic Indeterminacy' (1996) 16 *Oxford Journal of Legal Studies* 4, 667.

[17] See, eg, Pascal Frantz and Norvald Instefjord, *Rules vs Principles Based Financial Regulation* (London School of Economics and Economic Science, 2015); Ian MacNeil, 'Uncertainty in Commercial Law'(2009) 13 *Edinburgh Law Review* 68; Rosie Cooney and Andrew TF Lang, 'Taking Uncertainty Seriously: Adaptive Governance and International Trade' (2007) 18 *European Journal of International Law* 523.

[18] See also Rob van Gestel, *Wetgeven is vooruitzien* (Boom Publishers, 2008).

[19] In particular on the innovation principle, eee Aurelien, Portuese and Julien Pillot, 'The Case for an Innovation Principle: A Comparative Law and Economics Analysis' (2018) 15 *Manchester Journal of International Economic Law* 214. See, however, Kathleen Garnett, Geert van Calster and Leonie Reins, 'Towards an Innovation Principle: An Industry Trump or Shortening the Odds on Environmental Protection' (2018) 10 *Law, Innovation, Technology* 1. On the regulation of innovation and the promotion thereof by regulatory mechanisms, see also Stijn Smismans and Elen Stokes, 'Innovation Types and Regulation: The Regulatory Framing of Nanotechnology as "Incremental" or "Radical" Innovation' (2017) 8 *European Journal of Risk Regulation* 364; Nathan Cortez, 'Regulating Disruptive Innovation' (2014) 29 *Berkeley Technology Law Journal* 175.

the need to adopt a proactive rather than a reactive approach to lawmaking.[20] We argue nonetheless that a future-proof approach should be embraced with caution and should primarily entail that legislation takes into account the needs of future generations, remains adaptable and does not entrench politically sensitive policy programmes or institutions. This study of future-proofing law and policy-making engages with literature from different fields, including existing scholarship on sustainable development, resilience, legislative studies and better regulation.[21] It also draws on the framework developed by scholars in the field of regulatory innovation.[22] In this chapter, we discuss the potential of two instruments for the implementation of the proposed future-proof approach: experimental legislation and future-proof impact assessments.[23] This chapter is organised as follows. The first section introduces the concept of future-proofing from an interdisciplinary perspective. We then analyse the challenges of future-proofing law and discuss the traditional instruments used to incorporate flexibility in legislation. This chapter advances in the final section a new approach to the implementation of future-proofing.

II. Future-Proofing: Interdisciplinary Analysis

Future-proofing has been defined as 'the process of anticipating the future and developing methods of minimising the negative effects while taking advantage of the positive effects of shocks and stresses due to future events'.[24] The idea of future-proofing legislation and regulations stems from other fields. In design,

[20] There is a growing body of scholarship in different fields advocating for proactive approaches to regulation, see, eg, Sandra Fredman, 'Breaking the Mold: Equality as a Proactive Duty' (2012) 60 *American Journal of Comparative Law* 265; George J Siedel, 'Using Proactive Law for Competitive Advantage' (2010) 47 *American Business Law Journal* 641; Joshua Schoonmaker, 'Proactive Privacy for a Driverless Age' (2016) 25 *Information & Communications Law* 96.

[21] See, eg, *The Brundtland Report: World Commission on Environment and Development, Our Common Future* (Oxford University Press, 1987); Bruno S Silvestre and Diana M Tirca, 'Innovations for Sustainable Development: Moving Toward a Sustainable Future' (2019) 208 *Journal of Cleaner Production* 325.

[22] See, eg, Julia Black, Martin Lodge and Mark Thatcher (eds), *Regulatory Innovation: A Comparative Analysis* (Edward Elgar, 2005).

[23] See Sofia Ranchordás, *Constitutional Sunsets and Experimental Legislation. A Comparative Perspective* (Edward Elgar, 2014); Sofia Ranchordás, 'The Whys and Woes of Experimental Legislation' (2013) 1 *Theory and Practice of Legislation* 3; Sofia Ranchordás, 'Innovation-friendly Regulation: The Sunset of Regulation, The Sunrise of Innovation' (2015) *Jurimetrics* 55; Fabrizio de Francesco, Claudio M Radaelli and Vera E Troeger, 'Implementing Regulatory Innovations in Europe: The Case of Impact Assessment' (2012) 19 *Journal of European Public Policy* 4; Anne Meuwese, 'Regulatory Review of European Commission Impact Assessments' (2017) 19 *European Journal of Law Reform* 16.

[24] See Principles of Future-proofing: Research on Future-proofing the Built Environment, http://principlesoffutureproofing.com. See also FutureProof Techopedia, www.techopedia.com /definition/2204/future-proof ('Future proof is a buzzword that describes a product, service or technological system that will not need to be significantly updated as technology advances.').

future-proofing challenges consumer culture and permanent demand for change, ephemeral products and fashions, and echoes the need for long-lasting design. It encourages consumers to acquire fewer items and keep them for a longer period.[25] A similar perspective is common in the luxury design sector where consumers are convinced to invest in costly products as they are marketed as 'timeless items' which are expected to survive trends and hypes. More generally speaking, a future-proof approach in design aims to ensure that a product or system will not be easily replaced by the need for future capability.[26] In practice, the ability of a product to be future-proof requires designers to make sure that products remain engaged with their audience and their changing preferences, match their form to the function of the product, strike the right balance between aesthetics and functionality, and can be continuously improved. Future-proof design does not succumb to the latest fads: it endures.

Future-proofing has also been applied in the field of infrastructure management and ecology where it has been defined in the asset management context as 'the process of anticipating the distant future and taking actions to minimize risks and maximize opportunities for value realization from assets'.[27] Future-proofing is thus an overarching approach that seeks to anticipate change, prolong the longevity of products and systems, and ensure that they are resilient. Indeed, this approach is based on a number of key concepts: foresight, adaptability, transformability and resilience.[28] Resilience is a well-researched concept which has been defined as the capacity of ecosystems with alternative attractors to persist in the original state subject to perturbations.[29] The pursuit of resilience as a policy goal, particularly in ecological governance, is often accompanied by the adoption of adaptive management approaches which include dynamic and flexible legislation.[30] While resilience has been focused on ensuring the persistence of systems and infrastructure, future-proofing is a broader approach that can also be integrated in long-term thinking for the design of policies and regulation. In this section, we discuss the two central elements of the future-proof approach: foresight (or anticipation) and adaptability. Both features are key to the translation of future-proofing to law and policymaking and they reflect the primary idea underlying this approach. This translation to law and policymaking entails foreseeing future uses, needs and

[25] Kerr (n 5).

[26] Ur Rehman and Ryan (n 11) 715.

[27] N Shetty, 'ISO 55001 Framework for Futureproofing' (2nd CSIC Workshop on Infrastructure Futureproofing, Cambridge UK, 2014).

[28] Mary Thornbush, Oleg Golubchikov and Stefan Bouzarovski, 'Sustainable Cities Targeted by Combined Mitigation – Adaptation Efforts for Future-Proofing' (2013) 9 *Sustainable Cities and Society* 1.

[29] CS Holling, 'Resilience and Stability of Ecological Systems' (1973) 4 (1) *Annual Review of Ecology and Systematics* 1; Carl Folke et al, 'Resilience Thinking: Integratng Resilience, Adaptability, and Transformation' (2010) 15(4) *Ecology and Society* 20.

[30] Jaye Ellis, 'Crisis, Resilience, and the Time of Law' (2019) 32 *Canadian Journal of Law & Jurisprudence* 305.

events in order to prepare appropriately and design different and flexible solutions for what is yet to come.[31]

Foresight or anticipation is a key element of the future-proofing approach which in the last few years has permeated regulation and governance, for example, through the theory of anticipatory governance.[32] Anticipatory governance offers a forward-looking framework through which social, ethical and legal aspects of, for example, emerging technologies, may be assessed.[33] This framework has been adopted in the context of urban planning, climate change policy, and the regulation of emerging technologies (eg, nanotechnology). In terms of future-proofing, anticipatory governance places the emphasis on less rigid, less prescriptive instruments and prefers governance instruments that foster collaboration between policymakers and stakeholders, and are forward-looking rather than reactive.[34]

Foresight entails the ability to anticipate 'alternative' futures and predict the occurrence of future problems.[35] In the context of future-proof business, anticipation is essential to the long-term survival of firms. Forecasting multiple futures in order to ensure that a company is future-proof requires taking a number of steps and overcoming multiple hurdles: taking a decision on what should be forecasted, sidestepping 'uncertainty avoidance' and the tendency to prefer incremental innovations to disruptive ones, allowing for ambiguous decisions, and promoting action-based organisational learning rather than strategic decisions.[36] In this context, it is important to highlight that forecasting traditionally refers to the prediction of risks, that is, situations that have a high likelihood of occurring and can be predicted on the grounds of existing evidence.[37] Future techniques and future thinking seeks to develop instruments to accommodate uncertainty by exploring a wide variety of conceivable futures.[38]

Foresight includes not only predicting long-term effects of certain policies but also anticipating different types of impacts of a policy on a certain problem. The challenging question in this context will nonetheless be: how far in the future

[31] Masood, et al (n 7) 28, 29.

[32] Leon S Fuerth and Evan MH Faber, 'Anticipatory Governance: Winning the Future' (2013) 47 *Futurist* 42; Leon S Fuerth, 'Foresight and Anticipatory Governance' (2009) 11 *Foresight* 14, 31.

[33] BA Wender et al, 'Anticipatory Governance and Anticipatory Lifecycle Assessment of Single Wall Carbon Nanotube Anode Lithium Ion Batteries' (2012) 9 *Nanotechnology Law and Business* 201; See also David H Guston, 'Understanding "Anticipatory Governance" (2013) 44 *Social Studies Sciences* 218.

[34] Edward W De Barbieri, 'Urban Anticipatory Governance' (2018) 46 *Florida State University Law Review* 75, 108.

[35] Fuerth and Faber (n 32) 43.

[36] George Burt and Kees van der Heijden, 'First Steps: Towards Purposeful Activities in Scenario Thinking and Future Studies' (2003) 35 *Future Studies* 1011.

[37] For a thorough study of risks in regulation, see Mathew D Adler, 'Risk, Death, and Harm: The Normative Foundations of Risk Regulation' (2003) 87 *Minnesota Law Review* 1293; Vern R Walker, 'Risk Regulation and the Faces of Uncertainty' (1998) 9 *Risk* 27.

[38] 'Foresight, Exploring the Future: Tools for Strategic Futures Thinking. Horizon Scanning Centre, 2008.

should one be able to forecast to implement a future-proof approach? In the context of the design of buildings, the literature has distinguished three types of future impacts that one should be considerate of: 'long-duration' impacts which start early in the lifecycle and continue far into the future, for example the destruction of natural ecosystems during the building's construction; 'ongoing impacts' which occur while a product is in use, for example, during its construction; and 'far-future' impacts which arise long after the construction due to decommissioning (refurbishment, deconstruction or demolition) at some future date.[39]

In law and policymaking, flexibility and foresight are also interconnected elements. Laws that successfully resist change are not only those that were drafted with change in mind but also those that contain flexible and responsive elements. In the future-proofing approach, the idea of flexibility is conveyed by the design of adaptable systems and products. For example, in the IT-sector, cloud computing is regarded as future-proof 'as it enables companies to convert IT-based capital expenditure into a variable operating cost that can be easily adjusted and adapted to changing requirements'.[40] In several sectors (such as IT) interoperability, departing from traditional proprietary models, and reliance on open source have been promoted as important tools to implement future-proof approaches. Thus far, flexibility in law and policymaking has also been interpreted in the sense that legislators should create room for leeway and future adjustments.

The legal literature has discussed several ways in which a law or regulation can be rendered more flexible by using open-ended norms, goal-based regulation, framework legislation and principle-based rules.[41] Goal-regulation in particular embraces flexibility as an important tenet of legislation and governance.[42] Setting the goal to be achieved but not the means can be a way to engage with experts or others actors within the sector rather than limiting the decision-making power to policymakers or regulators. Directives in EU Law are well-known examples of goal-based regulation as they allow Member States to choose the national implementation methods as long as they achieve the goals defined by the EU legislator.

[39] Theopilus Hacking, *Assessment for Sustainable Development: Theoretical Framework and Mining Sector Case studies from Canada, Namibia, and South Africa* (Centre for Sustainable Development, Department of Engineering, University of Cambridge, 2006, PhD thesis); Maria Christina Georgadiou, Theophilus Hacking and Peter Guthrie, 'A Conceptual Framework for Future-Proofing the Energy Performance of Buildings'(2012) 47 *Energy Policy* 145, 147.

[40] Paul Taylor, 'Future Proofing: Being Adaptable and Flexible Means Using New Methods' *Financial Times*, 4th December 2012, available at www.ft.com/content/66a027c6-34a2-11e2-8986-00144feabdc0.

[41] See, eg, Rob van Gestel, 'The "Deparliamentarisation" of Legislation: Framework Laws and the Primacy of the Legislator' (2013) 9 *Utrecht Law Review* 106; Elizabeth Garrett, 'The Purposes of Framework Legislation' (2005) 14 *Journal of Contemporary Issues* 717; Julia Black, 'Forms and Paradoxes of Principles Based Regulation', LSE Legal Studies Working Paper No 13/2008, 1–36 < https://papers.ssrn.com/sol3/papers.cfm?abstract_id=1267722.

[42] Pauline C Westerman, 'Governing by Goals: Governance as a Legal Style' (2007) 1 *Legisprudence* 51, 57.

More complex examples of goal-regulation are present in other fields (eg, construction) where the legislator introduces flexibility in statutes to allow private actors to fill in dispositions with their expertise, for example on the most common standards of construction.[43]

The idea of anticipating different futures and enacting rules that are adaptable to uncertain circumstances may resonate with scholars working in the field of responsive regulation. The concept of responsive regulation was introduced by Ayres and Braithwaite in 1992.[44] Responsive regulation is adaptive because it develops around the idea of a constant interaction between the regulator and the regulatee. This interaction relationship reflects upon the knowledge of immediate requirements within the regulated sector.[45] The focus on the sector creates an adaptable and anticipative platform because of knowledge and experience within the sector.

Until now, the discussion on legislative flexibility has not been framed as a way to render legislation future-proof and it has not relied upon interdisciplinary insights from other disciplines that are more familiar with this perspective. A step in this direction was taken by more recent literature that has acknowledged the importance of relying on time considerations and temporary legislation (sunset clauses, sunrise clauses, experimental legislation) in the context of better regulation debates and, as the following section explains, as instruments of a future-proofing approach to law and policymaking.[46]

III. Future-Proofing Law

For centuries legal scholars have assumed that law was to a certain extent 'future-proof': general principles would resist the ravages of time and change, constitutions in several countries (eg, Germany, Portugal, Brazil, Israel) included eternity clauses to guarantee the protection of fundamental values against changing parliamentary democracies.[47] The long-lasting nature of legislation and constitutions was in reality explained by different political, social and economic factors and was not always connected to their ability to adapt to future challenges. In the last half

[43] For a thorough analysis of goal regulation, see Westerman, *Outsourcing the Law* (n 15).

[44] Ayres and Braithwaite (n 15).

[45] See Robert Baldwin and Julia Black, 'Really Responsive Regulation' (2008) 71 *Modern Law Review* 59, 73–76 http://doi.wiley.com/10.1111/j.1468-2230.2008.00681.x. See also: John Braithwaite, 'The Essence of Responsive Regulation' (2011) 44 *UBC Law Review* 475, 476.

[46] Rob van Gestel and Gijs van Dijck, 'Better Regulation through Experimental Legislation' (2011) 17 *European Public Law* 539.

[47] Ulrich K Preuss, 'The Implications of "Eternity Clauses": The German Experience' (2001) 44 *Israel Law Review* 429; Andrew Friedman, 'Dead Hand Constitutionalism: The Danger of Eternity Clauses in New Democracies' (2011) 4 *Mexican Law Review* 77, 80; Mila Versteeg and Emily Zackin, 'American Constitutional Exceptionalism Revisited' (2014) 81 *University Of Chicago Law Review* 1641, 1669.

century, the ability of legislation and regulation not only to last, for example, due to entrenchment clauses or legislative inertia but truly to respond to change has been more seriously researched in the literature and policymaking.[48]

In 2016, the European Economic and Social Committee (EESC) expressed its interest in exploring the development of future-proof legislation. This initiative obtained the support of a number of Member States, including Slovakia and the Netherlands.[49] In its report, the EESC presented future-proof legislation as an instrument to achieve more consistent legislation, boost innovation, and promote sustainable economic growth. Future-proofing emerges in the context of the REFIT programme and as a new framework or set of tools for the implementation of the controversial innovation principle.[50]

A. Potential

The design of future-proofing legislation is intrinsically connected to the long-term promotion of research and development and thus innovation. This argument has been used by the supporters of the so-called innovation principle. While this principle does not yet have the same importance and acceptance as other traditional legal principles such as proportionality, precautionary and subsidiarity, innovation is undoubtedly a pillar of the EU economy. The promotion of innovation also finds its basis in the Treaties (eg, Article 3(3) Treaty on the European Union) and it is mentioned in EU policy and strategy documents.[51] The innovation principle was initially advanced in 2013 by the European Risk Forum, a think tank supported by a number of industries. This think tank put forward this principle in order to draw attention to the serious impact of European regulation on innovation. The innovation principle entails taking into account the impact of legislation on research and innovation in the process of developing and reviewing regulation in

[48] A different debate is that on super-entrenchment, that is, statutes that are able to resist judicial challenges and legislative amendment attempts because of their general acceptability and quasi-constitutional character, see William N Eskridge, Jr and John Ferejohn, *A Republic of Statutes: The New American Constitution* (Yale University Press, 2010); Kathryn E Kovacs, 'Superstatute Theory and Administrative Common Law' (2015) 90 *Indiana Law Journal* 1207.

[49] European Economic and Social Committee (n 12) 1.7.

[50] REFIT is a part of the European Commission's better regulation agenda which ensures that 'EU laws deliver their intended benefits for citizens, businesses and society while removing red tape and lowering costs. It also aims to make EU laws simpler and easier to understand'. See European Commission, 'REFIT – Making EU Law Simpler and Less Costly', available at https://ec.europa.eu/info/law/law-making-process/evaluating-and-improving-existing-laws/refit-making-eu-law-simpler-and-less-costly_en.

[51] European Commission, 'Better Regulations for Innovation-Driven Investment at EU Level', Commission Staff Working Document, 2016, available at https://publications.europa.eu/en/publication-detail/-/publication/404b82db-d08b-11e5-a4b5-01aa75ed71a1/language-en/format-PDF/source-79728021; European Commission, 'Towards an Innovation Principle Endorsed by Better Regulation', European Political Strategy Center, 2016, 14, available at https://ec.europa.eu/epsc/file/strategic-note-14-towards-innovation-principle-endorsed-better-regulation_en.

all policy domains. The complex relationship between regulation and innovation, as well as the pacing gap between technological change and regulation tend to be underestimated.[52] The latter is often accused of stifling innovation and being at the outset of the so-called innovation deficit in Europe.

The innovation principle sets the promotion of innovation as a key goal in addition to other existing goals of the European Union. In this context, regulators are expected to balance the need for regulation with the importance of promoting research and development in the EU context. The EESC considers that future-proofing legislation could further the innovation principle and ensure that this principle is integrated throughout the regulatory lifecycle and expressed through different instruments.[53] Future-proof legislation is thus innovation-friendly, it seeks to accommodates change whatever this may be (disruptive or incremental) and is tune with future developments. The EESC also underlines in its report that future-proof legislation is not only characterised by an anticipatory and flexible character but also by the ability to 'remain true to its original objective' for a longer period of time.[54] In terms of future-proofing in other disciplines, future-proofing is a synonym of endured effectiveness in the legislative context. In practice, this means that future-proof legislation at the EU level should not be too detailed and it should limit itself to providing a framework.

The EESC posits a paradigm shift for legislation, defining future-proof legislation as legislation that not only responds to the current state of our information society but that also anticipates it, adapts to it, and accompanies it. The traditional model of reactive legislation, on the contrary, is set aside in this context as it is no longer thought to be effective in a society characterised by the fast pace of technology.[55]

In 2016, the Dutch Government included the concept of 'future-proof legislation' in a policy document and initiated a programme to explore the possibility of developing laws that are able to facilitate innovation.[56] The Dutch Government explained that achieving future-proof legislation would require enacting laws that rely on vague and indeterminate concepts, experimental legislation and principle-based regulation.[57] The future-proof programme was short-lived and terminated in 2017. The Dutch Government concluded that legislation already accommodates social requirements and accompanies change but the future-proof programme did make particular progress in stimulating innovation in novel policy fields

[52] See Susan Strange, *The Retreat of the State: The Diffusion of Power in the World Economy* (Cambridge University Press, 1996) 7 (noting that the 'technological factor' and, more specifically, its fast pace, is often highly overlooked in the analysis of the shift in the functions of the state).

[53] See also European Commission, 'Towards an Innovation Principle' (n 51).

[54] European Economic and Social Committee (n 12).

[55] See also Wulf A Kaal, 'Dynamic Regulation for Innovation' (2016) *Perspectives in Law, Business and Innovation*.

[56] *Kamerstukken II* 2015/16, 33009, nr. 12.

[57] *Ibid.*

such as the regulation of the sharing economy and blockchain technologies.[58] The popularity of the future-proofing legislative programme has remained fairly limited beyond the borders of the Netherlands and the legislative studies and regulation communities. Nevertheless, future-proofing appears to have great potential in environmental law and urban law and policy where this approach offers the perfect set of tools for the fight against climate change and taking into account the interests of future generations.[59]

Future-proofing has been primarily promoted in the context of environmental legislation, food safety and regulation, building regulations and standards and the regulation of emerging technologies. An example of the application of future-proofing can be found in the Code for Sustainable Homes.[60] Future-proofing also underlies many energy policies and legislative measures that seek to minimise the environmental impact of building solutions over their lifetime.[61] In the food industry, there are also increasing demands for the promotion of sustainable and future-proof solutions.[62] It is in this context that the literature has also turned to the defence of an adaptive and anticipatory legislative framework.[63] In the field of media regulation, future-proofing has also been discussed due to the need to re-evaluate the protection of minors at a time when media content is changing rapidly and the cooperation between public and private regulators appears to be a more sustainable, flexible and future-proof way of regulating audiovisual content than traditional command-and-control regulation.[64]

Thus far, the connection between future-proofing, environmental law and the innovation principle has seemed unlikely. The corporate origins of the innovation

[58] Kamerbrief Ministerie van Economische Zaken over afsluiting programma toekomstbestendige wetgeving, 26 June 2017, available at www.rijksoverheid.nl/documenten/kamerstukken/2017/06/26/kamerbrief-over-afsluiting-programma-toekomstbestendige-wetgeving. See, nonetheless, for a critical analysis of this future-proof programme, Gert Jan Veerman, 'Toekomstbestendige Wetgeving: Duurzaam? Wendbaar? Duurzaam Wendbaar?' (2016) 31 *RegelMaat* 82.

[59] See, eg, Richard L Revesz and Matthew R Shahabian, 'Climate Change and Future Generations' (2011) 84 *Southern California Law Review* 1097; Susan Martin, 'Climate Changes, Migration, and Governance (2010) 16 *Global Governance* 397.

[60] Department of Communities and Local Government (DCLG), '2009a. Code for Sustainable Homes: Technical Guide' May 2009, Version 2, available at www.communities.gov.uk/publications/planningandbuilding.

[61] Georgadiou, Hacking and Guthrie (n 39) 146.

[62] Luis Gonzalez Vaqué, 'Unfair Practices in the Food Supply Chain' (2014) 9 *European Food & Feed Law Review* 293; Carmen G Gonzalez, 'Climate Change, Food Security, and Agrobiodiversity: Toward a Just, Resilient, and Sustainable Food System' (2011) 22 *Fordham Environmental Law Review* 493; M Zidianaki, 'Sustainable Food Consumption in the EU: Filling the Gaps of the Legal Framework' (2013) 8 *European Food and Feed Law Review* 113; Carmen G Gonzalez, 'The Global Food Crisis: Law, Policy, and the Elusive Quest for Justice' (2010) 13 *International Legal Materials* 462.

[63] Gonzalez, 'The Global Food Crisis', *ibid* 475 ('Policy flexibility is necessary in order to rebuild the agricultural sector, to protect small farmers from devastating import surges, and to nurture higher value-added food processing industries.'); Luis Gonzalez Vaqué, 'Food Loss and Waste in the European Union: A New Challenge for the Food Law?' (2015) 10 *European Food and Feed Law Review* 20, 29–32.

[64] Madeleine de Cock Bunning, 'Towards a Future-Proof Framework for the Protection of Minors in European Audiovisual Media' (2014) 10 *Utrecht Law Review* 9.

principle have been used to argue that this principle consolidates the rhetoric against the precautionary principle. Future-proofing is nonetheless a broader approach that can be indeed invoked to promote an analysis focused on the promotion of economic growth but it can equally support a broader (and more social) interpretation of the innovation principle.

B. Objections to Future-Proofing Law

Future-proof legislation offers the promise of supporting economic development, combating climate change, and taking into account the interests of future generations. However, future-proofing legislation can be criticised from different legal perspectives.

First, the design of future-proof legislation entails making decisions for future generations. While this would not necessarily amount to the entrenchment of legislation, it could be regarded as a measure that binds future voters to what present-day legislators predicted. This objection refers as well to a potential deficit in the democratic legitimacy of future-proof legislation. Interestingly, in its report, the EESC argues that future-proof legislation should engage with the needs of businesses and citizens and would have a higher degree of democratic legitimacy because it would be effective for a longer period of time. In this context, it is important to break down this potential objection to the legitimacy deficit of future-proof legislation in two elements: first, the shift from a reactive to proactive regulation; second, the existence of different types of legitimacy.[65]

Regulation tends to be mostly reactive. Technology and regulatory assessments made to evaluate whether a product or system should be introduced on the market also tend to offer a retrospective analysis of socio-technical transformations. Reactive perspectives and tools fail to offer insights regarding future changes.[66] The idea of embracing an anticipatory or foresight approach has nonetheless strong implications for existing paradigms of regulation and governance.[67] A proactive and forward-looking approach can be received with strong opposition as it does not always serve current political majorities and their voters. Instead, it focuses on the long-term and unpredictable effects of policies in an attempt to serve future voters. These voters are nonetheless those who will have to deal with problems that can already be foreseen at the present time (eg, climate change, limited availability of drinking water). The legitimacy challenge of this shift can be addressed with a focus on the collaborative dimension of anticipatory governance and the

[65] See Douglas R Williams, 'Environmental Law and Democratic Legitimacy' (1994) 4 *Duke Environmental Law & Policy Forum* 1; Juliet Lodge, 'Transparency and Democratic Legitimacy' (1994) 32 *Journal of Common Market Studies* 343.

[66] De Barbieri (n 34) 86; K Grieger, A Baun and R Owen, 'Redefining Risk Research Priorities for Nanomaterials' (2010) 12 *Journal of Nanoparticle Research* 383.

[67] Gregory N Mandel, 'Regulating Emerging Technologies' (2009) 1 *Law, Innovation, and Technology* 75, 80.

future-proof approach.[68] Anticipating and accommodating future changes should entail the broader engagement of citizens as early on as possible. These collaborative and participatory dimensions are currently discussed in the context of urban law. Although planning for example is by definition anticipatory, urban law does not yet embrace a future-proof approach by seeking to anticipate future challenges and including citizens in the planning phase rather than only in hearings.[69] In the field of environmental law, the legitimacy of iterative and anticipatory tools has been justified by the deliberative democratic theory which supports the continuous process of dialogue between decision-makers.[70]

Moreover, although the idea of governing and regulating for the future may sound contrary to democratic legitimacy, it is important to underline that there are different types of legitimacy that can also be covered by future-proofing. In particular, the perception that future-proof policies and institutions are more effective could enhance other forms of legitimacy that are based on the trust that citizens have in an effective government.[71] In addition, the literature has showed that regulation in uncertain contexts or contexts characterised by rapid changes in technology and infrastructures requires forward-looking, dynamic and heterogeneous approaches to regulation.[72]

Second, the principle of legal certainty is another potential objection against the adoption of a future-proofing approach. It is well-known that the principle of legal certainty cannot be reduced to its most famous dimension of continuity.[73] On the contrary, legal certainty contains a number of other equally important dimensions such as predictability and clarity which encompass a dynamic interpretation. Legal certainty means that individuals are aware of their rights and duties in a specific situation and not in an abstract way, this certainty is dynamic and not static.[74] The proponents of future-proofing would then argue that a forward-looking approach to legislation can favour legal certainty as it allows the legislator to avoid scenarios

[68] See Neil Gunningham, 'The New Collaborative Environmental Governance: The Localization of Regulation'(2009) 36 *Journal of Law and Society* 145.

[69] De Barbieri (n 34) 79.

[70] Neil Craik, 'Deliberation and Legitimacy in Transnational Governance: The Case of Environmental Impact Assessments' (2007) 38 *Victoria University of Wellington Law Review* 381.

[71] See Jonathan Boston, John Wanna, Vic Lipski and Justin Richard (eds), *Future-Proofing the State: Managing Risks, Responding to Crises and Building Resilience* (Australian National University Press, 2014).

[72] Kenneth A Bamberger and Deirdre K Mulligan, 'New Governance, Chief Privacy Officers, and the Corporate Management of Information Privacyin the United States: An Initial Inquiry' (2011) 33 *Law & Policy* 477, 479–80. Deirdre K Mulligan and Kenneth A Bamberger, 'Saving Governance by Design' (2018) 106 *California Law Review* 697, 744–45.

[73] For an extensive analysis of the principle of legal certainty, see Patricia Popelier, 'Legal Certainty and the Principles of Proper Lawmaking' (2000) 2 *European Journal of Law Reform* 32; Patricia Popelier, 'Five Paradoxes on Legal Certainty and the Lawmakers' (2008) 2 *Legisprudence* 42; Patricia Popelier, *Rechtszekerheid als beginsel van behoorlijke regelgeving* (Intersentia, 1997).

[74] See the thorough analysis of the different dimensions of the principle of legal certainty and the expectations of citizens under specific circumstances, Harald A Oldenziel, *Wetgeving en rechtszekerheid* (Kluwer, 1998).

of deeply ineffective and obsolete laws. Moreover, it advances legal certainty in fast changing contexts and ensures that novel innovative technologies and discoveries fit more effortlessly into the existing system.[75]

Third, adopting future-proof regulation under uncertain conditions also raises numerous questions in the context of the principles of good administration. For example, anticipating legal decisions and regulations at a time when information is still scarce seems to be contrary to the duty to gather sufficient information, enact proportionate, well-founded and fair rules. This is a frequently heard objection both in Europe and the United States that often results in the postponing of decision-making to future legislatures. As David Super explains, 'while information typically becomes more plentiful over time, other inputs to legal decisions, particularly decisional resources, often become scarcer. Moreover, postponed legal decisions often have considerably less value than a decision made earlier'.[76] While this is not a plea for future-proofing law, it underlines the importance of taking regulatory control over situations when the first signs of problems start emerging.

Enacting fully future-proof legislation might be an impossible feat to achieve as legislators cannot fully predict what future majorities will decide and what the future problems will be. Nevertheless, the interdisciplinary future-proof approach described in this article (see section II) can inspire legislators to manage risks, promote innovation and ensure the resilience and effectiveness of legal systems. The adoption of vague and indeterminate concepts might not always be the answer to legislating future-proof as it creates interpretation challenges and it does not embrace a truly learning-by-doing approach that can help legislators prepare laws for future challenges. The following section offers a framework for the design of future-proof legislation.

IV. A Framework

In this chapter, future-proofing has brought multiple disciplines together, calling in particular for a comparison between law and design, two disciplines which in abstract terms would seem to have little in common. As Deirdre Mulligan and Kenneth Bamberger have recently argued, it is increasingly important for regulators to adopt a similar mindset to that of designers since 'regulation must be designed to (…) adapt as more information is learned (…) It must be an iterative process

[75] A similar perspective characterised by flexibility which can also favour similar results is the so-called 'regulatory sandbox', see Michael Huertas, 'The UK FCA's Regulatory "sandbox": Any Lessons for the EU?' (2018) 33 *Journal of International Banking Law and Regulation* 50; DW Arner, J Barberis and RP Buckley, 'FinTech, RegTech and the Reconceptualization of Financial Regulation' (2017) 37 *Northwestern Journal of International Law and Business* 373, 404–406.

[76] David A Super, 'Against Flexibility' (2011) 96 *Cornell Law Review* 1377, 1380.

whereby technology is invented and redesigned based on user interactions'.[77] Legislators can indeed learn a great deal from other sciences on how to design laws in order to extend their duration and effectiveness.[78] Nevertheless, lawmaking is also bound by constitutional limits which are unknown in the scientific, artistic and business worlds.[79] While designers should engage with their clients and the interests of the public in general, they are not their political representatives. Considering the objections raised in the previous section to a fully-fledged form of future-proofing, this section offers a framework for future-proofing legislation by delving in particular into two instruments: experimental legislation and future-proof impact assessments. These instruments can help legislators and regulators slowly adopt future-proof solutions without putting at stake longstanding principles of law (legal certainty, legality, transparency, proportionality).[80] In addition, experimentation promotes the implementation of iterative and trial-and-error approaches that can support the advancement of innovation policies.[81] Since innovators face various uncertainties in the different stages of the innovation process (eg risks of not recovering investment, risk of delay to the commercialisation of their new products), there is no 'one-size-fits-all' solution for the regulation of emerging technologies.[82]

Although experimental laws and regulations and regulatory impact assessments (RIAs) are well-known legislative tools in the legal literature, their implementation for the purpose of future-proofing has been underanalysed. In this section we contend that both instruments could ensure that new legislative measures are not only isolated responses to current political debates (eg, the refusal of many voters to acknowledge climate change and to accept the fiscal burden of climate-change policies) but are also fit-for-future challenges.

[77] Mulligan and Bamberger (n 72) 743.

[78] For a reflection on the role of interdisciplinary studies in law, see Mark Tushnet, 'Interdisciplinary Legal Scholarship: The Case of History-in-Law' (1996) 71 *Chicago and Kent Law Review* 909.

[79] See generally Peter Cane, 'The Constitutional and Legal Framework of Policy-Making' in C Forsyth and I Hare (eds), *The Golden Metwand and the Crooked Cord: Essays on Public Law in Honour of Sir William Wade* (Oxford University Press, 1998) 39.

[80] For a more extensive discussion of the relationship between innovation and experimentation, see Sofia Ranchordás, 'Innovation Experimentalism in the Age of the Sharing Economy' (2015) 19 *Lewis & Clark Law Review* 871; Cortez (n 19).

[81] Michael C Dorf and Charles F Sabel, 'A Constitution of Democratic Experimentalism' (1998) 98 *Columbia Law Review* 267; Justice Brandeis famously celebrated the value of policy experimentation and the learning-by-doing approach in *New State Ice Co v Liebmann*, 285 US 262, 311 (1932) (Brandeis, J, dissenting). For a more recent critical perspective of experimentation in the context of federalism, see David A Super, *Laboratories of Destitution: Democratic Experimentalism and the Failure of Antipoverty Law* (2008) 157 *University of Pennsylvania Law Review* 541, 546–48; Michael A Livermore, 'The Perils of Experimentation' (2017) 126 *Yale Law Journal* 636.

[82] See, eg, Wilfred Dolfsma and DongBack Seo, 'Government Policy and Technological Innovation – A Suggested Typology' (2013) 33 *Technovation* 173, 174; Jan Fagerberg, 'Innovation: A Guide to the Literature' in Jan Fagerberg, David C Mowery, Richard R Nelson (eds), *The Oxford Handbook of Innovation* (Oxford University Press, 2007).

A. Experimental Legislation

The term temporary legislation includes a number of legislative instruments that have a set termination date. Experimental laws, regulations and experimental clauses are examples of temporary laws that, contrary to sunset clauses, are not meant to expire, but test the effectiveness of a solution. If the tested measure produces positive results, then the experimental law will be converted in a permanent legislative instrument. Experiments are nonetheless temporary by definition because their key feature is to evaluate the effectiveness of a new policy or legislative disposition or approach.[83] In theory the future of an experiment depends on the result of its evaluation and the attainment of the goals for which the experiment was initiated.[84] Based on the evaluation, the experimental legislation will either be discontinued or remain in effect.

There are different types of experimental laws: on the one hand, experimental laws may consist of delegation clauses that allow for the temporary derogation of the general legislative system.[85] This common type of experiment means that a statute will delegate the power to experiment under conditions determined beforehand (eg, period of the experiment, geographical circumscription, evaluation). To illustrate, the legislature will establish a general framework for the production of electricity. In order to test novel forms of energy production, an experimental disposition may be inserted in the law according to which small and temporary experiments may be conducted to test solutions that deviate from the general one. If these solutions are successful, they may be adopted as permanent, if they are unsuccessful, the legislature may choose to terminate the experiment. On the other hand, in federal and highly decentralised systems, it is not uncommon for state legislators or local legislators to confer the power to experiment with new solutions in the context of shared or exclusive powers.[86] In this context, states are allowed to waive federal requirements and implement their own legislative and policy solutions that they may find fit to address state challenges.[87] Legislative experimentation in federal systems, particularly in the United States, has been captured by the well-known 'states-as-laboratories' metaphor.

Enacting future-proof legislation is a challenging task and shifting the paradigm of lawmaking from reactive to proactive or anticipatory would be difficult to

[83] van Gestel and van Dijck (n 46); Mader (n 13) 125.

[84] For a reflection on the evaluation process of temporary legislation, see Mader (n 13); Mandelkern Group on Better Regulation (Brussels, 2001) 11–13.

[85] Francoise Crouzatier-Durand, 'Reflexions sur le concept d'experimentation legislative (a propos de la loi constitutionnelle du 28 mars 2003 relative a l'organisation decentralisee de la Republique)' (2003) 56 *Revue Francaise de Droit Constitutionnel* 687.

[86] For a theoretical reflection on this distinction, see Michiel Heldeweg, 'Experimental Legislation Concerning Technological and Governance Innovation – An Analytical Approach' (2015) 3 *Theory and Practice of Legislation* 169.

[87] See Dorf and Sabel (n 81). See also Charles R Shipan and Craig Volden, 'Bottom-Up Federalism: The Diffusion of Antismoking Policies from U.S. Cities to States' (2006) 50 *American Journal Political Science* 825 (2006).

achieve in practice. A more realistic option is to experiment with laws and policies in fields that are subject to rapid social, ecological, and technological changes.[88] Experimental legislation embodies many of the features of the future-proof approach: first, it allows for anticipation. The design of an experiment in legislation requires reflecting upon the goals that need to be achieved and the different possible instruments that are available at a certain point in time. In order to protect the principle of legal certainty, the legislature needs to balance anticipation with legal certainty and define a framework that on the one hand creates legislative room for experimentation and on the other secures a minimum of clarity to individuals.[89]

Second, experimenting with legislation can advance the future-proof approach as it offers a flexible framework that can be altered as more information becomes available. Experimentation creates room for change as it relies on temporary dispositions that need to be evaluated every three to five years. Depending on the underlying legislative framework, the legislature may also allow for variations in the implementation of the experiment. As these dispositions are bound to expire after a certain period, legislation needs to be periodically evaluated and new measures introduced to replace ineffective ones.[90]

The interdisciplinary perspective of future-proofing also entails two features that can be translated into legal terms by experimental legislation: resilience and risk management. Experimental dispositions guarantee that the main pillars of a policy remain standing despite future changes as they distinguish between permanent and temporary elements. The idea of an experiment is to test different ways to achieve specific goals and gather more information as time elapses. Moreover, experimental legislation limits the implementation risks of a novel regulation and policy in time and in space.[91] Experiments are often only implemented for a specific number of municipalities and a short period of time (for example, a number of cities may be selected to test new rules over a three-year period). This form of risk management should however not put at stake the meaningful character of the experiment: while some experiments may deliver results after one or two years, in some sectors, the effectiveness of new policies and laws may only be visible after a decade.

B. Future-Proof Impact Assessments

The EESC stresses the importance of impact assessments at national and European levels, 'including the SME test, for all legislative or non-legislative measures,

[88] On a framework for experimental legislation, see Ranchordás, *Constitutional Sunsets* (n 23).

[89] For an analysis of the constitutional framework of experimental legislation, see Hans Detlev Horn, *Experimentelle Gesetzgebung unter dem Grundgesetz* (Duncker & Humblot, 1989).

[90] See Ranchordás, 'The Whys and Woes (n 23) 415.

[91] See Gabriel Domenech Pascual, 'Sharing Economy and Regulatory Strategies toward Legal Change' (2016) 4 *European Journal of Risk Regulation* 717.

so that political decisions are informed and based on specific data.'[92] Drawing on this suggestion, we argue that legislators should consider incorporating future-proofing in existing regulatory, technology and environmental impact assessments.[93] Regulatory impact assessments are conducted in a large number of Organisation for Economic Co-operation and Development (OECD) member countries and they involve an ex ante systematic appraisal of the social, economic and environmental impacts of proposed regulations and policy instruments.[94] These impact assessments consider the consistency of regulation, their ability to assist business environments, and also the proportionality of novel rules. Many of these evaluations already encompass a certain degree of future thinking as these impact assessments seek to predict the effect of a measure on the regulatory framework. However, a future-proof impact assessment would include future considerations in a consistent manner in order to guarantee that legislation is resilient and adaptable to what the future brings and remains engaged with future societies.

The introduction of some elements of the future-proofing approach, possibly as a last step in a regulatory impact assessment, could ensure that legislation remains effective for a longer period of time. Moreover, future-proofing could enhance the effectiveness, collaborative and legitimacy dimensions of lawmaking as this approach requires legislators and regulators to inquire whether the enacted legislation is likely to be in tune with future changes. In terms of future-proofing in other areas, a future-proof regulatory assessment should consider the short-term and long-term risks of a novel regulation or policy and create room for the adoption of flexible solutions.[95] To illustrate, if one accepts that climate change is an ongoing problem with future challenges that can be predicted for the short- and long-term, future-proof assessments could test the ability of new laws to defend the interests of future generations. These laws could be future-proof in the context of environmental law if they entrench certain measures that will help combat future challenges (eg, ban on plastic, large investments in reforestation) or if they disentrench longstanding policies that will likely be harmful in the future.

An additional step to future-proof assessments could include the adoption of the so-called 'lifecycle thinking' that is widely present in different sciences and arts that promote future-proofing approaches. 'Lifecycle thinking' would require

[92] European Economic and Social Committee (n 12) 1.14.

[93] Peter Caroll, 'Does Regulatory Impact Assessment Lead to Better Policy?' (2010) 29 *Policy and Society* 113.

[94] For a full appraisal of regulatory impact assessments, see Colin Kirkpatrick and David Parker (eds), *Regulatory Impact Assessments: Towards Better Regulation* (Edward Elgar, 2007); Claudio Radaelli and Claire Dunlop (eds), *Handbook of Regulatory Impact Assessment* (Edward Elgar, 2016).

[95] We draw here on the risk regulation literature, see, eg, Alberto Alemanno, Frank den Butter, André Nijsen, Jacopo Torriti, *Better Business Regulation in a Risk Society* (Springer, 2012); Elizabeth Fischer, *Risk Regulation and Administrative Constitutionalism* (Hart Publishing, 2010); Jonathan Wiener, Michael D Rogers, James Hammitt, Peter H Sand (eds), *The Reality of Precaution: Comparing Risk Regulation in the United States and Europe* (RFF Press, 2010); Stephen Breyer, *Breaking the Vicious Circle: Towards Effective Risk Regulation* (Harvard University Press, 1999).

that legislators consider the legislative process as a whole, from its very beginning until the far-reaching effects of a law. While doing so, legislators and policymakers would be expected to include a larger number of stakeholders when defining policies that could have an impact on future generations (for example, by including an ombudsman or other type of representative that could defend the interests of future voters. In conclusion, a future-proof impact assessment would add a new layer to existing ex ante evaluations by ensuring that new legislation is not only consistent with existing frameworks but also flexible, innovation-friendly and adaptable to future changes.

V. Conclusion

According to the EESC, European legislation is future proof if it is proactive and forward-looking and provides the maximum legal clarity and certainty [...] and citizens see it as legitimate.' Future-proof legislation is presented in the EESC report on this subject as adaptable legislation that 'is also able to look ahead'.[96] While this idea sounds attractive, it is not yet clear how legislators can fully transpose this future-proof perspective into law. Nevertheless, this chapter has shown that future-proofing regulation and governance is not a chimera either: legislators and policymakers can draw a number of important lessons from other sciences' experience with future-proofing.

In this chapter, we explained that future-proofing is not as novel as suggested in policy documents.[97] Rather, it is a perspective from other sciences and arts that can change the way in which legislators draft legislation and facilitate its adaption to social, economic and technological changes.

The adoption of a future-proof approach to lawmaking entails promoting a switch from a reactive to proactive perspective to law.[98] A proactive approach could be particularly useful for the regulation of emerging technologies, where a simple one-size-fits-all approach does not suffice.[99] However, the benefits of future-proofing will only be visible if the interests of future generations play a stronger role in the legislative process and the political debate. The active and compulsory participation of children or organisations representing them could be a way of making sure that a future-proof assessment takes place.[100] However, future-proofing implicates a broader shift and the adoption of new theories of regulation and governance that embrace experimentation, the study of lifecycles and the definition of key public values that societies hold dear.

[96] European Economic and Social Committee (n 12) 2.2., 2.5.

[97] European Economic and Social Committee (n 12).

[98] Rob A DeLeo, 'Anticipatory Policymaking in Global Venues: Policy Change, Adaptation, and the UNFCCC' (2016) 92 *Futures* 39, 44–46 http://dx.doi.org/10.1016/j.futures.2016.09.001.

[99] Mandel (n 67) 77.

[100] DeLeo (n 98) 46.

Concluding Remarks: Time, Law and Change: It Takes Three to Tango

LUC VERHEY

I. Introduction

Time, law and change – it is not easy to indicate how these three phenomena relate to each other. It is obvious that law must change over time because law must accommodate itself to evolving social, economic and political circumstances. Law that is outdated will lose its essential functions in society and will ultimately be ignored. However, the questions of when, how and by whom the law must be changed is much more difficult to solve. This seems to have become even more challenging, since technological developments occur more frequently nowadays.

In this concise contribution, I offer some general observations about the subject drawing on my expertise and practical experience as a former lawyer of the Dutch Department of Justice and as a current member of the Advisory Division of the Dutch Council of State. Through both roles, I have participated in the legislative process for almost 20 years. I will therefore limit my concluding remarks to the analysis of the relationship between change and legislation. My observations are neither exhaustive nor conclusive. They are meant as 'food for thought' and for further academic debate.

II. Change: A Matter of Urgency?

Legislation is an overarching term. Under its umbrella, one finds all sorts of regulations, such as those which are vague and general, or those which are elaborated and detailed. There are also rules that are sustainable and therefore can be left unchanged for decades, while there are rules that are strongly influenced by social and economic developments and thus have to be changed often and quickly. Further, there are laws that are made by the parliamentary legislator, the Government, an independent agency or regional and municipal authorities. Laws are so different that when it comes to changing the law you cannot formulate general guidelines on how and when to change them. There is simply no 'one-size-fits-all' solution.

Making sound legislation is often time-consuming. Sometimes, there is no time available due to particular circumstances. In emergency cases, it can be necessary to act immediately and put legislation in place as soon as possible. In this situation, it can even be necessary to deviate from constitutional rules of procedure. As a general idea, this is not controversial. In times of war or disaster, the Government needs room to manoeuver to take urgent measures. If strict requirements are met, these measures can depart from legislation and sometimes even from fundamental rights and freedoms. Similar powers to issue emergency rules and measures have been conferred on local authorities if public order is seriously threatened. Emergency rules are always temporary; they can only exist in so far and so long as they are strictly necessary. If they expire, regular legislation will be reintroduced.

In Western democracies, these emergency procedures have rarely been applied. As such, they are seldom subject to serious debate. This can be different if circumstances emerge that make particular emergency legislation necessary. A recent example in Dutch legislation concerns Brexit. The Government has sent a legislative proposal to Parliament which contains specific legal provisions to make the changeover to the situation after Brexit easier and more convenient. A highly controversial element in this bill was a general power conferred to the Government which, under certain conditions, makes it possible to issue emergency rules, departing from the regular parliamentary procedures. After severe criticism from academics and politicians, Parliament has vastly amended the bill. The question remains whether, due to the numerous guarantees that were added, a fair balance has been struck between the rule of law and the need for effective powers in case of emergency.

In daily practice, however, emergency legislation is not the most challenging issue. Instead, it is the social and political pressure to quickly update legislation. This pressure arises in every policy area. Politicians want to utilise the political momentum to change. As technology has developed increasingly fast, the pressure on the legislator seems to have increased. As a result, there seems to be less time to think about evidence-based legislation. The risk that we create costly and ineffective legal measures increases if we do not take enough time to think about whether new rules are necessary and effective. If they are, we must then consider which rules will work and which rules will not and how much time will be needed for a proper implementation. How should these matters be handled? In the next section, I will raise some issues based on my experience in daily practice.

III. Differentiation in Time Pressure

An important issue which has been often debated is how much time society gets to prepare itself for the new rules before they enter into force. This is especially important for rules which are highly complicated. If and how the legislator takes into account the needs of citizens, companies and executive organs in order to prepare them properly depends on the circumstances. Sometimes, the legislator leaves reasonable opportunity for implementing the necessary changes although time always seems to be too short. Notably, adjusting computer software or setting up entirely new computer systems has proved to be extremely time-consuming and can sometimes even have disastrous effects, preventing the original aims of the legislator from being achieved.

A Dutch example is the Environment and Planning Act. This Act contains a complete, very complex and radical change of environmental law. It replaces 26 Acts and intends to simplify and harmonise rules. Extensive parliamentary debates took place between 2014 and 2016. The Act was published on the 26th April 2016. However, it was decided that the Act and all other necessary regulations will probably enter into force only in 2021. The main reason for this is that, because of the radical changes the new rules will bring about, public authorities, companies and citizens need time to prepare themselves. Specifically, the introduction of an effective digital system proves to cost much time and effort. The Government made available 90 million euros in order to ensure that the new rules can be applied properly at the time they enter into force.

In other cases, however, time pressure is so high that legislation has to enter into force in the short-term, despite the general knowledge that society is simply not prepared. This happens, for example, when the Government has to cut down costs drastically due to financial reasons. An extreme instance in the Netherlands was the transition of public youth and social care to the municipalities. Far-reaching and complex legislation came into force on the 1st January 2015. Parliamentary debates took place under immense pressure. Albeit, despite serious doubts about how these radical changes should be implemented in practice, the rules were adopted and entered into force almost immediately. Municipal authorities and citizens would then see how to deal with it.

This does not mean that time pressure, as such, is a negative thing. People often need time pressure to achieve great outcomes. This surely applies in the legislative process. Narrow, but realistic, time schedules are necessary to keep operating at a certain speed in a process as complicated as lawmaking. In some cases, there is simply no time to completely develop evidence-based solutions. In our age, climate change is probably the most convincing example. Waiting until we know everything is in some cases the worst option. Sometimes, there is no other option than to take timely measures. This requires the courage to take risks and to accept that mistakes will be made.

IV. Delayed Drop Effect

An important issue is the follow-up in the event that there was a significant lack of time for implementing changes properly. What happened, for example, after the above-mentioned transition of the public youth and social care to the municipalities entered into force?

At the time, all warnings that this legislation would create chaos seemed unjustified at first. However, what happened was that actual implementation only started after the entering into force. As a result, the public and private bodies which have to deal with this and, more importantly, many citizens, are still faced with major difficulties as a direct effect of this legislation five years later. Therefore, this operation, which at the time seemed to be very successful, has now become increasingly controversial. Recently, the Dutch Association of Municipalities (VNG) has even written a letter to the Government, stating that if municipalities do not get more money for the implementation of the tasks which were conferred upon them in 2015, they intend to return these tasks to the Government. Of course, municipalities cannot decide this on their own; it needs a change of the law. But, it is meant to be a clear signal to the Government and the Parliament that things will go terribly wrong if radical change does not take place soon.

An important lesson of all this is that the apparent time gain at the beginning is illusionary. There is, what I would call, a 'delayed drop effect'; the serious problems faced by municipalities and citizens are a result of insufficient preparation time before the entry into force, which only gradually but becomes painfully visible long after political and public attention has faded away. Here, time has a distorting effect. Many citizens are confronted with problems which seem to be clearly related to the new law. However, the causes of the problems are not precisely clear because the possible effects were not thoroughly analysed at the appropriate time. In fact, this analysis is still needed to determine more accurately how problems could be solved.

The over-hasty introduction of new legislation has an important impact on the lifecycle of legislation. If new legislation has been put in place under time pressure, there is a substantial risk that its legal content and its impact on society would not have been carefully considered. If the new law is not fit for purpose, there can be negative consequences. As a result, major changes can become necessary shortly after the legislation has entered into force. The risk of over-hasty decision-making is then again at stake. The political pressure to take immediate action will be high and, therefore, there is once again the danger of miscalculating the effects of new changes or reparations of the law.

V. Speeding Up the Legislative Process

Nevertheless, new legislation is not only caused by the need for reparation of mistakes that are made because of a lack of time at earlier stages. In a broader sense, there is a tendency to speed up the legislative process. In some cases, changes of the law are introduced within a short period of time because a new political majority wants it or because an incident leads to political pressure to take action.

In those cases, one regularly seems to overestimate what the law can achieve. One neither asks whether an incident points to a structural problem that necessitates a change of the law, or whether a change of the law contributes to solving the problem and whether the change one wants could have negative effects which overshadow the expected positive impact. Politicians sometimes seem uninterested in the effects that legislation will have; they want to show to the public that they are acting immediately and quickly. Trying to reduce risks by introducing new rules, sometimes called the 'risk-rule reflex', is an intractable phenomenon in modern times. It is perhaps an understandable reaction to unfortunate outcomes and the publicity caused by such events. But, once again, hurried legislation can cause new problems soon afterwards and even make amendments to the law necessary.

Incidents are not the only reason why politicians take action. Other factors can also make quick responses of the legislator necessary. The speed of technological development is nowadays largely impactful. It can make legislation become outdated in a short period of time. Legislation can rapidly become meaningless when it is not updated. Continuous maintenance by changing detailed laws frequently is, however, not always an effective solution. Sometimes, open legal norms, which leave room for interpretation and incorporating new developments, are a better option. The recent European General Data Protection Regulation (GDPR) is a good example. Due to rapidly developing technology, it is simply impossible to have data protection law without open norms. But, the result of all this is that public bodies, companies, citizens, supervising authorities and, ultimately, the courts have much work to do in order to give data protection practical meaning. Some experts even have serious doubts whether data protection rules can be effective at all.

All the factors put together, the quick adoption of legislation could bring about an accumulation of consecutive changes to the law. This causes major difficulties for citizens, but also to companies, executive bodies and courts. As a result, the impression that the legislator lags behind events not only arises, but also seriously threatens the core functions of legislation.

VI. Democratic Legitimacy Deficit

Time effects as mentioned above can have a major impact on the democratic legitimacy of legislation and political decision-making in general. If legislation comes into society precipitously while politicians are saying that there will be no chaos and everything is working effectively, this can and probably will have damaging effects on the acceptance of what has been decided. Citizens can become disappointed and angry because expectations that were aroused during the legislative process were not met at a later stage.

This happened in the case previously discussed above, ie, the transition of the public youth and social care to the Dutch municipalities. Having citizens experience many serious problems with the delayed negative effects of the new law after its passing will cause understandable feelings of disappointment and anger, especially as during the legislative process there was no time to look seriously at its possible impact. These feelings will become even stronger if, soon after the law has entered into force, the law will change again because one wants to fix it. A long period of uncertainty, during which the level of social protection is not what it should be, will be the foreseeable result.

In terms of democratic legitimacy, the damage in this example is twofold. At the beginning of the decision-making process, there was not enough time to look at the impact of the far-reaching proposals and have serious input from experts and society as a whole in order to take into account all the relevant aspects of this highly complex legislation. This is the problem of lacking input-legitimacy. But, perhaps even more problematic is that politics has not delivered. It has failed to achieve satisfying results in an extremely sensitive area that deeply affects the daily lives of many citizens – citizens who often belong to the lower social classes. Failing to deliver on such a fundamental issue reduces trust in politics, a damage which is difficult to compensate.

These problems cannot be solved by simply making better legislation; they concern complicated changes in society as a whole, which arise in daily politics. We have to ask ourselves how we handle them. The answer is far from simple. We have to think more thoroughly about mechanisms which can create legislation which conforms to the practices and speed of our time. One of the mechanisms to achieve this could be experimental legislation (see also chapter 16 by Ranchordás and van 't Schip).

VII. Our Time Asks for Innovation

Within the agreement of the Dutch Coalition Government of 2017, experimental legislation plays an important role. The Government wants to start experiments in many policy areas to make innovation possible. Some of these experiments will make it necessary to deviate from existing legislation. The present guidelines on experimental legislation will probably be made more flexible. The Government's intentions are an illustration that, in the coming years, the debate on experimental legislation will intensify.

Indeed, our era asks for innovation. To prevent legislation from lagging too far behind social, economic and technological developments, the need for innovation also concerns our way of legislating. The classical methods seem to fall short of what is needed in present-day circumstances. We need more opportunity for experimental legislation. Legal experiments make it possible to test new solutions in cases for which we are not sure about the effects. In some cases, it will be necessary to experiment by derogation from present rules and procedures.

The crucial question is under which conditions derogations from the law are acceptable. Fundamental principles of the rule of law are at stake, such as the principle of legal certainty and the principle of equal treatment. Experimental legislation has to comply with these principles. For this reason, experimental legislation has to meet certain conditions. For example, Dutch government policy requires that, if a regulation as an experiment derogates from present legislation, this regulation specifically provides for:

- the goal of the experiment;
- the scope of the experimental regulation (for example the territorial and/or personal scope);
- the parts or clauses of the law from which can be derogated; and
- the maximum duration of the derogation.

Recently, this framework became the subject of discussion. It would be too restrictive in the face of the innovations that society needs. At the time of writing, a debate is ongoing as regards specific proposals for making this framework more flexible.

VIII. Experimental Legislation: Three Caveats

Indeed, a thorough debate on easing the conditions of experimental legislation is necessary. Society asks for more room for experiments of which the outcome cannot always be foreseen. Particularly, it cannot always be predicted which derogations from existing law are necessary in order to test solutions under consideration. Experiments can also prevent over-hasty legal reform, and can support preparing evidence-based legislation by testing what really works in practice. However, there at least three caveats one has to keep in mind.

First, one often underestimates what is already possible under valid legislation. For example, on the request of the Dutch Minister of Interior Affairs, Dutch municipalities proposed 75 experiments for which derogation from present legislation would be necessary. Closer examination demonstrates, however, that in only two of them such derogation perhaps would be necessary. In all other cases, solutions could be found without setting aside existent legislation. This makes clear that one must carefully consider what is really needed. In some cases, it also requires creativity and courage to think about and apply new interpretations of existing rules.

The second caveat is that experimental legislation can be misused. The idea of experimenting has become popular. As a result, the concept of an 'experiment' can easily become an all-purpose word which is used in cases for which it is not meant. Political pressure can lead to temporary legislation as a first step to get rid of rules that are seen as obstacles to economic growth or combatting terrorism or serious crime. This does not mean that there cannot be valid reasons to reconsider current legislation. However, in that case, the political debate must focus on the legislation itself in order to transparently and carefully consider if and how a new balance of interests has to be struck. In any case, experiments should not be used to undermine fundamental guarantees such as the individual rights and freedoms as laid down in constitutions and international treaties.

Lastly, even if the conditions for experimental legislation were made more flexible, it will remain crucial that experiments meet procedural and methodological requirements. After the experiment period has expired, one wants to draw conclusions – whether the experiment has been successful or not and what these conclusions could or should mean for future legislation. This means that, before an experiment starts, one has to determine the goals one wants to achieve and, in the face of this, the content of the experiment. This is necessary in order to carefully examine the effects of the experiment at a later stage and to conclude the extent to which the goal has been achieved. To draw authoritative conclusions to this end it will, last but not least, be necessary to think about when, how and by whom the experiment will be evaluated.

IX. Final Remark

We live in turbulent times where changes occur quickly. Sometimes, there is insufficient time to direct and control changes and transform them into acceptable results. In this context, legislation is both an indispensable tool to achieve our goals and a fundamental guarantee for the protection of our democracy and the rule of law. The ongoing debate illustrates that we have to think more thoroughly about mechanisms which can ensure that legislation fits the demands of our time.

As this book shows, time, law and change are inseparable. It thus takes three to tango. This brings me to my final remark. Problems can no longer be solved by using classical legal instruments. Innovative thinking about legislation is therefore not only a tool for lawyers. What is desperately needed is cooperation between disciplines and cooperation between academics and practitioners. Only by the efforts of several professionals with different backgrounds, experiences and views, will we be able to effectively address the great challenges of our time.

INDEX